Logic, Argumentation & Reasoning

Interdisciplinary Perspectives from the Humanities and Social Sciences

Volume 10

Series editor
Shahid Rahman

Logic, Argumentation & Reasoning explores the links between Humanities and the Social Sciences, with theories including, decision and action theory as well as cognitive sciences, economy, sociology, law, logic, and philosophy of sciences. It's two main ambitions are to develop a theoretical framework that will encourage and enable interaction between disciplines as well as to federate the Humanities and Social Sciences around their main contributions to public life: using informed debate, lucid decision-making and action based on reflection.

The series welcomes research from the analytic and continental traditions, putting emphasis on four main focus areas:

- Argumentation models and studies
- Communication, language and techniques of argumentation
- Reception of arguments, persuasion and the impact of power
- Diachronic transformations of argumentative practices

The Series is developed in partnership with the Maison Européenne des Sciences de l'Homme et de la Société (MESHS) at Nord - Pas de Calais and the UMR-STL: 8163 (CNRS).

Proposals should include:

- A short synopsis of the work or the introduction chapter
- The proposed Table of Contents
- The CV of the lead author(s)
- If available: one sample chapter

We aim to make a first decision within 1 month of submission. In case of a positive first decision the work will be provisionally contracted: the final decision about publication will depend upon the result of the anonymous peer review of the complete manuscript. We aim to have the complete work peer-reviewed within 3 months of submission.

The series discourages the submission of manuscripts that contain reprints of previous published material and/or manuscripts that are below 150 pages / 85,000 words.

For inquiries and submission of proposals authors can contact the editor-in-chief Shahid Rahman via: shahid.rahman@univ-lille3.fr or managing editor, Laurent Keiff at laurent.keiff@gmail.com.

More information about this series at http://www.springer.com/series/11547

Sven Ove Hansson • Gertrude Hirsch Hadorn
Editors

The Argumentative Turn in Policy Analysis

Reasoning about Uncertainty

Springer

Editors
Sven Ove Hansson
Department of Philosophy and History
Royal Institute of Technology
Stockholm, Sweden

Gertrude Hirsch Hadorn
Department of Environmental Systems Science
Swiss Federal Institute of Technology
Zurich, Switzerland

ISSN 2214-9120 ISSN 2214-9139 (electronic)
Logic, Argumentation & Reasoning
ISBN 978-3-319-30547-9 ISBN 978-3-319-30549-3 (eBook)
DOI 10.1007/978-3-319-30549-3

Library of Congress Control Number: 2016936269

Printed on acid-free paper

This Springer imprint is published by Springer Nature
The registered company is Springer International Publishing AG Switzerland

Preface

The history of this book goes back to a discussion that we had in December 2012 on recent developments in decision analysis. There is a long tradition of criticizing overreliance on the standard models of decision theory, in particular expected utility maximization. What we found to be new, however, is a more constructive trend in which new tools are provided for decision analysis, tools that can be used to systematize and clarify decisions even when they do not fit into the standard format of decision theory. Discussions with colleagues confirmed that we were on the track of something important. A new approach is emerging in decision research. It is highly pluralistic but it also has a common theme, namely the analysis of arguments for and against decision options. We decided that a book would be the best way to sum up the current status of this argumentative turn in decision analysis, and at the same time provide some impetus for its further development.

The book consists of an introduction, a series of chapters outlining different methodological approaches, and a series of case studies showing the relevance of argumentative approaches to decision analysis. The brief Preview provides the reader with an overview of the chapters, and an Appendix recapitulates some of the core concepts that are used in the book.

We would like to thank all the contributors for excellent co-operation and not least for their many comments on each other's chapters that have contributed much to the cohesion of the book. All the chapters were thoroughly discussed on a workshop in Zürich in February 2015 that has been followed by many e-mail exchanges. We would also like to thank Marie-Christin Weber for invaluable editorial help and the publisher and the series editors, Shahid Rahman and Laurent Keiff, for their support and their belief in our project.

Stockholm, Sweden
Zurich, Switzerland
September 24, 2015

Sven Ove Hansson
Gertrude Hirsch Hadorn

Contents

Contributors

Gregor Betz is professor in philosophy of science at the Karlsruhe Institute of Technology, Karlsruhe. In his publications, he develops argumentation-theoretic models of complex debates, reconstructs moral, political, philosophical and scientific controversies, defends the ideal of value-free science, simulates social opinion dynamics, vindicates the veritistic merit of plurality and critique, and assesses the predictive limits of climate science and economics. He is also contributing to the Argunet project (http://www.argunet.org), which seeks to promote a culture of reasoning. His books include: *Prediction or Prophecy? The Boundaries of Economic Foreknowledge and Their Socio-Political Consequences* (DUV 2006), *Theorie dialektischer Strukturen* (Klostermann 2010), *Debate Dynamics: How Controversy Improves Our Beliefs* (Springer 2012).

Georg Brun is a research fellow at the Institute of Philosophy at the University of Berne. Before that he was a research fellow at the Institute for Environmental Decisions at ETH Zurich, contributing to interdisciplinary projects on the analysis of policy arguments and decisions. His areas of research include epistemology, argumentation theory, philosophy and history of logic, metaethics and aesthetics. Book publications: *Die richtige Formel. Philosophische Probleme der logischen Formalisierung* [*The Right Formula: Problems of Logical Formalization*] (Ontos 2004), *Textanalyse in den Wissenschaften. Inhalte und Argumente analysieren und verstehen* [*Text Analysis in the Sciences: Analysing and Understanding Content and Arguments*] (as co-author, vdf 2014) and *Epistemology and Emotions* (as - co-editor, Ashgate 2008).

Neelke Doorn holds a master degree in civil engineering (MSc, cum laude) and philosophy (MA, cum laude) and a Ph.D. degree in philosophy of engineering and technology, with additional training in water and nature conservation law (LLB, cum laude). She wrote her Ph.D. thesis on moral responsibility in R&D networks. Dr. Doorn is currently an assistant professor at the School of Technology, Policy and Management of the Technical University Delft, Department of Values,

Technology and Innovation. Her research focuses on moral and distributive issues in water and risk governance. In 2013, she was awarded a prestigious personal Veni-grant for outstanding researchers from the Netherlands Organization for Scientific Research (NWO) for her project on the ethics of flood risk management. Dr. Doorn is Editor-in-Chief of *Techné*: *Research in Philosophy and Technology* (Journal of the Society for Philosophy and Technology).

Karin Edvardsson Björnberg is associate professor in environmental philosophy at KTH Royal Institute of Technology, Stockholm. Her research interests lie at the intersection of environmental philosophy and environmental policy analysis, where she pursues both normative and empirical questions. Among her more recent publications are articles published in *Ethics, Policy and Environment*, *Ethical Theory and Moral Practice*, *Energy Policy*, and *Futures*. She is currently leading two research projects, one on the ethical aspects of using biotechnology in agriculture and one on delay mechanisms in environmental policy.

Kevin C. Elliott is an associate professor in Lyman Briggs College, the Department of Fisheries & Wildlife, and the Department of Philosophy at Michigan State University, East Lansing. His major research areas include the philosophy of science, research ethics, and environmental ethics. In his recent work, he has focused especially on the roles of values in science, the management of financial conflicts of interest in scientific research, the ethical standards and practices of scientific teams, and ethical issues related to scientific and environmental communication. He is the author of *Is a Little Pollution Good for You? Incorporating Societal Values in Environmental Research* (Oxford University Press 2011) and more than 50 journal articles and book chapters.

Till Grüne-Yanoff is professor in philosophy at the Royal Institute of Technology (KTH) in Stockholm. His research focuses on the philosophy of science and on decision theory. In particular, he investigates the practice of modelling in economics and other social sciences, develops formal models of preference change and discusses the use of models in policy decision-making. Till is also a member of the TINT Finnish Centre of Excellence in the Philosophy of Social Science in Helsinki, sponsored by the Academy of Finland.

Armin Grunwald is professor of philosophy and ethics of technology and director of the Institute for Technology Assessment and Systems Analysis (ITAS) at Karlsruhe Institute of Technology (KIT). He is also director of the Office of Technology Assessment at the German Bundestag (TAB). His research includes contributions to the theory and methodology of technology assessment, ethics of technology, philosophy of science, and approaches to sustainable development. Currently he focuses on the hermeneutic side of technology assessment. Armin Grunwald is co-founder and member of the editorial board of the Journal of Responsible Innovation. He is member of several expert groups, e.g. the Commission for nuclear waste disposal of the German Bundestag and the Science

Committee of the international Future Earth research programme on sustainability. His recent book publications include *Responsible Nanobiotechnology. Philosophy and Ethics* (Panstanford Publishing 2012) and the *Handbuch Technikethik* (edited, Metzler 2013).

Sven Ove Hansson is professor in philosophy at the Department of Philosophy and History, Royal Institute of Technology, Stockholm. He is editor-in-chief of *Theoria* and member of the Royal Swedish Academy of Engineering Sciences. His research includes contributions to decision theory, the philosophy of risk, moral and political philosophy, logic, and the philosophy of science and technology. He is the author of around 300 refereed journals papers and books chapters. His recent books include *The Ethics of Risk. Ethical Analysis in an Uncertain World* (Palgrave Macmillan 2013), *Social and Ethical Aspects of Radiation Risk Management* (edited with Deborah Oughton, Elsevier 2013) and *The Role of Technology in Science. Philosophical Perspectives* (edited, Springer 2015).

Gertrude Hirsch Hadorn is an adjunct professor at the Department of Environmental Systems Science, Swiss Federal Institute of Technology, Zurich. She has worked in environmental ethics and in the philosophy of environmental and sustainability research with case studies in the fields of climate change and ecology. More recently, she has contributed to the methodology of transdisciplinary research, the analysis of values in science, the epistemology of computer simulations, and the analysis of uncertainty in decision-making. She is lead editor of the *Handbook of Transdisciplinary Research* (Springer 2008) and member of the Scientific Board of the interdisciplinary journal *GAIA*. She has acted as Vice President of the Swiss Academy of Sciences in 2001–2006.

Niklas Möller is associate professor at the Department of Philosophy and the History of Technology, Royal Institute of Technology (KTH), Stockholm. His research interest lies in normative and metanormative questions, mainly in political philosophy, moral philosophy, and the philosophy of risk. Möller received his Ph.D. in philosophy at KTH in 2009, after which he worked 2 years at Cambridge University as a post-doctorate researcher. Thereafter, he worked as a research scholar at the Department of Philosophy at Stockholm University, before returning to KTH. Möller has published numerous articles in international peer review journals such as *Philosophical Studies*, *British Journal of Political Science*, *Journal of Political Philosophy*, *Social Theory & Practice*, *Journal of Applied Philosophy*, *Ethical Theory & Moral Practice*, *Ethics*, *Policy & Environment*, *European Journal of Political Theory*, *Journal of Philosophical Research* and *Risk Analysis*.

Michael Schefczyk is professor of philosophy at Karlsruhe Institute of Technology. He is co-founder and editor-in-chief of *Moral Philosophy and Politics*. Recent publications include "Background Justice over Time: Property-Owning Democracy versus a Realistically Utopian Welfare State", in *Analyse & Kritik* 35 (1), 193–212; "The Financial Crisis, the Exemption View and the Problem of the Harmless

Torturer", in *Philosophy of Management*, Special Issue "Philosophical Lessons from the Global Financial Crisis", Volume 11 (1), 25–38; and "Neutralism, Perfectionism and Respect for Persons", in *Ethical Perspectives* 19 (3), 535–546.

Kristin Shrader-Frechette has degrees in mathematics, philosophy of science, and 3 NSF-funded post-docs, in biology, economics, and hydrogeology. O'Neill Professor, University of Notre Dame, in Philosophy and in Biological Sciences, she held professorships at University of California and University of Florida. Funded for 28 years by the US National Science Foundation, her research addresses models in biology/hydrogeology; default rules under mathematical/scientific uncertainty; quantitative risk analysis; and science and values/ethics. Translated into 13 languages, her work includes 15 books – such as *Tainted* (how flawed scientific methods influence policy); *What Will Work: Fighting Climate Change with Renewable Energy, Not Nuclear Power*; *Taking Action, Saving Lives*; *Method in Ecology*; and *Risk and Rationality*. Her 400+ journal articles appear in *Biological Theory, Philosophy of Science*, *Quarterly Review of Biology*, *Bulletin of the Atomic Scientists, Risk Analysis*, *Ethics*, and *Science* (3 pieces). She has served on many US Department of Energy, Environmental Protection Agency, and National Academy of Sciences boards/committees. Her pro-bono scientific/ethics work, to protect poor/minority communities from pollution-caused environmental injustice, have won her many awards, including the World Technology Association's Ethics Prize.

Part I
Introductory

Chapter 1
Preview

Sven Ove Hansson and Gertrude Hirsch Hadorn

Abstract This is a short summary of the multi-authored book that is the first comprehensive survey of the argumentative approach to uncertainty management in policy analysis. The book contains chapters that introduce various argumentative methods and tools for structuring and assessing decision problems under uncertainty. It also includes five case studies in which these methods are applied to specific policy decision problems.

Keywords Argumentation • Risk • Uncertainty • Rationality of decisions • Argumentative methods for decision support • Great uncertainty • Deep uncertainty • Expected utility • Policy analysis

The argumentative turn in policy analysis is a new approach that is currently developing out of many research efforts. It provides us with new tools for decision analysis that are based on methods and insights from philosophy and argument analysis. With these methods we can provide decision support in cases when traditional methods cannot be used due to their higher demands on the information input.

This book is the first comprehensive presentation of the argumentative turn. It contains an introductory chapter, a series of chapters proposing methods and tools for argumentative decision analysis, a series of chapters with case studies illustrating these methods, and a brief glossary of key terms.

S.O. Hansson (✉)
Department of Philosophy and History, Royal Institute of Technology, Stockholm, Sweden
e-mail: soh@kth.se

G. Hirsch Hadorn
Department of Environmental Systems Science, Swiss Federal Institute of Technology, Zurich, Switzerland
e-mail: hirsch@env.ethz.ch

S.O. Hansson, G. Hirsch Hadorn (eds.), *The Argumentative Turn in Policy Analysis*, Logic, Argumentation & Reasoning 10, DOI 10.1007/978-3-319-30549-3_1

1 Introduction

Conventional decision analysis, for instance in the form of risk analysis or cost-benefit analysis, is based on calculations that take the probabilities and values of the potential consequences of alternative actions as inputs. But often, we have to make decisions in spite of insufficient information even about what options are open to us and how they should be evaluated. In "Introducing the argumentative turn in policy analysis" *Sven Ove Hansson and Gertrude Hirsch Hadorn* show how methods from philosophical analysis and in particular argument analysis can be used to systematize deliberations about policy decisions under great uncertainty, i.e. when information is lacking not only about probabilities but also for instance about what the options and their potential consequences are, about values and decision criteria, and about how the decision relates to other decisions that will be made by others and/or at a later point in time.

The concept of argument analysis is wide and covers a large and open-ended range of methods and tools, including tools for conceptual analysis, structuring decisions, assessing arguments, and evaluating decision options. The use of these methods extends the rational treatment of decisions in at least two respects. First, argumentative methods can be used to clarify the grounds for applying the formal tools of traditional decision theory and policy analysis when these tools are useful. This can be done e.g. by analysing the decision frame. Secondly, when traditional tools are inapplicable or insufficient, the tools of argumentative decision analysis can replace or supplement them. For instance, such tools can deal with information gaps and value uncertainties that are beyond the scope of traditional methods. In this way, the argumentative turn in policy analysis provides a "widened rationality approach" to decision support. This is useful for all decision-makers, but perhaps in particular for those striving to make decisions that have democratic legitimacy. Such legitimacy has to be grounded in a social framework in which rational argumentation has a central role.

2 Part I: Methods

In policy debates, practical arguments – that is, arguments for or against some policy options – are often presented in incomplete and opaque ways. Important premises or steps of inference are not expressed explicitly, and their logical structure is intransparent. To make the argumentation perspicuous, argument analysis is needed. It specifies implicit premises and inference steps, represents the argument in a clear way, evaluates the validity of inferences, and clarifies the points of agreement and disagreement. In "Analysing practical argumentation" *Georg Brun and Gregor Betz* provide an introduction to methods of argumentation analysis with a special focus on their application to decisions under great uncertainty. The analysis of arguments is guided by a descriptive and a normative goal: on the one hand, reconstructing a given argumentation as clearly as possible and on the other hand, evaluating its validity. The more specific tasks, goals and uses of argument analysis are described and illustrated with examples. (More examples can be found in Michael Schefczyks's chapter "Financial markets: Applying argument analysis to the stabilisation task" in the second part of the book.) As a tool for structuring complex argumentation, Brun and Betz then introduce argument maps and exemplify their use with reference to a case study on ethical aspects of climate geoengineering. For the reconstruction and evaluation of different types of practical arguments, they suggest argument schemes which spell out various decision principles such as, for example, the Principle of Optimal Choice or the Principle of Absolute Rights Violation.

In "Evaluating the uncertainties" *Sven Ove Hansson* applies argument analysis to the task of evaluating and prioritizing among the large number of uncertainties pertaining to a complex decision. He begins by showing that many of the argument patterns that are commonly applied to such problems are in fact fallacies, since they programmatically disregard information that may be of crucial importance. Instead he proposes decision tools that are intended to ensure that no important factors are left without consideration. These tools are divided into three main groups: tools that help us find important uncertainties, tools for evaluating each uncertainty, and tools for comparing the uncertainties. The application of these tools requires a flexible and iterative process in order to account for new and unforeseen types of arguments. The chapter also contains a discussion of how ethical aspects of uncertainties should be dealt with. Hansson proposes as a moral starting-point that each person has a prima facie right not to be exposed by others to risks or dangers. This prima facie right can be overridden in cases of mutually advantageous, reciprocal risk exposures. Risks can be acceptable if they are part of a social system of reciprocal risk exposures that is beneficial to all members of society. This is a much stricter requirement than the usual impersonal criterion that the sum of all benefits (irrespective of whom they accrue to) should be larger than the sum of all expected detriments (irrespective of whom they accrue to).

In "Value uncertainty" *Niklas Möller* gives an overview of different sources of value uncertainty in decision problems. The chapter combines concepts and considerations from decision theory and moral philosophy. Four types of value uncertainty are discussed that a decision maker may face. First, she may be inconclusive about which values she is committed to. Secondly, she may be uncertain about the specific content of the values she is committed to, such as justice or freedom. Thirdly, she may be uncertain which values to apply to the problem at hand and fourthly, how to compare the different values in order to rank the options for choice. For dealing with value uncertainty in decision support Möller considers it imperative that the parameters of the problem are clearly specified. This is typically a non-trivial task since these parameters are often implicit, ambiguous, or vague. Making them explicit allows us to apply argument analysis for instance to conflicting values and unclear rankings. Techniques to specify value uncertainty include e.g. contextualization of the decision problem, making one's hierarchy of values explicit, and considering the strength of values or how the decision problem is embedded. The use of these methods can transform the decision problem into a more tractable one. However, it will rarely result in a single unanimous conclusion about how to decide. Möller recommends a search for a reflective equilibrium as a means to modify incompatible positions and achieve more coherence.

Often, decision problems are associated with uncertainties on factual knowledge that cannot be probabilistically characterized. This makes them inaccessible to the standard methods of decision analysis. In "Accounting for possibilities in decision making" *Gregor Betz* reviews arguments that may justify choices in view of merely possibilistic foreknowledge. He distinguishes between those conceptual possibilities that have been shown to be consistent with background knowledge and those that just have not been refuted. On this basis, he suggests how to extend standard argument patterns to reasoning under great uncertainty. Instructive examples from various policy fields are provided. To address the challenge of balancing the many and often conflicting reasons that speak for and against various options in a decision he proposes to use the methods described in "Analysing practical argumentation", especially the technique of argument maps.

We have goals on what we want to achieve. These goals regulate the decisions that we make in order to act in their direction. An agent could have a reason to revise her goals, for instance if it turns out to be difficult or entirely impossible to achieve or approach the goal to a meaningful degree. Emission targets to mitigate climate change would be a prominent case in question. However, goals need to have a certain stability to regulate action in a way that contributes to an agent's long-term interests and facilitates cooperation with others. In "Setting and revising goals" *Karin Edvardsson Björnberg* addresses the question when it is rationally justified to reconsider and potentially revise one's prior goals. By analysing an agent's argumentative chain, she identifies achievability- and desirability-related considerations that could provide a prima facie reason to reconsider the goal. Whether there is sufficient reason – all things considered – to revise the goal hinges on additional factors, such as pragmatic, moral and symbolic ones. She uses various examples from both public and personal decisions to show the importance and the challenges of investigating the reasons for and against revising a specified goal.

In "Framing" *Till Grüne-Yanoff* provides a concise introduction to the various aspects of framing. Decision framing in a narrow sense refers to how the elements of a decision problem such as the options or goals are formulated. Framing in a wide sense refers to how a decision problem is structured and how it is demarcated or embedded in a particular context. Grüne-Yanoff surveys some of the experimental evidence of the influence of framing on decision-making. He also describes the dominant descriptive theories and the main attempts that have been made to assess the rationality or irrationality of behaviour sensitive to framing. Two conclusions are especially important: First, different experimental designs elicit quite heterogenous phenomena, and the processes through which framing affects decision-making stay opaque. Secondly, it is not clear whether framing phenomena should be assessed as irrational. This depends on the status of the principle of extensionality as a rationality requirement, a topic that Grüne-Yanoff discusses in detail, using a distinction between semantic equivalence and informational equivalence.

He also points out three ways in which framing is relevant for policy making. First, framing introduces elements of uncertainty into a policy decision. Second, it is used to justify policy interventions intended to correct or prevent irrationality. Finally, framing effects are used to influence behaviour in a desired direction. All this combines to make the analysis of decision framing an important part of argumentative decision analysis.

It is not unusual to postpone decisions, to reconsider provisional decisions later on, or to partition decisions for taking them sequentially. In business for instance, strategies like delaying activities in the supply chain until customer orders have been received or the concept of real options for investments under uncertainty that adapts budgeting in accordance with new information are well-known. In public policy, we find strategies like the moratorium applied to nuclear energy, adaptive governance for ecosystems, and sequential climate policies. However, using these strategies is not always conducive to a rational decision. In "Temporal strategies for decision making" *Gertrude Hirsch Hadorn* discusses the conditions when these temporal strategies are appropriate means to learn about, evaluate, and account for uncertainties in decision making. She proposes four general criteria: the relevance of uncertainties for the decision, the feasibility of improving information on the relevant uncertainties, the acceptability of trade-offs related to the temporal strategy, and the maintenance of governing decision-making over time. These criteria serve as heuristics that need to be specified and weighted for systematically deliberating whether a certain temporal strategy will be successful in improving decision making.

3 Part II: Case Studies

In the case study "Reasoning about uncertainty in flood risk governance" *Neelke Doorn* explores the use in flood risk governance of argumentative strategies such as analysis of framing, temporal strategies, considering goal setting and revising, and making value uncertainty explicit. Flood risk governance is an interesting case of decision making under great uncertainty. There is a broad consensus that the probability and the potential impacts of flooding are increasing in many areas of the world, endangering both human lives and the environment. But in spite of this, the conditions under which flooding occurs are still uncertain in several ways. From the application of argumentative strategies she sketches a tentative outlook for flood risk governance in the twenty-first century, delivering important lessons concerning the distribution of responsibilities, the political dimension of flood risk governance, and the use of participatory approaches in order to achieve legitimate decisions.

The case study "Financial markets: applying argument analysis to the stabilisation task" by *Michael Schefczyk* applies the argument analysis techniques introduced in "Analysing practical argumentation" to Alan Greenspan's justification for the Federal Reserve's inactivity regarding the housing price boom between 2002 and 2005. During the chairmanship of Alan Greenspan, the Federal Reserve

Bank of the United States developed a new approach to monetary policy, which appeared to be highly successful at the time. This approach emphasised the crucial role of uncertainty in monetary policy. Schefczyk reconstructs the argumentative basis of Greenspan's so called "risk management approach". He examines whether monetary policy under Greenspan unduly relied on contested assumptions and whether the Great Recession was a foreseeable consequence of this overreliance, as some economists have argued. Scherczyk identifies more than ten arguments of relevance for this issue, which he structures with the help of argument maps. The central problem appears to be Greenspan's reliance on the stabilising effects of innovative financial instruments that were taken to make it unnecessary to uphold regulatory checks against the potential harmful effects of a housing price reversal. In this case study, argument analysis techniques are used in retrospect to put focus on dubious argumentation. Of course, these techniques may be even more useful in prospective policy analysis.

In the case study "Uncertainty analysis, nuclear waste, and million-year predictions", *Kristin Shrader-Frechette* analyses the information basis for decisions by American authorities on the clean-up of a former nuclear-reprocessing site, contaminated with large amounts of shallow-buried radioactive waste, including high-level waste, some in only plastic bags and cardboard boxes, all sitting on a rapidly eroding plateau. She shows how squeezing a decision under great uncertainty into the format of traditional risk assessment methods has led to biased and severely misleading information, which she calls "special interest science". The ensuing policy failure seems to be the result of faulty characterization, evaluation and management of both factual and value-related uncertainties.

Proposals have been made to deliberately manipulate earth systems, in particular the atmosphere, to cope with climate change. In "Climate geoengineering" *Kevin Elliott* shows how the issues that these proposals give rise to can be structured, analysed and assessed with argumentative methods. He highlights the weaknesses of framing climate geoengineering as an insurance policy or a form of compensation, but he finds the "technical fix" frame less misleading. He provides a structured overview of the ethical questions involved, highlighting the analytical work that is required to clarify them. For instance, he shows that the precautionary principle does not provide sufficient guidance without further specification, and that conceptualizing climate geoengineering as a moral hazard would need further analysis to clarify the precise meaning of that concept. Elliott argues for the use of argumentative strategies to identify the issues that need to be addressed as part of geoengineering governance schemes and to evaluate the procedures used for making governance decisions. For instance, it is not clear whether the concept of informed consent is appropriate for addressing a global issue of this sort.

Synthetic biology has given rise to public controversies long before specific technologies and their possible consequences are on the table for decisions on their use. This is not surprising, since technology shaping living systems, possibly up to creating artificial life, is an ethically sensitive issue. In "Synthetic biology: seeking for orientation in the absence of valid prospective knowledge and of common values" *Armin Grunwald* argues that important lessons can be learned from an analysis of the visionary narratives on synthetic biology. By studying these narratives we can gain a better understanding of the different ways in which the issue of synthetic biology is embedded in social contexts. By combining textual analysis with information on the social context of the narratives we can investigate the social structure of the communication among the various groups involved. All this can serve as a basis for assessing and reconstructing the arguments put forward in this debate. For instance, value uncertainties can be highlighted by making implicit parameters of the issue explicit. Such an analysis can contribute to preventing the fallacy of disregarding possible future consequences that cannot yet be detected.

4 Appendix

Several concepts are needed to characterize the methods proposed in the argumentative turn. In "Ten core concepts for the argumentative turn in policy analysis" *Sven Ove Hansson and Gertrude Hirsch Hadorn* provide short explanations of some of the most important of these concepts. References are given to the chapters where these concepts are introduced and discussed more extensively and used to develop methods and tools for policy analysis.

Chapter 2
Introducing the Argumentative Turn in Policy Analysis

Sven Ove Hansson and Gertrude Hirsch Hadorn

Abstract Due to its high demands on information input, traditional decision theory is inadequate to deal with many real-life situations. If, for instance, probabilities or values are undetermined, the standard method of maximizing expected values cannot be used. The difficulties are aggravated if further information is lacking or uncertain, for instance information about what options are available and what their potential consequences may be. However, under such conditions, methods from philosophical analysis and in particular argumentation analysis can be used to systematize our deliberations. Such methods are also helpful if the framing of the decision problem is contested. The argumentative turn in policy analysis is a widened rationality approach that scrutinises inferences from what is known and what is unknown in order to substantiate decision-supporting deliberations. It includes and recognises the normative components of decisions and makes them explicit to help finding reasonable decisions with democratic legitimacy.

Keywords Argumentation • Argumentative methods for decision support • Deep uncertainty • Expected utility • Fallacy • Great uncertainty • Risk • Uncertainty • Rationality of decisions • Policy analysis

1 A Catalogue of Uncertainties

If life were orderly and easy, making decisions would just be a matter of deciding what you want to achieve, finding out whether there is some way to achieve it and, in that case, choosing accordingly. But life is not orderly or easy. Much to the chagrin of orderly minded people, we have to make most of our decisions without knowing anywhere near what we would need to know for a well-informed decision.

S.O. Hansson (✉)
Department of Philosophy and History, Royal Institute of Technology, Stockholm, Sweden
e-mail: soh@kth.se

G. Hirsch Hadorn
Department of Environmental Systems Science, Swiss Federal Institute of Technology, Zurich, Switzerland
e-mail: hirsch@env.ethz.ch

S.O. Hansson, G. Hirsch Hadorn (eds.), *The Argumentative Turn in Policy Analysis*, Logic, Argumentation & Reasoning 10, DOI 10.1007/978-3-319-30549-3_2

This is true in our personal decisions, such as the choice of education, occupation, or partner. It applies equally to the decisions we make in small groups such as families and workgroups, and to the large-scale decisions in public policy and corporate management.[1] Let us briefly review the major types of lack in knowledge that affect our decisions.

First of all, we often have to make decisions without *knowing whether or not various possible future events that are relevant for our decisions will in fact take place* (Betz 2016). If you decide to spend 3 years in a vocational education programme, will you get the type of job it prepares you for? If you go to Norway on vacation next August, will there be rain? And if you go with your partner, will you quarrel? If the government increases public spending to cope with a recession, will the inflation go out of control?

But it is often even worse than that. In some decisions we are even unable to *identify the potential events that we would take into account if we were aware of them*. Choosing Norway for a vacation trip may have unexpected (both positive and negative) consequences. Perhaps you make new friends there, develop a new hobby, break your leg, or fall victim to swindlers that empty all your bank accounts. In a case like this we tend to disregard such unknown possible consequences since they can occur anytime everywhere.[2] However, there are decisions in which we take unknown possibilities into account (Hansson 2016). Many have moved from the countryside to large cities, more because of the wider range of positive options that they anticipated there than due to any particular, foreseeable such option. On the other hand, we buy insurance not only for protection against foreseeable disasters but also to protect ourselves against calamities we cannot foresee. In large-scale policy decisions, unforeseeable consequences often have a larger role than in private life. In a military context, it would be unwise to assume that the enemy's response will be one of those that one is able to think of in advance. We have considerable experience showing that emissions of chemicals into the environment can have unforeseeable consequences, and this experience may lead us to take measures of caution that we would not have taken otherwise. The issue of unknown consequences seems to be particularly problematic in global environmental issues. Suppose that someone proposes to eject a chemical substance into the stratosphere in order to mitigate the greenhouse effect. Even if all concrete worries can be assuaged, it does not seem irrational to oppose such a proposal solely on the

[1] We use "policy" to refer to "[a] principle or course of action adopted or proposed as desirable, advantageous, or expedient; esp. one formally advocated by a government, political party, etc." (http://www.oed.com; meaning 4d). However, we do not restrict the use of "policy" to public policies only. In this chapter we neither distinguish between "policy analysis" and "decision analysis" nor between "policy/decision analysis" and "policy/decision support". Decisions on policies are normative decisions on whether a course of action is e.g. permissible or mandatory. Therefore, in philosophy, policy decisions are analysed as practical decisions, which means that practical arguments which use normative principles are required in order to justify them (Brun and Betz 2016).

[2] This is a case of the "test of alternative causes", see (Hansson 2016).

ground that it may have consequences that we have not even been able to think of (Betz 2012; Ross and Matthews 2009; Bengtsson 2006). The term "unknown unknowns" for this phenomenon was popularized by the former U.S. Secretary of Defense Donald Rumsfeld (Goldberg 2003).

In most scholarly discussions of decision-making it is assumed that we base our decisions on values or decision criteria that are well-defined and sufficiently precise. In practice that is often not the case; we have to *make decisions without knowing what values to base them on, or how the alternatives for choice compare all things considered* (Möller 2016). For instance, suppose that you are looking for a new flat to rent, and you have several options to choose among. Even if you know everything you wish to know about each of the apartments, the decision may keep you awake at night since you do not know how to weigh different factors such as a quiet location, closeness to public transportation, travel time to your present workplace, a modern kitchen, a large living-room, generous storage facilities, prize, etc. against each other. The situation is similar in many large-scale decisions. For instance, in major infrastructure projects such as the building of a new road there are a sizeable number of predicted consequences, including health effects from air pollution, deaths and injuries from traffic accidents, losses of species due to environmental effects, gains in travel time, economic costs and gains etc. In decisions like these, the uncertainty for many of us is so fundamental that it cannot be decreased by making values explicit and reconstructing them as a coherent system to determine which decision is best. Such a procedure often results in an unreliable ranking not doing justice to the range of values at stake (Sen 1992). Instead, we may face "hard choices" that have to be made in spite of unresolved conflicts between the multiple values involved (Levi 1986).

Not only the consequences, but also the options that we can choose between may be unknown to us. Of course there are decisions with only two or very few options. For instance, a marriage proposal will have to be answered with a "yes" or a "no". But there are also decisions with *(potential) options that are so many or so arduous to evaluate that you could not possibly find and evaluate all of them.* Suppose that you are looking for a nice, small Italian village for a vacation week. A good guidebook will provide you with quite a few alternatives, but of course there are many more. If you want to make sure that you choose the very best village for your purposes, you will probably have to spend much more time in choosing the destination than in actually holidaying there. In this case, the disadvantages of a perfectly well-prepared decision (the "decision costs" in economists' parlance) tend to be so large that we will in practice base the decision on much less information. Similar problems arise in many large-scale decisions. There are many ways to dispose of nuclear waste, and the evaluation of any such method is time- and resource-consuming. Therefore, any proposal for nuclear waste management can be met by demands that it should be further investigated or that additional alternative proposals should be developed and investigated. Such demands may of course be eminently justified, but if repeated indefinitely they may lead to protracted storage in temporary storage facilities that are much more risky than any of the proposed alternatives for permanent disposal. So, while a decision on the embedding of the decision problem is needed to determine the options to be decided on, it is typically uncertain how to appropriately draw the demarcation.

Despite having agreed on the embedding and demarcation of the decision problem, it may be *uncertain how to properly phrase the options for choice* (Grüne-Yanoff 2016). For instance, if a 70 % post-surgery survival chance is re-described as a 30 % risk of dying from this surgery, you may change your mind about undergoing that medical treatment. Or, your attitude to a new technology may depend to some extent on whether this technology is proposed with the goal of "maximizing profit" or with the goal of "increasing efficiency". How the components of a decision problem are formulated or how the problem is presented for choice may have an influence on which of the available options you will go for. Therefore, different ways of framing a decision problem are a further source of uncertainty about policy decision problems.

The structure of the decisions that we have to make should not be taken for given (Hirsch Hadorn 2016). Often we can influence it to a high degree. In particular, we can *divide the mass of decisions we have to make into individual decisions in different ways*. In a restaurant you can decide before the meal exactly what you are going to eat and drink throughout the whole meal. Alternatively, you can first choose an entrée and a main course, and then decide on a dessert only after the main course. There are obvious advantages and disadvantages with both methods (Hammond et al. 1999). As individuals we tend to deal with this flexibility in the delimitation of decisions in different ways; some of us prefer to make plans whereas others tend to improvise as they go along. The same type of issue arises in social decisions. For instance, should a parliament decide on the national budget on a single occasion? Or should it make piecemeal decisions: one decision for each budget area, one decision for each tax or other income? There are usually both advantages and disadvantages attached to different ways to divide a decision complex into individual decisions. Sometimes it can make a big difference if we merge or split up decisions.

We often have to make *a whole series of decisions concerning the same or related subject-matter* (Edvardsson Björnberg 2016; Hirsch Hadorn 2016). When making one of the decisions in such a series (other than the last) we have to make up our minds on how to treat our own future decisions and in particular whether or not we are able to make them in advance and then stick to what we have decided (Hansson 2007; Rabinowicz 2002; Spohn 1977). Suppose that you have made up your mind to go to the gym twice a week the following year. The cheapest way to pay for the exercise is to buy a 12 months gym membership. Paying per month or per visit would be much more expensive. Therefore, at first sight it would seem self-evident that you should buy the 12 months membership. However, you have many more decisions to make concerning your fitness activities: In each of the coming weeks you will have to decide whether to carry out your previous commitment and actually go to the gym. If you end up going there seldom or not at all, it will be much cheaper to buy a ticket for each visit. But on the other hand, paying in advance for the whole year may be a way to bind yourself to your resolution to exercise twice a week. But then, does paying in advance really make a difference in that respect? The decision turns out to be quite complex. Similar complications arise in many other contexts. Often it is an advantage to be able to make a decision once and for all and just carry it through as if the future decision points were not really decision points – this is usually what it takes to stop smoking or carry through a tedious exercise programme. But there are also situations when such resoluteness can lead us wrong. Perseverance in "saving a relationship" has ruined many a woman's life.

Unless you live the life of an eremite, *the effects of most of your decisions are combined in unforeseeable ways with those of others*. There are basically two ways to deal with this: We can try to influence the decisions that others make, and we can try to foresee and adjust to them. Often, we combine both strategies, and so do the other agents who are involved. If you want to make friends with a person, then your success in doing so will depend on a complex interplay of actions by both of you. The same applies if you want to achieve a desired outcome in a negotiation, or if you try to arrange a vacation trip so as to make it agreeable to all participants.

An important class of multi-agent decisions are those in which the agents have contradictory goals (Edvardsson Björnberg 2016). Excellent examples can be found in team sports: How will the other team respond if our team tries to slow down the game at the beginning of the second half time? In the area of security more ominous examples are legion. How vulnerable is the city's water supply to sabotage? Will measures to improve it be counter-productive by spurring terrorists to attack it? If a country improves its air defence, will its potential enemies compensate for this for instance with anti-radiation missiles and stealth technology? In cases like this both sides try both to figure out and to influence how the other side reacts to various actions that they can take themselves. There is no limit to the entanglement.

2 Classifying Uncertainties

In order to develop strategies to deal with this profusion of uncertainties, we need a terminology to distinguish between different types of uncertainties. Perhaps unsurprisingly, there is a fairly standardized terminology for some cases that are reasonably close to the ideal case when the decision-maker has all the information needed for the decision. The terminology for more information-poor decisions is much less clear. Let us begin at the end where we have a standardized terminology.

The case when we have all the relevant information, including what our options are and what outcome will follow after each of them is called decision-making *under certainty*. Obviously there is no full certainty in the real world, but some decisions are so close to it that we can in practice treat them as performed under certainty. The consequences of climbing into the cage of a hungry tiger are known almost for certain, and so are important consequences of watering a dry lawn, pouring an egg into a hot frying pan, or disconnecting a TV from the wall socket.

Table 2.1 Five common meanings of the word "risk" (from Hansson 2011)

Definition of "risk"	Example
An unwanted event which may or may not occur	"Lung cancer is one of the major risks that affect smokers."
The cause of an unwanted event which may or may not occur	"Smoking is by far the most important health risk in industrialized countries."
The probability of an unwanted event which may or may not occur	"The risk that a smoker's life is shortened by a smoking-related disease is about 50 %."
The statistical expectation value of an unwanted event which may or may not occur	"The total risk from this nuclear plant has been estimated at 0.34 deaths per year."
The fact that a decision is made under conditions of known probabilities	"If you choose to place a bet at a roulette table, then that is a decision under risk, not under uncertainty."

The case traditionally counted as closest to certainty is that in which at least some of our options can have more than one outcome, and we know both the values and the probabilities of these outcomes. This is usually called decision-making *under risk*. This terminology is well established but may be somewhat confusing since the word "risk" has several other, more common meanings (See Table 2.1). A more instructive term would be "decision-making under known probabilities", or even better: "decision-making under specified probabilities". A typical case would be gambling at the roulette table. If we have no reason to believe that the wheel has been tampered with, then we can assume that we know the probabilities of each of the outcomes that can follow after each bet that we make. (The term "decision-making under risk" is used irrespectively of how the probabilities are interpreted; they may for instance be taken to be objective probabilities, subjective estimates of objective probabilities, or entirely subjective degrees of belief.)

The next step downwards in information access differs from the previous case only in that we do not know the probabilities, at least not all of them. This is usually called decision-making *under uncertainty*. The distinction between risk and uncertainty is commonly attributed to Frank Knight ([1921] 1935) and J. M. Keynes (1921). In principle the distinction is simple – it is just a matter of whether or not we know the probabilities involved. However, although uncertainty and risk are usually defined in decision theory as two mutually exclusive concepts, it is in practice common to use "uncertainty" as a general term for lack of knowledge, regardless of whether it can be characterized probabilistically or not. (This practice is followed in the IPCC's guidance note on treatment of uncertainty, Mastrandrea et al. (2010)). So, the phrase "uncertainty" is used instead of "risk or uncertainty" (Eisenführ et al. 2010). When we wish to make it clear that we do *not* take "uncertainty" in this broad sense, we can use the phrase "non-probabilistic uncertainty".

In many cases when we do not know (exact) probabilities we nevertheless have some meaningful information about probabilities or likelihood. Even if you do not know the probability that it will rain in London tomorrow you may be confident that it is more likely that it will rain than that it will not, and perhaps you will be sure that the probability of rain is between 60 % and 95 %. We can describe this as a case of partially probabilistic uncertainty, also called "imprecise probability". Cases when we know nothing about how likely the possible outcomes are (more than that their probabilities are above zero) are sometimes called decision-making under *ignorance* (Alexander 1975). (However, some authors reserve the term "ignorance" for decisions where some possible outcomes are unknown (Betz 2010)).

Let us now turn to the terminology for more information-poor decisions, i.e. decisions in which more information is lacking than that about the probability of options. As we saw in the previous section, this may imply several types of information shortage, such as unidentified consequences, undecided values, unidentified options, undetermined demarcation of the decision, unclear connections with later decision on the same subject-matter, and, unforeseeable dependence on decisions by others. In spite of this diversity, all decisions with any of these features are commonly merged into a single category.[3] However, the terminology for that category differs, and so does its more precise definition (if any). In what follows we will consider five of the terms that have been used.

The tools of standard decision theory have been developed for reasonably well-defined problems that are assumed to have a clear solution. In the early 1970s, some scholars in operations research criticised the application of these tools to less well-defined policy problems. They saw the latter type of problems as entirely different and called them "wicked problems" in contrast to the traditional types of problems which they called "tame problems" (Rittel and Webber 1973). They listed ten characteristics of wicked problems:

1. There is no definitive formulation of a wicked problem.
2. Wicked problems have no stopping rule.
3. Solutions to wicked problems are not true-or-false, but good-or-bad.
4. There is no immediate and no ultimate test of a solution to a wicked problem.
5. Every solution to a wicked problem is a "one-shot operation"; because there is no opportunity to learn by trial-and-error, every attempt counts significantly.
6. Wicked problems do not have an enumerable (or an exhaustively describable) set of potential solutions, nor is there a well-described set of permissible operations that may be incorporated into the plan.
7. Every wicked problem is essentially unique.
8. Every wicked problem can be considered to be a symptom of another problem.
9. The existence of a discrepancy representing a wicked problem can be explained in numerous ways. The choice of explanation determines the nature of the problem's resolution.
10. The planner has no right to be wrong (Rittel and Webber 1973).

[3] Some attempts have been made to subdivide this large category. However many of these attempts are philosophically unsatisfactory since they unsystematically mix different criteria for subdivision, such as the source of lack of knowledge and the type of knowledge that is uncertain. "Model uncertainty", for instance, refers to the type of information that is uncertain, namely in this case the model of the decision problem. A model or parts of it could be uncertain for various reasons. One kind of source could be lack of information regarding e.g. parameterizations, the temporal and spatial grid, how to set up the model equations, etc. Another kind of source could be the problem itself, in cases when it is conceived as a system with intrinsic variability as in the case of modeling climate change. For details on model uncertainty in decision support see e.g. Walker et al. (2003).

However, it was not made clear which of these characteristics have to be satisfied in order for a problem to be classified as wicked. The term is poorly defined, and it is also confusing since the primary sense of the word "wicked" refers to an inclination towards wilful wrong-doing, and intentionality cannot be ascribed to problems. What can be considered morally objectionable is treating wicked problems as if they where tame ones (Rittel and Webber 1973; Churchman 1967), since decision makers may be misled by taking such results as solutions to policy problems.

The term "great uncertainty" has been used in various meanings at least since the eighteenth century (E.g.: Locke 1824:xii). In Hansson (1996) an attempt was made to delineate it more precisely. It is essentially a negative term since it refers to cases in which the information required in decision-making under uncertainty, in the usual sense, is not available. The following types and subtypes of great uncertainty were listed:

Uncertainty of demarcation

- Unfinished list of options
- Indeterminate decision horizon

Uncertainty of consequences

- Unknown possibilities

Uncertainty of reliance

- Disagreement among experts
- Unclear who are experts
- General mistrust of experts

Uncertainty of values (Hansson 1996)

In a later paper, "great uncertainty" was defined more succinctly as a situation in which other information than the probabilities needed for a well-informed decision is lacking (Hansson 2004a). Lack of information may include, for instance, unidentified consequences, undecided values, unidentified options, undetermined demarcation of the decision, unclear connections with later decision on the same subject-matter, or unforeseeable dependence on decisions by others.

More recently, the term "deep uncertainty" has been used for decisions on complex problems for which important information about factors other than probabilities is lacking. The concept of deep uncertainty is framed from a systems analysis perspective, and it is used in decision support for topics such as climate change (e.g. Swart et al. 2009; Kandlikar et al. 2005; Lempert et al. 2004). Deep uncertainty covers different sources of uncertainty such as missing or imprecise information, but also disagreement on information, unreliable information and untrustworthy information sources. Furthermore, deep uncertainty also refers to issues that go beyond information about outcomes:

> Deep uncertainty exists when analysts do not know, or the parties to a decision cannot agree on, (1) the appropriate models to describe the interactions among a system's variables, (2) the probability distributions to represent uncertainty about key variables and parameters in the models, and/or how to value the desirability of alternative outcomes. (Lempert et al. 2003:3f)

> Second, climate change is associated with conditions of deep uncertainty, where decision-makers do not know or cannot agree on: (i) the system models, (ii) the prior probability distributions for inputs to the system model(s) and their interdependencies, and/or (iii) the value system(s) used to rank alternatives. (Lempert et al 2004:2)

In the literature it is emphasized that deep uncertainty refers to situations in which the conventional decision-theoretical models are difficult to apply and may not correspond to the needs of decision-makers. Instead, it is proposed that decision-makers will need "adaptive, evolving strategies" (Lempert 2002).

It should be clear from the above that the terms "great" and "deep" uncertainty refer to roughly the same preconditions for decisions, namely those in which the available information is too incomplete for the standard definition of decision-making "under uncertainty". However, there is a difference in emphasis. "Deep uncertainty" has its focus on uncertainties that come into view in attempts to construct models of complex real-world system, whereas "great uncertainty" has more emphasis on uncertainties pertaining to the situation of the decision-maker her- or himself.

Since the beginning of the twenty-first century, the term "black swan" has been used in descriptions of situations partly coinciding with those covered by the terms "great" and "deep" uncertainty. The term was introduced in two books by Nassim Nicholas Taleb (2001, 2007). However, by a "black swan" is not meant a type of decision but a type of events that is difficult to take into account in decision-making, namely events with large effects that come as a surprise but will be believed in hindsight to have been predictable. The notion of a black swan is somewhat related to that of unidentified potential events that was mentioned above in Sect. 1. "Black swans" is a popular terminology, in particular in discussions on the financial sector for which it was first developed. However, we need to consider a broader category of decision situations, including but not limited to unpredictable events with large consequences.

Recently, the term "radical uncertainty" has been proposed to cover various uncertainties that cannot be characterized probabilistically:

> With the notion of radical uncertainty we might mean a number of things. For one, we could be referring to a state of utter cluelessness, in which we have no language to express what we are uncertain about. We can also mean a state of, what may be called, model uncertainty, in which we doubt our modelling assumptions but have insufficient means in the model to express alternative assumptions. And radical uncertainty may refer to an epistemic state in which we have insufficient grasp of our uncertainty regarding a distinct set of propositions. (Romeijn and Roy 2014:1222)

The term has been introduced for new formal approaches that go beyond probability to characterize something like a degree of uncertainty. However, as in the case of deep uncertainty, the emphasis is not on accounting for the range of uncertainties pertaining to the situation of the decision-maker her- or himself.

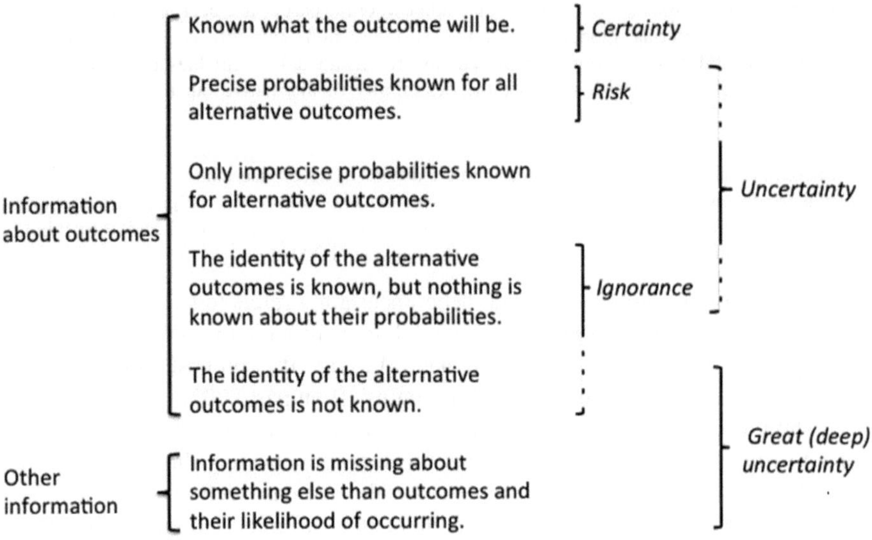

Fig. 2.1 The major types of lack of knowledge in decision-making

So, for the purpose of this book, "radical uncertainty" is not useful as a general term for considering uncertainties.

The terminologies for types of decisions that we have reviewed in this section are summarized in Fig. 2.1. Three of the terms used for uncertainty exceeding that of standard "decision-making under uncertainty", namely "wicked problem", "black swan" and "radical uncertainty" are not included in the figure since they do not demarcate types of decisions. Two of these terms are also unsuitable for philosophical analysis: "wicked problem" is explained in terms of a set of criteria several of which are ill-defined or irrelevant, "black swan" is too limited in scope since it only refers to unforeseen events, and both terms are linguistically misleading. As already indicated, the terms "great uncertainty" and "deep uncertainty" are approximately synonymous. Linguistically we prefer the former term since "deep" connotes something like a one-dimensional extension or high degree, which is unfortunate due to the multidimensionality of the types of uncertainty that we wish to capture. Also "radical uncertainty" does not capture this multidimensionality.

It is important to recognize that there are many types of great uncertainty. The use of a single term to cover them all is of course an oversimplification. Different types of uncertainty may require very different treatments in decision-making practice. Therefore it is often useful and sometimes imperative to distinguish between different types of great uncertainty. We propose that this is best done by reference to the type of decision-relevant information that is lacking: uncertainty about values, uncertainty about demarcation, uncertainty about control etc.

In addition to this classification in terms of what information is lacking, other characterizations of uncertainties can be useful. For instance, it is often helpful to clarify whether different uncertainties in a decision problem can be removed, if the time and resources needed to do so are available, or whether they are irreparable. Some uncertainties can be reduced or eliminated through the collection of more information, whereas others cannot, often since they concern issues that the inherent indeterminacy of complex systems makes inaccessible to human knowledge. Some uncertainties, for instance about values or decision framing may be eliminable through decisions or negotiations, whereas others are not. Therefore, knowing about the sources of uncertainty could be important for decision makers, for instance when considering whether a temporal decision strategy would be appropriate for the decision problem at hand (Hirsch Hadorn 2016). However, information about sources is often missing, for instance, if only a degree of uncertainty is communicated: be this a classification that ranges from "exceptionally unlikely" to "virtually certain" or the assignment of a numerical probability as for example in IPCC's uncertainty assessment for policy-makers that uses both metrics (Mastrandrea et al. 2010).

3 The Reductive Approach

Decision theory is dominated by what can be called a reductive approach to the wide range of information deficiencies and other indeterminate factors that characterize real-life decision-making. The reduction consists in disregarding most types of uncertainties in order to make the decision accessible to a particular type of (elegant and often efficient) formal analysis. It is almost universally assumed in decision theory that the problem to be treated consists in making a single well-determined decision, that the available options and the outcomes that can follow them are well-defined, and that well-determined valuations of the outcomes are available. In combination these assumptions ensure that the decision problem can be represented in the standard formal format of decision theory, namely decision matrices. Furthermore, it is commonly assumed that in some way, all the relevant probabilities are available, which means that the decision problem can be squeezed into the format of decision-making under risk or under specified probabilities, properly speaking.[4]

This approach has an important attraction that should not be underestimated: Once we have managed to express a decision problem as a decision under risk, we have access to an efficient decision-making method that always tells us which

[4] We call the traditional approach of decision theory and policy analysis a reductive approach, because this approach has to disregard most types of uncertainties in order to make the decision accessible to a specific type of formal analysis. The traditional approach is also called "probabilism" (Betz 2016) because it assumes that all relevant probabilities are available.

option is optimal (in a fairly reasonable sense of optimality), given the values that we have incorporated into our description of the problem. The method in question is the *maximization of the expectation value*, also called expected value maximization or expected utility maximization. The term "expected" is statistical jargon for "probability-weighted". What we should maximize, according to this method, is the probability-weighted value of the outcome.

For a very simple example, suppose that monetary outcomes are all that matter. You have won a competition, and as a winner you can choose between two options: Either € 500 in cash, or a lottery ticket that gives you 1 chance in 10,000 of winning € 5,000,000 and 5 chances in 10,000 of winning € 50,000 (and then of course 9994 chances in 10,000 of winning nothing). The expected gain if you choose the cash is of course € 500. The expected gain if you choose the lottery ticket is, in euros:

$$1/10{,}000 \times 5{,}000{,}000 + 5/10{,}000 \times 50{,}000 + 9994/10{,}000 \times 0 = 525$$

According to the maxim of maximizing the expectation value you should choose the lottery ticket. (We assume here, for simplicity, that the value to you of a sum of money is proportionate to that sum. Otherwise, the calculation will be more complex, but the principle is the same.)

In probabilistic risk assessment, this approach is applied to negative outcomes such as fatalities. Since risks are negative events, their expected occurrence has to be minimized instead of maximized, but of course that makes no essential difference. The standard procedure is to determine for each possible outcome both a measure of its disvalue (in other words its severity) and its probability. These two are multiplied with each other, and the values thus obtained are added up for each option in order to determine the risk that is associated with it. Perhaps surprisingly, the number of deaths in an accident is often used as a measure of its severity, thus non-fatal injuries are either disregarded or (more plausibly) assumed to occur in proportion to the number of fatalities. For a concrete example, suppose that two major types of accidents are anticipated if a chemical factory is constructed in a particular way: one type with a probability of 1 in 20,000 that will kill about 2000 persons and another type with a probability of 1 in 1000 that will kill 10 persons. The expected number of fatalities (often confusingly called "the risk") for that factory can then be calculated to be

$$1/20{,}000 \times 2000 + 1/1000 \times 10 = 0.11$$

i.e. 0.11 fatalities.

Another area in which expectation values are used is cost-benefit analysis (CBA). This is a decision-aiding methodology (or rather, set of methodologies). It is based on the fairly uncontroversial assumption that when preparing a decision, we should weigh advantages against disadvantages. It is also based on the more controversial assumption that this should be done by assigning monetary values to all potential outcomes. In a typical CBA, two or more options in a public decision are compared to each other by adding up the monetary values assigned to their respective consequences. The value of an uncertain outcome is obtained as an expectation value, thus a chance of 1 in 100 of saving € 1,000,000 is treated in the same way as a certain gain of € 10,000. If the loss of a life is assigned the value of € 10,000,000, then a risk of 1 in 1000 that two persons will die corresponds to a loss of

$$1/1000 \times 2 \times €10{,}000{,}000 = €20{,}000,$$

and this is then often taken to be the highest economic cost that is defensible to avoid such a risk. Cost-benefit analysis is much more comprehensive than probabilistic risk assessment. It can be applied to in principle any social decision, as long as we can identify the possible outcomes and assign both probabilities and monetary values to all of them.

4 Problems with the Reductive Approach

Given the immense complexity of many human decisions, we need to simplify and to prioritize among the aspects involved, and it will often be necessary to leave out some aspects in order to focus more on others. This is what the reductive approach does, and in principle it is also what it should do. However, for many purposes it does not do it well enough. Each of the aspects discussed in Sect. 1 is of paramount importance in some decisions but easily negligible in others. Therefore we need mechanisms to pick out the important aspects, which are different in different decisions. The reductive approach always selects the same few aspects and always neglects all the others even in cases in which they are of paramount importance (Hansson 2016). In this section we are going to show how this can create problems for decision-makers.

4.1 Unknown Probabilities

In order to calculate useful expectation values, we need reasonably reliable probability estimates. In some cases, these estimates can be based on empirically known frequencies. As one example, death rates at high exposures to asbestos are known from epidemiological studies. In most cases, however, the basis for probability estimates is much less secure. This applies for instance to the failure probabilities of new technological constructions, and also to the probabilities of most societal events. When probabilities cannot be estimated from empirically known frequencies, the standard method is to instead use experts' estimates of probabilities. The reliability of decision analysis will then depend on the assumption that there is a good correlation between objective probabilities and experts' estimates of these probabilities.[5]

However, this assumption is not correct. Experimental studies indicate that the probability estimates of experts (and of everyone else) are biased and highly unreliable. Like the rest of us, experts tend to underestimate the difference between small probabilities (Tversky and Kahneman 1986). We tend to "see" (or correctly estimate) the difference between a 10 % probability of disaster and a 1 % probability of disaster, but are much less adroit at distinguishing between the probabilities or likelihoods of events whose real probabilities are for instance 0.001, 0.0001, 0.00001, or 0.000001. Obviously, this tendency to treat small probabilities as all the same can be seriously misleading. If we base a decision on an expert's estimate that the probability of some serious type of accident is 0.000001, but it is really 0.0001, then we find ourselves with a 100 times higher expected damage than what we expected. Furthermore, experts tend to be overconfident, i.e. they believe that their own probability estimates are more reliable than what they really are (Morgan 2011). In summary, the practice of treating expert's estimates of unknown probabilities in the same way as the probabilities we actually know (approximatively) from actual experience can lead us to systematically make the wrong decisions (Hansson 2016).

The common tendency in the decision-supporting disciplines to proceed as if reasonably reliable probability estimates were available for all possible outcomes has been called the tuxedo fallacy (Hansson 2009a). It consists in treating all decisions as if they took place under epistemic conditions analogous to gambling at the roulette table, rather than under the conditions of the jungle, in which you do not know beforehand what situations you may find yourself in, what you can do, or what consequences you need to take into account, not to speak of the probabilities of those consequences.

4.2 *Counterproductive Probabilities*

In some cases it can be counterproductive to think in terms of numerical probabilities since doing so can make us deemphasize important aspects of the decision. For instance, in many countries about half of all marriages end in divorce (Salmela-Aro

[5] Many attempts have been made to represent uncertainties in somewhat more resourceful formal structures such as probability intervals, second-order probabilities etc. Some of these methods provide a better representation of some aspects of (epistemic) uncertainty than what classical probabilities can do. However, they obviously cannot capture the many other indeterminate factors in complex decisions such as uncertainties about values, about the demarcation of the decision and about its relationship to other decisions by the same or other agents. There is also a trade-off: the richer a formal representation is and the more it deviates from traditional probability functions, the more difficult is it to use it in unequivocal decision rules such as (adapted versions of) expected utility maximization.

et al. 2006; Raley and Bumpass 2003). If all spouses in the country based their degree of commitment to the marriage on this probability, then the frequency of divorce might well be still higher. For a person wanting to avoid divorce, an attempt to improve the odds might be more useful than a strategy that takes the probability for given. The same applies to many other decisions. When making plans for a joint vacation, it does not seem advisable to make probability estimates of your companions' reactions to different proposals. It would be more useful to interact with them with the purpose of finding a plan that is agreeable to all of you. The participants in formal negotiations for instance between companies or governments are often in a similar situation. There is an abundance of situations in which a successful decision-maker will not be one who takes it for given what her own options are and how other agents are inclined to act, and estimates the probabilities of various outcomes, based on that information. Instead we should expect the successful decision-maker to be one who tries to change the initial conditions of the decision, for instance by developing new and better options and by communicating gainfully with others in order to influence the ways in which they will act (Edvardsson Björnberg 2016; Hirsch Hadorn 2016).

4.3 Undetermined Values

In order to calculate expectation values, we also need to have well-determined values for all the relevant outcomes. This is usually easy for single-criterion decisions. A physician may be looking for the treatment that gives the patient the largest chances of survival. The evaluation of different treatments can then be based exclusively on the expected number of remaining life years. A person wanting to reduce her weight can make a decision on what to eat for dinner based only on the aim of weight reduction. And of course a CEO may have such a strong focus on profits that monetary income is all that counts in the evaluation of different options for the company.

However, many, arguably most, decisions that we make have a more complex value structure (Möller 2016). In the parlance of decision theory they are multi-dimensional and cannot easily be reduced to a single dimension. This can easily be exemplified with decisions in our private lives. When you decide where to go for a vacation, or choose what apartment, car, computer, or sofa to buy, you typically have a mental list of requirements that you want to satisfy (and almost invariably, keeping down the costs is one of these criteria). During the decision process, you may add new items to the list and remove others. Your priorities among the items are typically vague and can also change during the decision process. The situation is very similar in large societal decisions. As mentioned in Sect. 1, a decision on an infrastructure project such as the building of a new road or railway will involve a large number of aspects such as reductions in travel time, environmental damage, lives gained (or lost) due to changed accident rates, economic costs and gains, and many others. The reductive approach requires that we reduce (translate) all of these aspects into one and the same category or dimension, and furthermore that this dimension allows for numerical measurement. In practice that dimension is always money, and consequently the unit of measurement is some monetary unit such as dollars or euros. When this reduction has been performed, all conflicts between different aspects can be solved by comparisons in terms of monetary cost or gain. To achieve such a reduction, conversion factors that express the values of human lives, the preservation of species, etc. in monetary terms are determined. It is assumed that these conversion factors should be the same for all decisions within a jurisdiction. This means for instance that the relative weights assigned to reductions in travel time and reduced death toll in traffic are decided beforehand for the different decisions to be made in the transport sector. It also means that the same "value of life" is used in all areas of decision-making.

Unfortunately these conversion factors have no tenable ethical foundations (Hansson 2007; Heinzerling 2000, 2002). Strong arguments can be made that for instance human lives and monetary gains or losses are incommensurable, i.e. they cannot be measured in the same unit. If a hi-fi system has a monetary price, then this means that you can buy it at that price and then do what you want with it, for instance destroy it. If a monetary value is assigned to the loss of a human life, then that does not imply that someone can buy that person, or the right to kill her, for that price. In short, these "life values" are not prices in the economic sense. Unfortunately, no fully satisfactory answer seems to be available to the question what these monetary values represent when they do not represent prices. A common answer is that they represent willingness to pay, but they can only do so in an idealized way that does not seem to have direct empirical correlates.

4.4 Counterproductive Values

Just as for probabilities, it can in some instances be counterproductive to think in terms of predetermined values for all decision outcomes. The reason for this is that doing so may engender thought patterns that have negative consequences. This has commonly been said about the assignment of monetary value to human lives. Even though we cannot pay an indefinite amount of money to save a human life, assigning a precise sum of money to it may send a message that can be conceived as desecrating the value of life (Hampshire 1972). This was exemplified when Working Group III of the Intergovernmental Panel on Climate Change claimed in their second report that human lives differ in their monetary "value", since national circumstances including opportunity costs differ greatly between developing and developed countries. This proposal was strongly contested with the argument that differences in the "values" of human lives are not morally acceptable (Brun and Hirsch Hadorn 2008).

There are also cases when it may be morally inadvisable to know beforehand what values one would apply in a decision. For an example from private life, consider a father who has two children. It is far from unthinkable that he may one day find himself in a terrible situation: Both their lives are threatened and he can save one but not both of them. It is to be hoped that if this happens, he will manage to choose one of them rather than letting them both die. However, it does not seem to be an advantage for him to know beforehand whom he would choose. Such knowledge might be an indication of emotional problems in relation to the child he would not save. (The example is based on William Styron's novel *Sophie's Choice*, Styron 1979). This is an individual predicament, but similar arguments can be made about social decisions in extreme situations. It is conceivable that in a disastrous pandemic, a country's healthcare system would have to deprioritize certain groups (such as the very old). But in a normal situation, members of these groups have the same priority as everyone else. A prior decision about which groups to deprioritize in an extreme emergency could most likely have a negative social impact. This is a reason not to make such decisions until they are really needed (Hansson 2012). In conclusion, we have good reasons not to base all decisions on predetermined values. In many decisions, the development of values and decision criteria is an essential part of the decision process up to its very end. It does not seem to be an advantage to replace that process by decision-making based on values that were developed before the specific decision arose (Hansson 2016).

4.5 Interpersonal Valuation Issues

In a traditional probabilistic risk assessment, it makes no difference how risks are distributed. A case in which 500 persons are subjected to a risk of 1 in 1000 of dying has to be treated in the same way as one in which 5000 persons are subjected to a risk of 1 in 10,000 of dying. In both cases, the expected number of additional fatalities is 0.5. Similarly in a cost-benefit analysis, all costs and all benefits are combined in one and the same balance. (Both in probabilistic risk assessment and cost-benefit analysis, supplementary distributional analyses are sometimes performed, but the total summing up is still the primary approach.) This means that a disadvantage affecting one person can be fully compensated for by an advantage for some other person. According to this type of reasoning it would make no difference if you expose yourself to a risk in order to obtain an advantage for yourself or instead expose someone else to the same risk in order to obtain the same advantage for yourself. In this way, both the main versions of the reductive approach are impersonal; persons do not matter other than as carriers of the goods and evils that are summed up the moral calculus (Hansson 2004c). Both methods aim at determining whether a disadvantage is acceptable per se, rather than whether it is acceptable to expose the persons to it who are actually to be exposed. An alternative approach would be to treat the actual risk-exposures of individual persons, rather than an abstract sum of their effects, as the primary issue for ethical deliberation (Hansson 2003, 2013).

The reductive approach conforms with a utilitarian way of thinking, but our moral thinking does not necessarily have to follow utilitarian patterns, which, for instance, have difficulties in accounting for moral rights or requirements of fair distribution. It usually makes a big difference for our moral evaluation who will receive the advantages of a decision and who will receive the disadvantages. If there is a group that would receive large disadvantages, without receiving any share of the advantages, then that could be reason enough to reject the proposal, without paying much attention to the total sum of advantages accrued to others. This would apply in particular if the disadvantaged group has a morally important right that is violated by the proposal. For instance, an infrastructure project with serious negative effects on the reindeer husbandry of the aboriginal Sami people in Sweden could legitimately be rejected on the basis of their indigenous rights, without much consideration of its potential advantages for other parts of the country's population.

4.6 The Choice of a Decision Rule

As should now be obvious, in many cases we lack the information about options, outcomes, probabilities and values that would be needed to calculate and maximize expectation values. But in the cases when we have that information, or acceptable proxies for it, should we then maximize expectation values? There are at least two strong reasons why this need not always be the case. One of these reasons is that we sometimes have to give priority to the interests and rights of individual persons who are particularly affected by a decision. For example, suppose that we have to choose, in an acute situation, between two ways to repair a serious gas leakage in the machine-room of a chemical factory. One of the options is to send in the repairman immediately. (There is only one person at hand who is competent to do the job.) He will then run a risk of 0.9 to die due to an explosion of the gas immediately after he has performed the necessary technical operations. The other option is to immediately let out gas into the environment. In that case, the repairman will run no particular risk, but each of 10,000 persons in the immediate vicinity of the plant runs a risk of 0.001 to be killed by the toxic effects of the gas. The maxim of maximizing expectation values requires that we send in the repairman to die. But it would be difficult to criticize a decision-maker who refrained from maximizing expectation values (minimizing expected damage) in this case in order to avoid what would be unfair to a single individual and infringe the rights of that person (Hansson 1993:24).

The other reason is that it cannot be taken for granted that the moral impact of a potential outcome is proportionate to its probability. In policy discussions the avoidance of very large catastrophes, such as a nuclear accident costing thousands of human lives, is often given a higher priority than what is warranted by the statistically expected number of deaths. Critics have maintained that serious events with low probabilities should be given a higher weight in decision-making than what they receive in a model based on the maximization of expectation values (Burgos and Defeo 2004; O'Riordan et al. 2001; O'Riordan and Cameron 1994). Such risk-averse or cautious decision-making has strong popular support, not least in environmental issues. Furthermore, reasonable arguments have been given why risk aversion can be evolutionarily advantageous (Okasha 2007, 2011). It does not seem to be a good idea to choose a framework for decision support that excludes risk aversive decision-making.

To all this we can add further features of actual decision-making discussed in Sect. 1 that we cannot account for with the reductive approach: repeated decisions, uncertainty about one's present control over one's own future decisions, indeterminate delimitation of the decision, and combination effects with decisions by other decision-makers. In many if not most of the decision problems of real life other factors than probabilities and values are so important that they need to be taken into account.

5 Introducing the Argumentative Turn

We hope to have shown that traditional decision theory, with its high demands on information input, is inadequate to deal with many real-life decisions since they have to be based on much less information. Does this mean that we have no means to decision support in such cases? No, it is not quite as bad as that. There is help to get, but it comes from somewhat surprising quarters. Recently philosophers have shown how methods from philosophical analysis and in particular argumentation analysis can be used to systematize discussions about policy issues involving great uncertainty. This is a "widened rationality approach",[6] that scrutinises inferences from what is known and what is unknown for the decision at hand. It recognises and includes the normative components and makes them explicit. This is what we mean by the *argumentative turn* in decision support and uncertainty analysis.

The argumentative turn includes a large and open-ended range of methods and strategies to tackle the various tasks that come up with the analysis of a decision problem. It comprises tools for conceptual analysis and for structuring procedures as well as for the analysis and assessment of arguments. Compared to the reductive approach, the argumentative approach is pluralistic and flexible, since it does not squeeze a decision problem into a standard format in order to make a particular type of calculation possible. The argumentative approach is a rational approach in a wider sense, since the analytical tools are used to clarify and assess reasons for and against options (Brun and Betz 2016).

Argumentative methods and strategies extend the rational treatment of decisions in traditional decision theory in two respects. Firstly, they can be used to clarify the grounds for the application of formal methods of traditional decision theory and policy analysis. In this way, argumentative methods provide justificatory

[6] Since we use "rationality" in a wider sense for decisions under great uncertainty and not in the restricted sense of traditional decision theory, we also use terms like "reasonable" and "sound" for the normative assessment of decisions.

prerequisites for the application of the reductive approach when it is appropriate (Hansson 2013:74–80). Secondly, when the reductive approach is inapplicable or in need of supplementation, argumentative methods and strategies can be used to replace it or to cover the aspects that it leaves out.

The argumentative approach goes beyond traditional approaches to policy analysis since it includes a pluralistic analysis of the normative issues involved in a decision (Hansson 2016; Möller 2016) as well as criteria for rational goal setting and goal revision (Edvardsson Björnberg 2016). For instance, argumentative methods analyse what would follow from the application of various decision principles for the problem at hand (Brun and Betz 2016; Betz 2016). In contrast, traditional approaches to policy support are restricted to descriptive information from empirical investigation or computer simulations. This restriction also holds for investigations of which values and norms are held by whom in society (see e.g. Walker et al. 2003). Obviously, such treatments are often essential components of decision support, and they provide indispensable inputs to argumentative analysis (Schefczyk 2016; Shrader-Frechette 2016; Doorn 2016), but normative analysis is also a necessary part of the deliberations that should precede a difficult decision (Elliott 2016; Grunwald 2016; Doorn 2016; Shrader-Frechette 2016).

The argumentative approach to policy analysis has to be distinguished from discourse analysis. Discourse analysis is defined as "the study of language in use" (Gee and Handford 2012:1). It includes a family of heterogenous approaches used in linguistics and various social sciences for "studying language in the context of society, culture, history, institutions, identity formation, politics, power and all the other things that language helps us to create and which, in turn render language meaningful in certain ways and able to accomplish certain purposes." (Gee and Handford 2012:5) A first direction in discourse analysis takes the position of an outsider to comment on positions and their interactions in a policy debate. For a synthesis of discourse analyses of environmental politics see e.g. Hajer and Versteeg (2005). A second direction in discourse analysis approaches policy analysis from the normative perspective of communicative ethics, elaborating on criteria for participation and deliberation. Dryzek, for example, argues in this line:

> The defensibility of policy analysis, and planning depends on the conditions in which arguments are made, received, and acted upon. I therefore conclude with a discussion of the radicalization of the argumentative turn which involves a rational commitment to free democratic discourse. (Dryzek 1993:214)

However, neither of these directions in discourse analysis enters into assessing the arguments to substantiate a rational argumentation about the issue at hand (Hajer and Versteeg 2005:175). Therefore, they cannot replace the methods of argumentative analysis. But they can certainly provide useful inputs to the reconstruction of arguments, e.g. by informing about how the decision problem is embedded. Furthermore, they may help to set up and guide procedures for deliberation on decision problems, which have been structured and analysed with argumentative methods.

The value of systematizing normative discussions about policy issues is perhaps most easily seen in the many cases when argumentation analysis can be used to reveal fallacies in reasoning about risk and uncertainty (Hansson 2004b, 2016). Indeed, such fallacies are quite common, and exposing them can be an important step towards more intellectually supportable decisions. We will mention just a few examples:

- It has sometimes been argued that exposure to a pesticide should be accepted since the ensuing risk is smaller than the risk of being hurt by a meteorite falling down on one's head, a risk that we in practice accept. However, we do not have a choice between pesticides and meteorites, and neither do have a reason to accept all risks that are smaller than some risk that we cannot avoid. Therefore, this argument is a fallacy.
- It is often argued that various risks are acceptable since they are natural. This is a petitio principii that uses the ambiguity of the term "natural" as both a descriptive and a positively evaluative term. It does not take much reflection to see that many natural risks (in the descriptive sense of the term) are far from acceptable, and that major endeavours are justified to avert them or mitigate their effects.
- Expert opinion or expert consensus is often taken to be the criterion for acceptability of risks. This is a fallacy in two respects: First, experts may be mistaken. Secondly, scientific expertise does not cover all aspects of risk acceptability, in particular not the ethical aspects (Hansson 2013).

But argumentative analysis in decision support can take us further than that. It is not only a means for the negative task to uncover fallacies, but also a means for the positive task to indicate what is needed for better substantiating decisions. We can use argumentation analysis for instance to better understand the uncertainties involved in decisions, to prioritize among uncertain dangers, to determine how decisions should be framed, to clarify how different decisions on interconnected subject-matter relate to each other, to choose a suitable time frame for decision-making, to analyse the ethical aspects of a decision, to systematically choose among different decision options, and not least to improve our communication with other decision-makers in order to co-ordinate our decisions. We believe that argumentation analysis is particularly useful in democratic decision-making. Democracy works not only by voting, but also requires rational communication, negotiations, compromises and active participation in order to achieve its purpose (Hansson and Oughton 2013). Therefore the goal of decision support should be to help making reasonable decisions with democratic legitimacy. Democratic legitimacy of decisions requires that arguments and their conclusions are reasonable from more than one perspective. To be reasonable from more than one perspective, democratic legitimacy cannot result simply from an aggregative approach, but requires deliberative procedures (Peter 2009). Argumentative analysis is a means for better substantiating deliberation to achieve democratic legitimacy of decisions.

This book is the first comprehensive survey of the argumentative approach to decision analysis and uncertainty management. It contains chapters discussing the various components of that approach, including its normative aspects.[7] In addition it presents a series of case studies in which these kinds of methods are applied to policy decision problems. We would like to conclude this introduction with a plea for pluralism in decision analysis. Our purpose is not to replace one attempted panacea by another but to open up for a wide range of decision-guiding methodologies. Needless to say, methods not treated in this book, such as mathematical representations of uncertainty, can also contribute to decision support (Hansson 2008, 2009b). One of the advantages of the argumentative turn is that argumentation is a wide enough concept to cover a plurality of approaches to decision support.

References

Alexander, E. R. (1975). The limits of uncertainty: A note. *Theory and Decision, 6*, 363–370.

Bengtsson, L. (2006). Geo-engineering to confine climate change: Is it at all feasible? *Climatic Change, 77*, 229–234. doi:10.1007/s10584-006-9133-3.

Betz, G. (2010). What is the worst case? The methodology of possibilistic prediction. *Analyse & Kritik, 32*, 87–106.

Betz, G. (2012). The case for climate engineering research: An analysis of the "arm the future" argument. *Climatic Change, 111*, 473–485. doi:10.1007/s10584-011-0207-5.

Betz, G. (2016). Accounting for possibilities in decision making. In S. O. Hansson & G. Hirsch Hadorn (Eds.), *The argumentative turn in policy analysis. Reasoning about uncertainty* (pp. 135–169). Cham: Springer. doi:10.1007/978-3-319-30549-3_6.

Brun, G., & Betz, G. (2016). Analysing practical argumentation. In S. O. Hansson & G. Hirsch Hadorn (Eds.), *The argumentative turn in policy analysis. Reasoning about uncertainty* (pp. 39–77). Cham: Springer. doi:10.1007/978-3-319-30549-3_3.

Brun, G., & Hirsch Hadorn, G. (2008). Ranking policy options for sustainable development. *Poiesis & Praxis, 5*, 15–30. doi:10.1007/s10202-007-0034-y.

Burgos, R., & Defeo, O. (2004). Long-term population structure, mortality and modeling of a tropical multi-fleet fishery: The red grouper epinephelus morio of the Campeche Bank, Gulf of Mexico. *Fisheries Research, 66*, 325–335. doi:10.1016/S0165-7836(03)00192-9.

Churchman, C. W. (1967). Wicked problems. Guest editorial. *Management Science, 14*, B141–B142.

Doorn, N. (2016). Reasoning about uncertainty in flood risk governance. In S. O. Hansson & G. Hirsch Hadorn (Eds.), *The argumentative turn in policy analysis. Reasoning about uncertainty* (pp. 245–263). Cham: Springer. doi:10.1007/978-3-319-30549-3_10.

Dryzek, J. S. (1993). Policy analysis and planning: From science to argument. In F. Fischer & J. Forrester (Eds.), *The argumentative turn in policy analysis and planning* (pp. 213–232). London: University College London Press.

Edvardsson Björnberg, K. (2016). Setting and revising goals. In S. O. Hansson & G. Hirsch Hadorn (Eds.), *The argumentative turn in policy analysis. Reasoning about uncertainty* (pp. 171–188). Cham: Springer. doi:10.1007/978-3-319-30549-3_7.

Eisenführ, F., Weber, M., & Langer, T. (2010). *Rational decision making*. Berlin: Springer.

Elliott, K. C. (2016). Climate geoengineering. In S. O. Hansson & G. Hirsch Hadorn (Eds.), *The argumentative turn in policy analysis. Reasoning about uncertainty* (pp. 305–324). Cham: Springer. doi:10.1007/978-3-319-30549-3_13.

[7] The normative aspects are most extensively discussed in Brun and Betz (2016), Hansson (2016), Möller (2016), Betz (2016), and Edvardsson Björnberg (2016).

Gee, J. P., & Handford, M. (2012). Introduction. In J. P. Gee & M. Handford (Eds.), *The Routledge handbook of discourse analysis* (pp. 1–6). London: Routledge.
Goldberg, J. (2003). The unknown. The C.I.A. and the Pentagon take another look at Al Qaeda and Iraq. *The New Yorker*. http://www.newyorker.com/magazine/2003/02/10/the-unknown-2. Accessed 21 May 2015.
Grüne-Yanoff, T. (2016). Framing. In S. O. Hansson & G. Hirsch Hadorn (Eds.), *The argumentative turn in policy analysis. Reasoning about uncertainty* (pp. 189–215). Cham: Springer. doi:10.1007/978-3-319-30549-3_8.
Grunwald, A. (2016). Synthetic biology: Seeking for orientation in the absence of valid prospective knowledge and of common values. In S. O. Hansson & G. Hirsch Hadorn (Eds.), *The argumentative turn in policy analysis. Reasoning about uncertainty* (pp. 325–344). Cham: Springer. doi:10.1007/978-3-319-30549-3_14.
Hajer, M., & Versteeg, W. (2005). A decade of discourse analysis of environmental politics: Achievements, challenges, perspectives. *Journal of Environmental Policy & Planning, 7*, 175–184. doi:10.1080=15239080500339646.
Hammond, J. S., Keeney, R. L., & Raiffa, H. (1999). *Smart choices. A practical guide to making better decisions*. Boston: Harvard Business School Press.
Hampshire, S. (1972). *Morality and pessimism*. Cambridge: Cambridge University Press.
Hansson, S. O. (1993). The false promises of risk analysis. *Ratio, 6*, 16–26. doi:10.1111/j.1467-9329.1993.tb00049.x.
Hansson, S. O. (1996). Decision-making under great uncertainty. *Philosophy of the Social Sciences, 26*, 369–386.
Hansson, S. O. (2003). Ethical criteria of risk acceptance. *Erkenntnis, 59*, 291–309.
Hansson, S. O. (2004a). Great uncertainty about small things. *Techne, 8*, 26–35.
Hansson, S. O. (2004b). Fallacies of risk. *Journal of Risk Research, 7*, 353–360. doi:10.1080/1366987042000176262.
Hansson, S. O. (2004c). Weighing risks and benefits. *Topoi, 23*, 145–152.
Hansson, S. O. (2007). Philosophical problems in cost-benefit analysis. *Economics and Philosophy, 23*, 163–183. doi:http://dx.doi.org/10.1017/S0266267107001356.
Hansson, S. O. (2008). Do we need second-order probabilities? *Dialectica, 62*, 525–533. doi:10.1111/j.1746-8361.2008.01163.x.
Hansson, S. O. (2009a). From the casino to the jungle. Dealing with uncertainty in technological risk management. *Synthese, 168*, 423–432. doi:10.1007/s11229-008-9444-1.
Hansson, S. O. (2009b). Measuring uncertainty. *Studia Logica, 93*, 21–40. doi:10.1007/s11225-009-9207-0.
Hansson, S. O. (2012). The trilemma of moral preparedness. *Review Journal of Political Philosophy, 9*, 1–5.
Hansson, S. O. (2013). *The ethics of risk. Ethical analysis in an uncertain world*. Basingstoke: Palgrave Macmillan.
Hansson, S. O. (2016). Evaluating the uncertainties. In S. O. Hansson & G. Hirsch Hadorn (Eds.), *The argumentative turn in policy analysis. Reasoning about uncertainty* (pp. 79–104). Cham: Springer. doi:10.1007/978-3-319-30549-3_4.
Hansson, S. O., & Oughton, D. (2013). Public participation – potential and pitfalls. In D. Oughton & S. O. Hansson (Eds.), *Social and ethical aspects of radiation risk management* (pp. 333–346). Amsterdam: Elsevier Science.
Hansson, S, O. (2011). Risk. *Stanford Encyclopedia of Philosophy*. http://plato.stanford.edu/entries/risk/. Accessed 21 May 2015.
Heinzerling, L. (2000). The rights of statistical people. *Harvard Environmental Law Review, 24*, 189–207.
Heinzerling, L. (2002). Markets for arsenic. *Georgetown Law Journal, 90*, 2311–2339.
Hirsch Hadorn, G. (2016). Temporal strategies for decision making. In S. O. Hansson & G. Hirsch Hadorn (Eds.), *The argumentative turn in policy analysis. Reasoning about uncertainty* (pp. 217–242). Cham: Springer. doi:10.1007/978-3-319-30549-3_9.

Kandlikar, M., Risbey, J., & Dessai, S. (2005). Representing and communicating deep uncertainty in climate-change assessments. *Comptes Rendus Geoscience, 337*, 443–455. doi:10.1016/j.crte.2004.10.010.

Keynes, J. M. (1921). *A treatise on probability*. London: Macmillan.

Knight, F. H. ([1921] 1935). *Risk, uncertainty and profit*. Boston: Houghton Mifflin.

Lempert, R. J. (2002). A new decision sciences for complex systems. *PNAS, 99*, 7309–7313.

Lempert, R. J., Popper, S. W., & Bankes, S. C. (2003). *Shaping the next one hundred years. New methods for quantitative, long-term policy analysis*. Santa Monica: Rand.

Lempert, R. J., Nakicenovic, N., Sarewitz, D., & Schlesinger, M. (2004). Characterizing climate-change uncertainties for decision-makers. An editorial essay. *Climatic Change, 65*, 1–9.

Levi, I. (1986). *Hard choices. Decision making under unresolved conflicts*. Cambridge: Cambridge University Press.

Locke, J. (1824). *The works of John Locke in nine volumes* (12th ed., Vol. 7). London: Rivington.

Mastrandrea, M. D., Field, C. B., Stocker, T. F., Edenhofer, O. Ebi, K. L., Frame, D. J., et al. (2010). Guidance Note for Lead Authors of the IPCC Fifth Assessment Report on Consistent Treatment of Uncertainties. Intergovernmental Panel on Climate Change (IPCC). http://www.ipcc.ch. Accessed 20 Aug 2014.

Möller, N. (2016). Value uncertainty. In S. O. Hansson & G. Hirsch Hadorn (Eds.), *The argumentative turn in policy analysis. Reasoning about uncertainty* (pp. 105–133). Cham: Springer. doi:10.1007/978-3-319-30549-3_5.

Morgan, M. G. (2011). Certainty, uncertainty, and climate change. *Climatic Change, 108*, 707–721. doi:10.1007/s10584-011-0184-8.

Okasha, S. (2007). Rational choice, risk aversion, and evolution. *The Journal of Philosophy, 104*, 217–235.

Okasha, S. (2011). Optimal choice in the face of risk: Decision theory meets evolution. *Philosophy of Science, 78*, 83–104. doi:10.1086/658115

O'Riordan, T., & Cameron, J. (Eds.). (1994). *Interpreting the precautionary principle*. London: Earthscan.

O'Riordan, T., Cameron, J., & Jordan, A. (Eds.). (2001). *Reinterpreting the precautionary principle*. London: Cameron May.

Peter, F. (2009). *Democratic legitimacy*. New York: Routledge.

Rabinowicz, W. (2002). Does practical deliberation crowd out self-prediction? *Erkenntnis, 57*, 91–122.

Raley, R. K., & Bumpass, L. L. (2003). The topography of the divorce plateau: Levels and trends in union stability in the United States after 1980. *Demographic Research, 8*, 245–260.

Rittel, H., & Webber, M. (1973). Dilemmas in a general theory of planning. *Political Science, 4*, 155–169.

Romeijn, J.-W., & Roy, O. (2014). Radical uncertainty: Beyond probabilistic models of belief. *Erkenntnis, 79*, 1221–1223. doi:10.1007/s10670-014-9687-9.

Ross, A., & Matthews, H. D. (2009). Climate engineering and the risk of rapid climate change. *Environmental Research Letters, 4*, 045103. doi:10.1088/1748-9326/4/4/045103.

Salmela-Aro, K., Aunola, K., Saisto, T., Halmesmäki, E., & Nurmi, J.-E. (2006). Couples share similar changes in depressive symptoms and marital satisfaction anticipating the birth of a child. *Journal of Social and Personal Relationships, 23*, 781–803. doi:10.1177/0265407506068263.

Schefczyk, M. (2016). Financial markets: The stabilisation task. In S. O. Hansson & G. Hirsch Hadorn (Eds.), *The argumentative turn in policy analysis. Reasoning about uncertainty* (pp. 265–290). Cham: Springer. doi:10.1007/978-3-319-30549-3_11.

Sen, A. (1992). *Inequality reexamined*. Harvard: Harvard University Press.

Shrader-Frechette, K. (2016). Uncertainty analysis, nuclear waste, and million-year predictions. In S. O. Hansson & G. Hirsch Hadorn (Eds.), *The argumentative turn in policy analysis. Reasoning about uncertainty* (pp. 291–303). Cham: Springer. doi:10.1007/978-3-319-30549-3_12.

Spohn, W. (1977). Where Luce and Krantz do really generalize Savage's decision model. *Erkenntnis, 11*, 113–134.

Styron, W. (1979). *Sophie's choice*. New York: Random House.

Swart, R., Bernstein, L., Ha-Duong, M., & Petersen, A. (2009). Agreeing to disagree: Uncertainty management in assessing climate change, impacts and responses by the IPCC. *Climatic Change, 92*, 1–29. doi:10.1007/s10584-008-9444-7.

Taleb, N. N. (2001). *Fooled by randomness: The hidden role of chance in life and in the markets*. London: Texere.

Taleb, N. N. (2007). *The black swan: The impact of the highly improbable*. New York: Random House.

Tversky, A., & Kahneman, D. (1986). Rational choice and the framing of decisions. *The Journal of Business, 59*, S251–S278.

Walker, W. E., Harremoës, P., Rotmans, J., van der Sluijs, J. P., van Asselt, M. B. A., Janssen, P., & Krayer von Krauss, M. P. (2003). Defining uncertainty. A conceptual basis for uncertainty management in model-based decision support. *Integrated Assessment, 4*, 5–17. doi:10.1076/iaij.4.1.5.16466.

Part II
Methods

Chapter 3
Analysing Practical Argumentation

Georg Brun and Gregor Betz

Abstract Argument analysis is a powerful tool for structuring policy deliberation and decision-making, especially when complexity and uncertainty loom large. Argument analysis seeks to determine which claims are justified or criticized by a given argumentation, how strong an argument is, on which implicit assumptions it rests, how it relates to other arguments in a controversy, and which standpoints one can reasonably adopt in view of a given state of debate. This chapter first gives an overview of the activities involved in argument analysis and discusses the various aims that guide argument analysis. It then introduces methods for reconstructing and evaluating individual arguments as well as complex argumentation and debates. In their application to decisions under great uncertainty, these methods help to identify coherent positions, to discern important points of (dis)agreement, as well as to avoid spurious consensus and oversimplification.

Keywords Practical reasoning • Argument analysis • Reconstruction • Argument mapping • Uncertainty • Argumentation schemes

1 Introduction

When experts derive policy recommendations in a scientific report, they set forth arguments for or against normative claims; they engage in practical reasoning – and so do decision-makers who defend the choices they have made, NGOs who argue against proposed policy measures and citizens who question policy goals in a public consultation. Practical reasoning is an essential cognitive task that underlies policy making and drives public deliberation and debate.

Unfortunately, we are not very good at getting practical arguments right. Intuitive practical reasoning risks to suffer from various shortcomings and fallacies as

G. Brun (✉)
Institute of Philosophy, University of Bern, Bern, Switzerland
e-mail: Georg.Brun@philo.unibe.ch; Georg.Brun@ethik.uzh.ch

G. Betz
Institute of Philosophy, Karlsruhe Institute of Technology, Karlsruhe, Germany
e-mail: gregor.betz@kit.edu

S.O. Hansson, G. Hirsch Hadorn (eds.), *The Argumentative Turn in Policy Analysis*, Logic, Argumentation & Reasoning 10, DOI 10.1007/978-3-319-30549-3_3

soon as a decision problem becomes a bit more complex – for example in terms of predictive uncertainties, the variety of outcomes to consider, the temporal structure of the decision problem, or the variety of values that bear on the decision (see Hansson and Hirsch Hadorn 2016). Hence we need to analyse policy arguments and to make explicit which scientific findings and normative assumptions they presume, how the various arguments are related to each other and which standpoints the opponents in a debate may reasonably hold.

Although argumentation does not provide an easy route to good decisions in the face of great uncertainty, the *argumentative turn* builds on the insight that substantial progress can be made with the help of argument analysis.[1] Consider, for example, the following text which is listed as an argument against "nuclear energy" in *Pros and Cons. A Debater's Handbook*:

> In the 1950s we were promised that nuclear energy would be so cheap that it would be uneconomic to meter electricity. Today, nuclear energy is still subsidised by the taxpayer. Old power stations require decommissioning that will take 100 years and cost billions. (Sather 1999:257)

It is unclear which claim(s) this professed argument is supposed to attack or support, and maybe even more so, in which way it is supposed to do so. Analysis is needed to make the reasoning more specific and to reveal its hidden assumptions. In general, we expect that argument analysis can help us understand which aspects of a decision challenge are crucial, and in what respects and why we disagree. Does a disagreement concern the truth or the relevance of some premises? Or rather which conclusion they support or about how strong the argument is? Clarity in such matters is important, not least because there is always a danger that policy debates lead to a spurious consensus on an ill-defined position all parties interpret in favour of their own views.[2]

If argument analysis should be of help in answering the questions mentioned and provide the desired clarity, it must provide reconstructions. It must start with the arguments that are put forward in a debate and try to represent them as clearly as possible in a form which allows for an effective evaluation. This is a task which differs not only from scientific research into the subject matter of the debate, but also from discourse analysis; that is, from empirical research which aims at describing and structuring the views and arguments different people put forward or subscribe to in a debate. As a reconstructive enterprise, argument analysis has both a descriptive goal, inasmuch as it deals with the arguments people actually use, and a normative perspective. This means that reconstructions of arguments are

[1] An "argumentative turn" in policy analysis and planning had first been proclaimed by Fisher and Forester (1993), who called for putting more emphasis on deliberative and communicative elements in decision making (see also Fischer and Gottweis 2012). We conceive of our chapter, and this book in general, as a genuinely normative, argumentation-theoretic contribution to – and extension of – the programme of an argumentative turn, which was so far mainly shaped by the perspectives of political science and empirical discourse analysis.

[2] For examples, see Singer (1988:157–9).

guided by the goal of making the given argumentation as clear as possible and by standards for evaluating arguments: premises can be right/true or wrong, arguments can be valid or invalid, strong or weak.

As a reconstructive enterprise, argument analysis is also not opposed to traditional decision theoretic reasoning. Quite the contrary, what has been said about argument analysis is true of applied decision theory as well: it is essentially a method for reconstructing and evaluating practical reasoning. But traditional decision theory is confined to problems which exhibit only a very limited range of uncertainty, namely unknown or not precisely known probabilities of outcomes (see Hansson and Hirsch Hadorn 2016). And it is restricted to a specific though important type of reasoning, so-called consequentialist arguments. Relying on traditional decision theory therefore also means *systematically* ignoring other kinds of practical arguments that may be set forth in order to justify policy conclusions. For this reason we suggest to conceive of argument analysis as the more general, more unbiased and hence more appropriate method for decision analysis, which incorporates the insights of traditional decision theory just as far as consequentialist arguments are concerned and the preconditions for its application are met.

In Sect. 2, we start with a brief survey of the various tasks involved in argument analysis, the aims guiding argument analysis and the uses to which argument analysis may be put. Section 3 then introduces the basic techniques for analysing individual arguments and discusses the most common problems. On this basis, we sketch an approach to analysing complex argumentation and debates in Sect. 4, while Sect. 5 addresses strategies for dealing with the specific challenges of analysing reasoning involving practical decisions under uncertainty.

Argument analysis is a lively field of research and the argumentative turn is no systematic, monolithic theory, but includes a plurality of approaches and methods. We therefore add the caveat that this chapter is neither a presentation of textbook-methods nor an overview of the available approaches, it is rather an opinionated introduction to analysing practical reasoning.[3]

2 Tasks, Aims and Uses of Argument Analysis

This section sets the stage for further discussion by giving a overview of argument analysis. We identify a range of tasks involved in argument analysis, give an account of the aims guiding argument analysis, and then briefly comment on the various uses which may be made of argument analysis. On the basis of this general overview, the subsequent sections discuss the individual tasks in more detail and with reference to examples.

[3] We freely draw on our earlier work, specifically Brun (2014), Brun and Hirsch Hadorn (2014), Betz (2013), and Betz (2010).

2.1 *Tasks of Argument Analysis*

Argument analysis, understood in a wide sense, involves two basic activities: reconstruction and evaluation of argumentation and debates.

Reconstruction of argumentation and debates comprises a range of tasks which take argumentative texts as inputs and return various representations as outputs. Roughly, one can distinguish the following activities of reconstruction:

- Text analysis: extract debates and arguments from texts.
- Debate Analysis: determine how the argumentation of different proponents relate to each other.[4] For example, does A's argument support or attack B's argument or position?
- Argument analysis in a narrow sense: break down complex argumentation into individual arguments and their relations. For example, identify attack and support relations between arguments, or distinguish "hierarchical" argumentation, in which one argument supports a premise of another argument, from "multiple" argumentation, in which several arguments support the same conclusion.[5]
- Analyse individual arguments and recast them in standardized form as inferences[6]: determine which premises and which conclusion are given; reformulate unclear, incomplete and nonuniform sentences; supply missing elements.

In this chapter, we discuss these tasks in reverse order and we take the analysis of debates and complex argumentation together since on a basic level debates and complex argumentation are analysed in the same way.

Each of these tasks not only involves the identification of some argumentative structure but also its representation in a form which supports the goals of the reconstruction, especially the aim of enhancing clarity. For both, analysis and representation, a broad range of tools are available, ranging from informal guidelines to formal languages and software support (see the resources listed at the end of this chapter).

It is important to note that the above list of reconstructive tasks is not to be read as implying that the activity of reconstructing has a simple sequential structure. Although the list can be used as a rough guide to reconstructing, the various tasks constitute neither a linear and nor a uniquely determined sequence of steps. They are rather (partly) interdependent, and backtracking and looping strategies will frequently be called for. One reason is that, in general, several competing reconstructions may be on offer in each and every step of analysis. This constantly

---

[4] We use "debate" in a sense which does not necessarily involve more than one person. One can "internalize" proponents of various positions and explore how they can argue against each other.

[5] Sometimes "serial" or "subordinate" are used in place of "hierarchical", and "convergent" in place of "multiple". See Snoeck Henkemans (2001) a survey on terminology and basic structures of complex argumentation.

[6] We use "inference" as a technical term for completely explicit and well-ordered arguments.

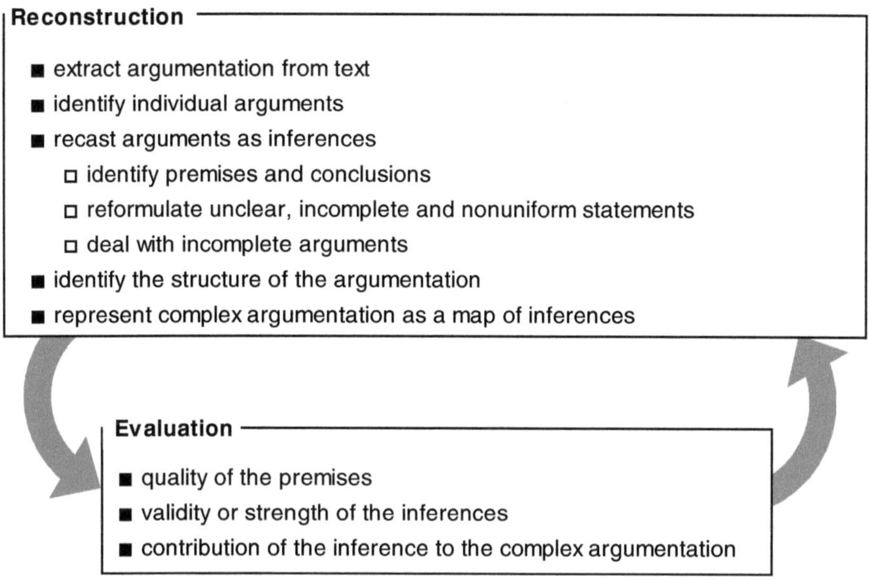

Fig. 3.1 Interplay of reconstruction and evaluation in argument analysis (Adapted from Brun and Hirsch Hadorn 2014:209)

requires taking decisions which need to be made with a perspective to the other reconstructive tasks. Another reason is that each subsequent step of reconstruction will identify additional structure, which may prompt us to revise or refine an "earlier" step. If, for example, the analysis of individual arguments uncovers ambiguities, this will often motivate exploring alternative reconstructions of the overarching complex argumentation. As we will shortly see, the reconstruction of an argumentation is also intertwined with its evaluation. The practical upshot is that reconstructing requires a strategy of trial and error, going back and forth between reconstruction and evaluation as well as between reconstructing individual arguments and more complex structures (see Fig. 3.1). Since all this requires creativity rather than following a predefined procedure, new ideas are always possible and consequently, the analysis of a realistically complex argumentation is an open-ended undertaking.

Speaking of "reconstruction" should also help to avoid, right from the beginning, the misunderstanding that argument analysis is just a matter of uncovering a given but maybe hidden structure. As the discussions below will make clear, argument reconstruction is an activity based on and relative to some theoretical background, it involves creative and normative moves, and it aims at coming up with representations of arguments that meet certain standards the original texts typically fail to comply with, for example, full explicitness. This fits well with the term "reconstruction", which refers to a construction guided by a pre-existing object or situation, in our case an argumentation.

Let us now turn from reconstruction to evaluation. A comprehensive evaluation of arguments and complex argumentation involves assessing a whole range of qualities. The following may be distinguished:

- Truth and acceptability of the premises of individual arguments.
- Validity or strength of individual arguments: does the truth of the premises guarantee or at least provide good reasons for the truth of the conclusion? Valid arguments with true premises are called "sound".
- Overall evaluation of a complex argumentation: is the argumentation valid or strong in view of the validity or strength of its component-arguments? Does the argumentation contain "gaps"?
- Contribution of arguments to a complex argumentation, debate, discussion or inquiry ("dialectical relevance").
- Coherence of a position (as characterized by an argumentation).
- Contribution of argumentation and debates to solve a problem, for example, a decision task.

Not all of these aspects can be addressed by argument analysis alone. Most importantly, assessing the truth of the claims involved is subject to other kinds of research in, for example, empirical sciences or ethics.

For some of these evaluations, extensive theoretical treatments are available. Logical theories, for example, make it possible to prove validity, the theory of dialectical structures can be used to effectively assess which position can be consistently adopted in a debate, and argumentation theory provides extensive treatments of fallacies; that is, of common patterns of invalid, weak, irrelevant, misleading or otherwise problematic arguments. Using some of these resources requires taking additional, non-trivial steps of reconstruction, such as formalizing inferences in order to prove their validity with the help of some logical theory.

2.2 *Aims and Guiding Perspectives*

Argument analysis may be done in the service of all kinds of practical or theoretical goals, but it always operates between two pulls. On the one hand, argument analysis is an interpretational undertaking dealing with some given argumentation, which it is therefore committed to take serious. On the other hand, argument analysis aims to represent the argumentation at hand as clearly as possible, evaluate it, and identify problems and potential for improvement. These two orientations open up a spectrum from exegetical to exploitative argument analysis (Rescher 2001:60), from argument analysis which aims at understanding as accurately as possible an author's argumentation to argument analysis which seeks to find the best argumentation that can be constructed following more or less closely the line of reasoning in some given argumentative text.

The exegetical aspect implies that reconstructions must answer to hermeneutic principles, especially accuracy (sometimes called "loyalty"[7]) and charity. "Accuracy" means that a reconstruction must be defensible with respect to the argumentative text, in particular its actual wording and the available information about its context. Charity calls for reconstructing an argumentation under the defeasible presumption that it performs well with respect to validity, soundness and the other evaluative dimensions mentioned above. In particular, charity is a "tie-breaker" if there are alternative, equally accurate interpretations. It requires, other things being equal, to select the most favourable interpretation. This makes sure that an unfavourable evaluation of an argument is not merely the result of interpretative mistakes of even malevolence. Charity is also a basic reason why reconstruction and evaluation are intertwined in argument analysis.

However, reconstruction is also guided by the fundamental aim of clarification. This ideal comprises three core aspects: explicit, precise and transparent representation. Explicitness not only requires that the relation between individual arguments in a complex argumentation be represented explicitly, but also that the individual arguments are framed as inferences, which implies that all premises and the conclusion are made explicit and formulated as self-contained statements. "Precision" is not used in its numerical sense, but means that argument reconstruction needs to address ambiguity, context-dependence and vagueness in a way which makes sure that they do not lead to misevaluation of the arguments at hand. Transparency, finally, calls for representing debates, complex argumentations and individual arguments in a way that makes it easy to grasp their structure and get an overview.[8]

In short, reconstruction means representing argumentation in a form which ensures that its structure is represented explicitly, precisely and transparently. Since these aspects of clarity as well as the hermeneutic principles of accuracy and charity may be partly antagonistic, trade-offs are often inevitable. And in such cases, deciding whether a proposed reconstruction is adequate requires judgement rather than applying a purely formal procedure. And in many cases more than one resolution of conflict, favouring different reconstructions, may be plausible.

2.3 *Uses of Argument Analysis*

The core function of arguing is to provide reasons for a claim, but arguments – even the same argument – may be put to very different uses. One may strive to identify supporting reasons as a means to, for example, support some statement, attack a position, resolve whether to accept a controversial claim, reach consensus on some

[7] See Walton (1996:211–6); for a more comprehensive discussion of hermeneutical principles in the context of argument analysis see Reinmuth (2014).

[8] On various aspects of clarification see also Morscher (2009:1–58) and Hansson (2000).

issue, shake an opponent's convictions or explore the consequences of adopting a certain position. Argument analysis by itself does not directly realize such aims, neither does it necessarily lead to better arguments. However, it may prove effective as a means to

- reflect on one's own reasoning and that of others; for example, by becoming more clearly aware of all the premises involved, of the exact relations between the constituents of a complex argumentation, or of the strengths and weaknesses of an argumentation;
- identify promising revisions of a position; for example, eliminate problematic premises or strengthen an argument by resorting to a weaker conclusion or by adding supporting premises;
- identify promising moves in a debate; for example, identify premises that could be used to support a position, finding arguments that may force an opponent to modify her position or identify arguments that can help to find a consensus.

3 Analysing Individual Arguments

In this section, we illustrate many aspects of argument analysis with the help of an argument from Singer's *Animal Liberation* and a passage from Harsanyi, in which he criticizes John Rawls's appeal to the maximin principle in *A Theory of Justice* (Rawls 1999). For the sake of exposition, we give comparatively meticulous reconstructions for these two untypically transparent examples (square brackets are used for cross-references and to indicate important changes to the original text):

> [**Singer**] So the researcher's central dilemma exists in an especially acute form in psychology: either the animal is not like us, in which case there is no reason for performing the experiment; or else the animal is like us, in which case we ought not to perform on the animal an experiment that would be considered outrageous if performed on one of us. (Singer 2002:52)

(1.1) Either the animal is not like us or else the animal is like us.
(1.2) If the animal is not like us, there is no reason for performing the experiment.
(1.3) If the animal is like us, we ought not to perform on the animal an experiment that would be considered outrageous if performed on one of us.
(1.4) [There is no reason for performing the experiment or we ought not to perform on the animal an experiment that would be considered outrageous if performed on one of us.]

> [**Harsanyi**] Suppose you live in New York City and are offered two jobs at the same time. One is a tedious and badly paid job in New York City itself, while the other is a very interesting and well paid job in Chicago. But the catch is that, if you wanted the Chicago job, you would have to take a plane [. . .]. Therefore there would be a very small but positive probability that you might be killed in a plane accident. [. . .]

> [3.2] The maximin principle says that you must evaluate every policy available to you in terms of the *worst possibility* that can occur to you if you follow that particular policy. [. . .] [2.1] If you choose the New York job then the worst (and, indeed, the only) possible outcome will be that you will have a poor job but you will stay alive. [. . .] In contrast, [2.2] if you choose the Chicago job then the worst possible outcome will be that you may die in a plane accident. Thus, [2.4/3.1] the worst possible outcome in the first case would be much better than the worst possible outcome in the second case. Consequently, [3.3] if you want to follow the maximin principle then you must choose the New York job. [. . .]
>
> Clearly, this is a highly irrational conclusion. Surely, if you assign a low enough probability to a plane accident, and if you have a strong enough preference for the Chicago job, then by all means you should take your chances and choose the Chicago job. (Harsanyi 1975:595)

(2.1) The worst possible outcome of the option *New York* is having a poor job.
(2.2) The worst possible outcome of the option *Chicago* is a dying in a plane accident.
(2.3) [Having a poor job is much better than dying in a plane accident.]
(2.4) The worst possible outcome of [the option *New York*] is much better than the worst possible outcome of [the option *Chicago*].
(3.1) The worst possible outcome of the option *New York* is much better than the worst possible outcome of the option *Chicago*. [=2.4]
(3.2) [Given two options, the maximin principle says that you must choose the one the worst possible outcome of which is better than the worst possible outcome of the other.]
(3.3) [The maximin principle says that] you must choose the option *New York*.

3.1 Basics of Reconstruction

A reconstruction of an individual argument takes an argumentative text as its input and aims at delivering an inference as its output. The guiding principles are the hermeneutic maxims of accuracy and charity as well as the ideal of clarity with its aspects of explicitness, precision, and transparency. In principle, the reconstruction proceeds by employing four basic types of operations: elements which do not contribute to the argument, for example, digressions and purely rhetoric embellishments, are deleted, unclear statements are reformulated, premises and conclusion are rearranged into a standard form, and missing elements, such as (parts of) a premise or the conclusion are added.

The first task is to find argumentative elements in a text. In argumentative passages, one or more statements are treated as providing a reason for a further statement (and this in turn may be done in the service of any of the many uses to which arguments can be put; see Sect. 2). Hence, the criterion which decides whether some element of a text is part of an argument is functional. Being a premise or a conclusion is not a matter of the form or the content of a sentence, but a role a statement can play, just like being an answer. Identifying arguments in a text therefore presupposes at least a rough understanding of the structure of the text. A well-tested strategy is to start by sketching the main argument(s) in a passage in one's own words and as succinctly as possible. For [Harsanyi] that could be (of course, many other formulations are equally plausible at this stage of analysis):

(4) The worst possible outcome of the option *Chicago* (dying in a plane accident) is much worse than the worst possible outcome of the option *New York* (a poor job). Therefore, according to the maximin principle you must choose the option *New York*.

One can then turn to the analysis of individual arguments, and tackle the problem of identifying the premises and the conclusion. In practice, this is not just a matter of applying formal techniques. "Indicator words" such as "therefore", "thus", "because" and many more are certainly worth paying attention to, but they cannot be used as simple and reliable guides to an argument's structure. It is usually best to try to identify a conclusion (which may not be stated explicitly) and then actively search for premises, also with the help of hypotheses about what would make for a good argument. A functional perspective provides the guide for this search: what would fit what we already have found out or assumed about the argument at hand? What makes sense in light of the complex argumentation or the debate the argument is part of? (Betz 2010:§ 99; Sect. 4 below). In [Harsanyi], we know (from the context) that Harsanyi wants to attack Rawls's use of the maximin principle and specifically the claim that one should take the New York job. Hence the conclusion of (4) is a good starting point.

Once some premises or a conclusion are identified, they must typically be reformulated for the sake of clarity. Explicitness requires that all premises and the conclusion must be specified as a complete, independently comprehensible sentence. This is of special importance if more than one element of an argument are given in one sentence. In extracting individual premises or a conclusion from such sentences, the result must be spelled out as a full sentence, which usually means that some anaphoric expressions (expressions used in such a way that their interpretation depends on the interpretation of other expressions, e.g. relative pronouns, or "first case" and "second case" in 2.4) must be replaced by expressions which can be independently interpreted.

A second aspect of clarity is precision. Eliminating ambiguity, context-dependence and vagueness altogether is neither realistic, nor necessary for the purposes of argument analysis. But certain problems call for reformulation. Concerning ambiguity and context-dependence, premises and conclusions must firstly be represented in a way which avoids equivocation; that is, the use of corresponding instances of the same expression with different meanings. In [Singer], for example, an equivocation would result if "is like us" did not refer to the same aspects of likeness in its two occurrences; reconstruction (1) assumes that this is not the case. Some of these problems can be tackled by employing, or if necessary introducing, a standardized terminology (e.g. restricting "risk" to known probabilities; see Hansson and Hirsch Hadorn 2016). Secondly, syntactical ambiguity needs to be resolved, for example, different readings of scope ("Transportation and industry contribute 20 % to the US greenhouse gas emissions."). Thirdly, context-dependent, for example, indexical ("I", "this", "here", "now", ...) and anaphoric ("Harsanyi quotes Rawls before *he* criticizes *him*."), expressions, must be replaced if there is a danger that their interpretation might not be clear in the resulting representation of the argument. In practice, the necessary reformulation of premises and conclusion is often greatly facilitated by introducing notational aids such as brackets or subscripts (e.g. "$risk_1$" for known probabilities of outcomes, "$risk_2$" for unwanted outcomes).

Argument analysis will also sometimes uncover vagueness; that is, expressions for which there are "borderline-cases" cases in which it is unclear whether the expression applies although the meaning of the expression is clear. Vagueness is a pervasive and to a large extent unproblematic feature of natural language expressions, but it can have the undesired effect that the truth of a sentence featuring a vague expression cannot be assessed. However, if reducing vagueness is necessary, this task cannot be handled with the resources of argumentation theory alone. Deciding in which way statements should be made more exact is rather a matter of considerations relating to the subject matter of the argument at hand.

The goal of transparency, the third aspect of clarity, means that it should be easy to recognize the meaning of every sentence in an inference as well as its logical structure and, more generally, any structure relevant to argument evaluation with respect to, for example, the strength of individual arguments or the coherence of a position. Key factors of transparency are abbreviation, simplicity and uniformity of expression, easily graspable symbols and a direct correlation between features of the representation and features of the argument which are relevant to its evaluation. In practice, all this boils down to representing debates, argumentations, inferences and individual sentences in standardized forms which are easily grasped.

Transparency is therefore to a considerable degree a matter of selecting appropriate tools for representing inferences. Examples range from the format *premises – inference bar – conclusion* (as in 1–3) and visualizations (e.g. Fig. 3.2) to logical languages (e.g. $\neg p \vee p$; $\neg p \rightarrow q$; $p \rightarrow r \Rightarrow q \vee r$ for (1)[9]). While the former are readily graspable, logical formulas become cognitively efficient only after some initial training.

On an informal level, streamlining formulations is nearly always of pivotal importance. This includes eliminating superfluous elements (e.g. purely illustrative examples), simplifying needlessly complex phrasing, introducing informal abbreviations, introduce standard expressions for marking out logical structure

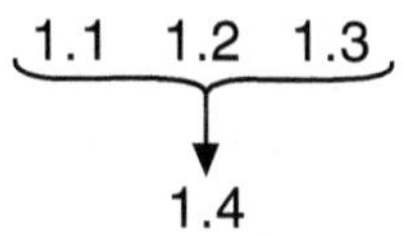

Fig. 3.2 Alternative representation of inference (1) reconstructed from [Singer]

[9] With *p* corresponding to "the animal is like us", *q* to "there is no reason for performing the experiment" and *r* to "we ought not to perform on the animal an experiment that would be considered outrageous if performed on one of us."

(e.g. "and" instead of "but", "not acceptable" instead of "inacceptable") and especially eliminating stylistic variations, for example, by replacing expressions which are synonymous in the context at hand by one and the same. In the examples (1)–(3), the most extensive reformulation is (3.2), which replaces Harsanyi's casual formulation of the maximin principle by a more precise one.

3.2 Dealing with Incomplete Arguments

A certain type of incomplete arguments, so called enthymemes, are responsible for notorious problems of argument reconstruction. Enthymemes are arguments which are weak in the form in which they have been put forward, but merely because a premise or the conclusion has been "left implicit". Such arguments are extremely common because efficient transmission of information with the help of relatively few explicit expressions is a basic trait of natural language communication. This favours leaving unexpressed what can be assumed as easily understood anyway. Enthymemes are arguments which exploit this feature of natural language communication by not explicitly stating a premise or the conclusion.[10] Accordingly, not all incomplete or otherwise weak arguments count as enthymemes, but only those which can more or less readily be completed in a way which can be assumed to go without saying in the context at hand.

In what follows, we introduce the traditional approach to deal with incomplete arguments by supplying premises or a conclusion.[11] This approach is motivated by the goal of explicitness and guided by the hermeneutic principles of accuracy and charity, which, however, are antagonistic in this context. Charity speaks in favour of reconstructing an inference that can be positively evaluated and accuracy in favour of respecting the actual wording of an argument. Adding a premise or a conclusion will therefore have a price in accuracy even if it is charitable.[12] Importantly, charity and accuracy come in degrees, can be traded off against each other, and often more than one candidate for completing an argument will remain

[10] Of course, reconstructing enthymemes does not rest on the highly dubious idea that *all* implicit information should be made explicit. Even complete arguments virtually always involve a great deal of presuppositions. That the premise "The 2-degree-target can no longer be achieved", as well as its negation, imply "Reaching the 2-degree-target is not impossible at every point in time" does not mean that the latter sentence should be reconstructed as an additional premise.

[11] In fact, missing conclusions are often neglected in the literature.

One alternative to the traditional approach relies on argument schemes and adds the elements needed to turn the argument at hand into an instance of such a scheme (Paglieri and Woods 2011). Another idea is to interpret arguments against the background of a belief-state ascribed to its author and deal with "incomplete" arguments by revising the ascribed belief state (Brun and Rott 2013).

[12] This presupposes that charity is interpreted as a presumptive principle, not merely a tie-breaker.

As Jacquette (1996) has pointed out, adding a premise is in some cases less charitable than strengthening a premise or weakening the conclusion.

plausible. Exercising judgement rather than applying a formal procedure is needed for assessing the alternative suggestions and deciding which one to select.

Both, the notion of an enthymeme and the appeal to charity are linked to the evaluation of arguments. Hence reconstruction and evaluation are intertwined in dealing with enthymemes. Considerations of deductive validity or non-deductive strength (to be discussed below) go into judging whether an argument counts as an enthymeme and in which ways it may be completed.

When reconstructing enthymemes by adding a conclusion, the leading consideration is whether a sentence can be found which turns the given enthymeme into a strong argument and which suits the conclusion's role in its dialectical context. Specifically, the argument resulting from adding a conclusion should fit into the complex argumentation, which it is part of according to the analysis in progress. If, for example, an argument is thought to constitute an attack on another argument, its conclusion may be expected to be incompatible with one of the latter's premises; if it is thought to be part of a hierarchical complex argumentation, its conclusion is expected to be effective as a premise of another argument (e.g. 2.4 and 3.1). In the example [Singer], the context in *Animal Liberation* strongly suggests a conclusion which speaks against experimenting on animals. In practice, the search for prospective conclusions can be facilitated by checking out whether the given premises fit an argumentation scheme; that is, a commonly used pattern of arguing (see Walton et al. 2008). For example, the reconstruction (1) and specifically the added conclusion (1.4) are guided by the idea (suggested by Singer) that this argument can be reconstructed as instantiating one of the standard schemes of dilemmas. For practical arguments, the decision principles discussed in Sect. 5 can be used as a heuristic guide.

For adding premises, the leading consideration is that one or more sentences need to be found which yield a strong argument and which can be defended as acceptable and more or less obvious *relative to their dialectical context*. The question is not whether the author of the argument or of the reconstruction actually finds the prospective premise acceptable or obvious, but whether it can be assumed to have these qualities in the context in which the argument at hand is supposed to provide a reason for its conclusion. This may well be a position an author is attacking or discussing, rather than endorsing herself. For example, since Harsanyi refers to Rawls's position, the added premises (2.3) needs to be acceptable to Rawls in the described fictional situation, not to Harsanyi. As a practical strategy (see van Eemeren and Grootendorst 2004:3, 117), one may start with the "logical minimum" as a candidate for the additional premise. For deductive arguments, this is a sentence of the form "If [the given premises], then [the given conclusion]". For non-deductive arguments, two strategies are available. One can either try to find a weakest[13] premise which yields a non-deductively strong argument, or one can convert the argument at hand into an equivalent deductive one with a

[13] Sentence S is logically stronger than sentence T (and T is logically weaker than S) just in case S implies T but not vice versa.

weakened premise and investigate which additional premises are needed for such a conversion. For both strategies, argumentation schemes may be used as a heuristic tool.

Once a candidate for a reconstruction has been found, one has to decide whether the supplementary premises can plausibly be ascribed to a proponent of the relevant position. This may not be the case for two reasons. If the premise is inacceptable to the proponent because it is too strong, the argument cannot be dealt with as an enthymeme, but must be evaluated as weak. However, a premise can also be implausible because it is too weak. Typically this is due to problematic implicatures; that is, claims not implied but suggested by the prospective premise in virtue of communicative principles (van Eemeren and Grootendorst 1992:ch. 6). In such cases, a stronger premise may yield a more adequate reconstruction. The logical minimum for (3) in [Harsanyi], for example, would be (3.2*), which is much less plausible than (3.2) as a premise expressing the maximin principle:

(3.2*) If the worst possible outcome of the option *New York* is much better than the worst possible outcome of the option *Chicago*, then the maximin principle says that you must choose the option *New York*.

Two important general points need be noted. The hypothesis that an argument is an enthymeme is, of course, defeasible. Hence, reconstructing incomplete arguments can take different routes. Either a complete inference can be reconstructed which can be defended in light of the hermeneutic principles and the specific considerations discussed, or else one may conclude that the argument presented is just weak, or even resolve that it is unclear what it is supposed to be an argument for. Secondly, there may be several ways in which an enthymeme can be reconstructed as a complete inference, each fitting into a different reconstruction of the complex argumentation at hand. Selecting a best reconstruction is than a matter of an overall judgement.

3.3 *Evaluation of Arguments*

Arguments can be evaluated in (at least) three respects: the quality of their premises, the strength of the relation between premises and conclusion, and the argument's contribution to the complex argumentation which it is part of. In this section, we focus on the first two perspectives; the third is discussed in Sect. 4. All these evaluations address inferences, and therefore presuppose that at least a tentative reconstruction of the argument at hand has been carried out.

With respect to the quality of the premises, the question whether they are true is obviously of central interest. In general, it cannot be answered by argument analysis but calls for investigation by, for example, perception, science or ethics. The main exceptions are inconsistencies that can be detected by logical or semantical analysis which shows that the logical form or the meaning of a set of premises guarantees that they cannot all be true.[14] Inferences involving an inconsistent set of premises are negatively evaluated since they cannot perform the core functions of arguments; they provide no reason in favour of the conclusion. However, arguments with an inconsistent set of premises are relatively seldom found. Much more common are inconsistencies arising in the broader context of a complex argumentation, when a proponent endorses an inconsistent set of sentences (see Sect. 4). Plainly, truth and consistency must be distinguished from acceptability since we do not live in a world in which people accept all and only true sentences (in such a world, there would be little need for arguments). Premises must therefore also be evaluated with respect to whether they are acceptable in their dialectical context. If, for example, an argument is supposed to convert an opponent or to undercut[15] its position (as in Harsanyi's argumentation against Rawls), its premises must be acceptable to the opponent, irrespective of whether they are acceptable to the proponent or the author of the argument. Again, this is a matter that needs to be assessed in the course of analysing the broader argumentative context.

The second perspective from which arguments are evaluated focuses on the relation between the premises and the conclusion. The leading perspective is that a good argument should lead from true premises to a true conclusion: does the truth of the premises guarantee the truth of the conclusion or does it at least provide strong support? Two standards are commonly distinguished, deductive validity and non-deductive strength. If an inference is evaluated for deductive validity, the question is whether the conclusion *must* be true if the premises all are. If evaluated for non-deductive strength, the question is whether the premises provide a strong reason, if not an absolute guarantee, for the truth of the conclusion.[16]

Deductive validity is conceived as a maximally strong link between premises and conclusion in the following sense: it guarantees (in a logical sense to be explained below) that the conclusion is true *if* the premises are. This leaves room for deductively valid inferences with premises or conclusions that are false; it only excludes the possibility that we could be confronted with true premises and a false conclusion. Hence a deductively valid inference can be put to two basic uses: showing that the conclusion is true, given that the premises are true; or showing that at least one premise is false, given that the conclusion is false (this is Harsanyi's overall strategy of argumentation). Another important consequence is that for showing an inference to be deductively invalid, it suffices to point out one situation in which the premises are true but the conclusion false. Showing that an inference is

[14] Other inconsistencies, e.g. inconsistency of a premise with known facts of science, are just a reason for assessing the premise in question as false.

[15] In an undercut argument, the proponent (who puts forward the argument) uses premises which the opponent accepts to infer a conclusion which the opponent denies. See Betz (2013) for a typology of dialectical moves.

[16] The distinction between deductive and non-deductive primarily applies to standards of evaluation and only derivatively to arguments. An arguments can then be called "deductive" either because it is meant or taken to be evaluated by deductive standards, or because it performs well with respect to deductive standards. (Skyrms 2000:ch. II.4).

deductively valid is more ambitious insofar as referring to just one case will not do. We rather need a general argument which shows that there cannot be a case in which the premises are true and the conclusion false.

Such arguments can be given in basically two ways, which correspond to two varieties of deductive validity. The first is called "formal" validity[17] and covers arguments which are valid in virtue of one of their logical forms. Logical forms are constituted by features of inferences which are relevant to their validity and "topic neutral" such as the way inferences can be analysed into constituents of logically relevant categories (e.g. sentences, predicates and singular terms) and logical expressions such as "and", "all" and "if . . . then". The core idea of formal validity is that some inferences are valid solely in virtue of such structural features and regardless of the meaning of the non-logical expressions they involve. The notion of logical form is relative to a logical theory (of, e.g. zero- or first order logic), and such a theory is also needed to actually show that an inference is formally valid. The basic structure of a proof of formal validity involves two steps. First, the inference at hand must be formalized. One of its logical forms must be represented by means of a formula; that is, a schematic expressions of the formal language which is part of the logical theory. Secondly, the logical theory can be used to prove that every inference which has a logical form represented by the scheme in question is valid. Well-known techniques for such proofs include truth tables and natural deduction. In this way, the validity of the example [Singer] can be shown by proving $\neg p \vee p$; $\neg p \rightarrow q$; $p \rightarrow r \Rightarrow q \vee r$.

The second form of deductively valid inferences are "materially" valid inferences (also called "semantically" or "analytically" valid), the validity of which is due to a logical form and the meaning of (some of) the non-logical expressions they contain (e.g. "Option *New York* is better than option *Chicago*. Therefore *Chicago* is worse than *New York*."). One way of dealing with materially valid inferences employs a strategy of treating such inferences as enthymematic counterparts of formally valid inferences. If a premise expressing the conceptual relationship responsible for the materially valid inference is added to the original, a formally valid inference results. The inference at hand is then materially valid just in case the resulting inference is formally valid and the added premise expresses a conceptual truth. In reconstruction (2) of [Harsanyi], for example, one could add (2.5) as a premise and then get (2.6) as a conclusion (in line with 4):

(2.5) x is much better than y just in case y is much worse than x.
(2.6) The worst possible outcome of the option *Chicago* is much worse than the worst possible outcome of the option *New York*.

Non-deductive strength is an attribute of inferences which are deductively invalid, but the premises of which nonetheless provide good reason for their conclusions. Three characteristics distinguish non-deductive strength from logical validity: non-deductive strength is compatible with the conclusion being false even

[17] In this chapter, we use "validity" simpliciter as an abbreviation for "deductive validity"; in the literature it often also abbreviates "formal validity".

if all the premises are true, it comes in degrees, and it is nonmonotonic; that is, adding premises can yield a stronger or weaker argument. An immediate consequence is that even if a strong non-deductive argument supports some conclusion, there can still be a counter-argument which shows that this conclusion is false. Evaluating the non-deductive strength of arguments is a much more heterogeneous business than assessing deductive validity. In the literature, a range of different types of non-deductive inferences are analysed. Examples include inferences based on probability ("inductive" inferences), analogies, inferences to the best explanation and inferences involving causal reasoning or appeal to the testimony of experts. It is debated how the various types of non-deductive inferences can best be analysed, whether they can be reduced to a few basic theoretical principles and whether they admit of a uniform and maybe even formal treatment. Some also defend a deductivist strategy of systematically correlating (some types of) non-deductively strong arguments to deductively valid ones with additional premises and a weaker conclusion. Again, argumentation schemes can be used as a heuristic tool for identifying candidates for additional premises.[18] One particular idea is to include premises which express that there are no valid or strong counter-arguments. We critically elaborate on this approach in Sect. 5, which also includes a range of examples.

Invalid and non-deductively weak inferences pose a particular challenge to the analyst. If she fails to show that an inference is valid or strong, this may be her fault rather than a deficit of the inference. For invalidity, there is the simple case mentioned above, in which we find that an inference has true premises and a false conclusion in some possible situation. But unless we can refer to such a direct counter-example, showing formal invalidity amounts to showing that the inference has *no* valid logical form, and there is, strictly speaking, no general way of conclusively showing that we have investigated *all* the inference's logical forms (see Cheyne 2012). All we can do, is making plausible that an inference has no valid form, and for this, we need to rely on the assumption that we have considered all formal features of the inference which may be relevant to its validity. So any verdict of invalidity is at most as plausible as this assumption. And similar considerations apply in case of material invalidity and non-deductive weakness. Still, verdicts of invalidity or non-deductive weakness can often be argued convincingly, for example, by pointing out a confusion about necessary and sufficient conditions.

Many more defects of arguments are systematically studied under the label "fallacies". In general, fallacies are arguments that are irrelevant or misleading, especially because they are presented as being valid or strong although they are in fact invalid or weak, or as performing a dialectical function they in fact do not perform. The first type, traditionally called *non sequitur*, has just been discussed. The second type is exemplified in problems of dialectical irrelevance such as

[18] Lumer (2011) explains how argumentation schemes can be exploited for deductivist reconstructions.

arguments which do not support the thesis they are presented as supporting (*ignoratio elenchi*) or arguments which attack a position the opponent does not in fact defend ("straw-man").[19] In this way, Harsanyi's undercut seems to miss the point because he includes assumptions about probabilities although Rawls intends maximin as a principle only for some situations which involve "choice under great uncertainty" (Rawls 1999:72); that is, choice situations, "in which a knowledge of likelihoods is impossible, or at best extremely insecure" (Rawls 1999:134).[20]

3.4 Practical Arguments

So far, our discussion has not been specifically tailored to practical arguments. The basic characteristic of practical argumentation is that it leads to a "normative" conclusion. In this chapter, we focus on normative sentences which qualify an action with some deontic modality; that is a phrase such as "it is forbidden to ...", "... must not do ..." or "... ought to ...".[21] On the one hand, there are many more such expressions which are commonly used. On the other hand, not all normative premises and conclusions are normative sentences, because they can have a normative meaning in the context at hand even if they do not contain an explicitly normative expression (e.g. "Boys don't cry."). A first task of reconstruction is therefore formulating the normative premises and the normative conclusion explicitly as normative sentences. One possibility is to qualify directly acts (e.g. "Agent A ought to do X" etc.), another is to is to rely on standard qualifiers for sentences ("It is obligatory that Agent A does X"), which are studied in deontic logic (see McNamara 2010):

As an example, we get the following standard formulation for the conclusion of inference 3:

(3.3*) The maximin principle says that it is impermissible that you choose the option *New York*.

Importantly, the relations depicted in Fig. 3.3 only hold if the various modalities relate to the same normative perspective. What is obligatory from a legal point of view is not merely optional from this point of view even if it is morally optional. Reconstructions therefore must make the normative perspective explicit unless all

[19] There is a rich literature on fallacies; see section *Resources*. For specific fallacies in argumentation about risk, see Hansson (2016).

[20] Harsanyi offers further considerations which may dispel the straw-man worry in the text that follows what we quoted as [Harsanyi].

[21] This is a restricted perspective since there are other types of non-descriptive sentences as well, for example those which include evaluative terms ("good", "better"). For a more precise and sophisticated discussion (using a different terminology), see Morscher (2013).

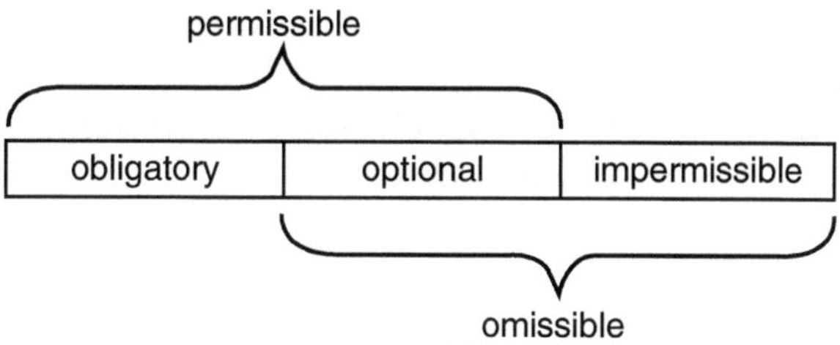

Fig. 3.3 Deontic modalities and their logical relations (e.g. everything optional is permissible)

explicit normative phrases in an argumentation relate to the same normative perspective.

A second challenge for reconstructing practical arguments arises in connection with the fact that there are no valid practical inferences without any normative premises.[22] Practical arguments are frequently enthymematic in this respect, and normative premises must then be supplied in reconstruction. For the purpose of a systematic study of practical arguments, it will be convenient to rely on inferences with a certain standard form that can be expressed with the help of a decision principle. This is a sentence which claims that a certain option for acting has a certain deontic status under some descriptive and normative conditions. Such principles can then be used as a premise which joins further premises stating the mentioned conditions with a conclusion expressing the normative status of the relevant option. In Sect. 5, we will discuss a selection of examples of decision principles.

Another cluster of problems which regularly arises in the analysis of practical arguments is the following. If an option or a decision problem can be assessed with reference to more than one action-guiding principle, one faces the question of how these principles relate to each other. Are they lexicographically ordered (e.g. moral considerations trump prudential ones)? Or can the principles be weighted against each other in some other way? And how can such information be accounted for in argument analysis? Furthermore, premises of practical arguments will often include so-called *prima facie* (or *pro tanto*, or defeasible) reasons or obligations (cf. Hansson 2013:99). These are normative claims which are stated without any restrictions, but may be overridden in specific cases of application nonetheless (e.g. "Lying is impermissible" may not apply to cases in which an insignificant lie can save the life of many). We suggest to deal with these challenges as problems of acquiring coherent positions in a complex argumentation (see Sect. 4.2).

[22] Strictly speaking, this is only true for practical arguments in which every premise and the conclusion either is *entirely* in the scope of a deontic modality or does not contain any deontic modality. The situation is much more complex if for practical arguments which include "mixed" sentences; that is, sentences only part of which are in the scope of a deontic modality. See Morscher (2013) for an accessible discussion.

4 Analysing Complex Argumentation

4.1 Reconstructing Complex Argumentation as Argument Maps

We have so far studied methods for analysing individual arguments. Virtually every policy debate and practical deliberation contains however multiple, typically conflicting arguments (see, e.g. Schefczyk 2016 on the monetary policy debate). If the *argumentative turn* aspires to represent an alternative to traditional risk analysis, it has to solve the problem of aggregating and compounding opposing arguments; at least, it has to *suggest* methods for balancing conflicting reasons.

Balancing reasons is a fundamental reasoning task we all perform regularly in a more or less systematic way. The basic tool we use to structure this task is a pro/con list. Still, such a pro/con list is insufficient for aggregating conflicting arguments. It may at best serve as a starting point for a more thorough analysis and should be seen as a mere heuristic one may use when nothing important is at stake (e.g. in many everyday decisions). The problem is that policy deliberation and analysis does frequently not go beyond giving a pro/con list. (And *if* it does, it uses highly questionable methods, e.g. cost benefit analysis.) There is a striking mismatch between the efforts our societies put into (a) getting the factual statements our policy analysis relies on right and (b) drawing the right conclusions from these factual statements in view of our complex normative beliefs. Put bluntly: where we find that a back-of-the-envelope-calculation is not good enough to establish the facts, we should not draw policy conclusions merely relying on pro/con lists, either.

But why precisely is a pro/con list not enough? There are three major issues with such lists:

1. *Macro structure*. It is unclear how exactly the different arguments relate to each other. Even worse, such lists wrongly suggest that all *pro* arguments (respectively *con* arguments) are related to the central thesis in a similar way.
2. *Micro structure*. The internal structure of the individual arguments remains unclear.
3. *Aggregation*. The plain juxtaposition of *pros* and *cons* suggests improper aggregation methods, such as simply counting (weighted) *pros* and *cons*.

Let us illustrate these points with an example. Consider the thesis:

[T] The global use of nuclear power should be extended.

The following list of arguments is drawn from the 18th edition of *Pros and Cons: A Debater's Handbook* (Sather 1999:255–7); the items have only been shortened (as indicated) and re-labelled. The fact that many of the descriptive claims made are false (as of today) does not prevent the example from being instructive.

Pro	Con
[Pro1.1] The world faces an energy crisis. Oil will be exhausted within 50 years, and coal will last less than half that time. It is hard to see how 'alternative' sources of energy will fulfil growing power needs. [Pro1.2] It is estimated, for example, that it would take a wind farm the size of Texas to provide for the power needs of Texas. [...]	[Con1.1] The costs of nuclear power stations are enormous, especially considering the stringent safety regulations that must be installed to prevent disaster. [Con1.2] Alternative energy, however, is only prohibitively expensive because there is no economic imperative to develop it when oil and gas are so cheap. [...]
[Pro2.1] The Chernobyl disaster, widely cited as the reason not to build nuclear power plants, happened in the Soviet Union where safety standards were notoriously lax, and often sacrificed for the sake of greater productivity. [...]	[Con2.1] It is simply not worth the risk. Nuclear power stations are lethal time-bombs, polluting our atmosphere today and leaving a radioactive legacy that will outlive us for generations. [Con2.2] Chernobyl showed the potential for catastrophe [...]. [...]
[Pro3.1] The problems of the nuclear energy programme have been a result of bureaucracy and obsessive secrecy resulting from nuclear energy's roots in military research. These are problems of the past. [...]	[Con3.1] In the 1950s, we were promised that nuclear energy would be so cheap that it would be uneconomic to meter electricity. Today, nuclear energy is still subsidised by the taxpayer. [...]

Now consider:

1. *Macro structure*. For example, does argument [Con3.1] back up [Con1.1], does it question [Pro1.1], or does it criticize the central claim [T]? – Maybe it even does all three things at the same time. That is just not transparent.
2. *Micro structure*. None of the arguments is fully transparent in terms of assumptions and validity. It is for example unclear to which implicit premises the argument [Pro1.1] appeals in order to justify the central thesis [T].
3. *Aggregation*. It is tempting to count how many *pros* and *cons* one accepts in order to balance the conflicting arguments. We will see that this would be irrational.

So, how can we improve on this? As a first step, we have to get a better understanding of the *structure* of complex argumentation in general.

Arguments exhibit an internal premise-conclusion structure. The logico-semantic relations between the statements arguments are composed of determine the "dialectic" relations between arguments, the relations of support and attack.[23]

[23] Pollock (1987:485) distinguishes two further dialectic relations. An argument *rebuts* another argument if the arguments possess contradictory (or at least contrary) conclusions; an argument *undercuts* another argument if it questions the validity or applicability of an inference scheme applied in the latter. (Note that this is another use of "undercut" than in footnote 15.) The undercut relation is, however, not directly relevant in the framework we propose here. Validity of the individual arguments is guaranteed qua charitable reconstruction. Rather than using controversial inference schemes for the reconstruction, we suggest to add corresponding general premises that can be criticized. Pollock's undercut-relation hence effectively reduces to the attack relation.

- An argument *supports* another argument if the conclusion of the supporting argument is identical with (or at least entails) a premise of the supported argument.
- An argument *attacks* another argument if the conclusion of the attacking argument negates (or at least contradicts) a premise of the attacked argument.

We can now state more precisely the shortcomings of pro/con lists. They suggest that all *pro* (*con*) arguments possess the same conclusion, which is identical with the central thesis (respectively its negation). Typically some *pro* arguments do however support other *pro* arguments, rather than the central thesis directly; or they attack *con* arguments. These exact dialectic relations remain obscure in mere pro/con lists.

Attack- and support-relations between arguments can be visualized as a network, a so-called argument or debate map. (Note that "argument map" sometimes refers to the visualization of the internal structure of a single argument, too.) Argument maps visualize the dialectical structure of a complex argumentation. It is convenient to display central theses besides arguments in such a map. This allows one for example to visually express so-called rebuttals without introducing an extra relation in the argument map; argument A rebuts argument B in case A supports a thesis that B attacks.

Conceptually, the micro-structure of arguments determines the macro-structure of a debate. Methodologically, i.e. in terms of reconstruction procedure, the reverse order of analysis has turned out to be practical. Accordingly, we suggest to *sketch* the dialectical structure first before reconstructing individual arguments in detail, which may (and typically does) lead to a revision of the original sketch. Sketching the dialectical structure essentially means to lay out the explicitly intended and intuitively hypothesized support- and attack-relations between arguments. The starting point of such a sketch may be a pro/con list.

Figure 3.4 shows a sketch of the debate about nuclear power, based on the pro/con list given above (solid arrows represent support, dashed arrows attack

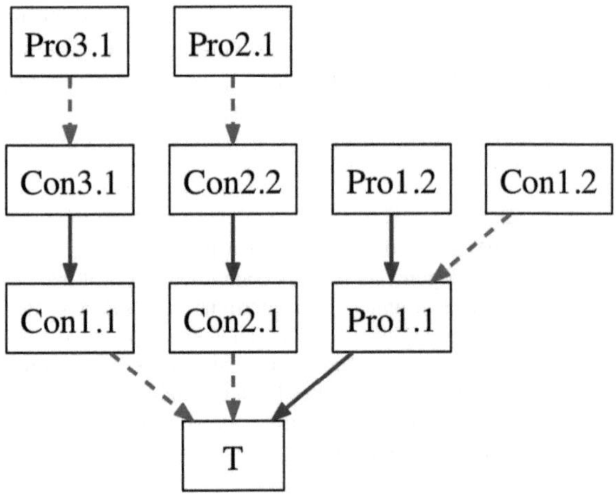

Fig. 3.4 Argument map visualizing support (*solid arrows*) and attack (*dotted arrows*) relations between arguments and theses (*boxes*) in the illustrative debate about nuclear power

relations between the arguments, and theses). The map is basically a hypothesis about the debate's dialectical structure, which has to be probed through detailed reconstructions of the individual arguments. At the same time, this hypothesis may guide the further reconstruction process, namely through suggesting constraints for (i) adding premises and (ii) modifying premises and conclusions in arguments.

We next present detailed reconstructions of two arguments mentioned in the illustrative pro/con list and the argument map above, the argument [Pro1.1] in favour of the global expansion of nuclear energy and the argument [Con2.1] against it.

[Pro1.1]

(1) If the global use of nuclear energy is not extended and the growing power need will be met nonetheless, then fossil fuels will fulfil growing power needs or 'alternative' sources of energy will do.
(2) It is impossible that fossil fuels will fulfil growing power needs (because of limited resources).
(3) It is impossible that 'alternative' sources of energy will fulfil growing power needs.
(4) **Thus** (1–3): The global use of nuclear energy is extended or growing power needs will not be met.
(5) The global energy crisis must be resolved, i.e. growing power needs must be met.
(6) *Practical-Syllogism*-Principle [cf. below].
(7) **Thus** (from 4–6): The global use of nuclear power should be extended. [T]

[Con2.1]

(1) The probability of accidents in nuclear power stations with catastrophic environmental and health impacts is non-negligible.
(2) Nuclear power stations pollute our atmosphere and leave a radioactive legacy that will outlive us for generations.
(3) If a technology exhibits a non-negligible likelihood of catastrophic accidents, pollutes the atmosphere and generates long-lasting, highly toxic waste, then its continued use – and a fortiori its expansion – poses severe environmental and health risks for current and future generations.
(4) **Thus** (1-3): The continued use of nuclear energy – and a fortiori its expansion – poses severe environmental and health risks for current and future generations.
(5) Any measure that poses severe environmental and health risks for current and future generations should not be implemented.
(6) **Thus** (4,5): The global use of nuclear power should not be extended. [N.B. entails non-T!]

These two reconstructions corroborate the dialectic relations as presumed in the preliminary argument map (cf. their conclusions).

4.2 Argument Maps as Reasoning Tools

Let us now suppose that all arguments have been reconstructed like [Pro1.1] and [Con2.1] above, and that the dialectic relations as visualized in Fig. 3.4 do really obtain, i.e. the debate's macro-structure dovetails with the micro-structure of the arguments. In addition, we assume that all individual arguments have been reconstructed as deductively valid (and non-redundant).[24] How can we evaluate such a debate?

It is important to understand that the reconstruction itself is not prescriptive. It neither decides on who is right or wrong nor on who has the final say in a debate. Hence argument analysts do not teach scientists or policy-makers what they should believe or do, and for what reasons. Essentially the reconstruction itself entails only if-then claims: if certain statements are true, then certain other statements that occur in the debate must also be true. The argument map does not reveal which statements are true; it is thus neutral and open to different evaluations (depending on which statements one considers to be true, false or open). In other words, the argument map identifies the questions to be answered when adopting a position in the debate, and merely points out the implications of different answers to these questions. Because of this, a thesis that is supported by many arguments is not necessarily true. And, by the same token, a thesis that is attacked by many arguments is by no means bound to be false. This applies equally to arguments. An attack on an argument does not imply that the very argument is definitely refuted. (It may be, for example, that the attacking argument itself draws – from an evaluative perspective – on premises that can easily be criticized by adding further arguments).

But then, again: how *can* we reason with argument maps? How do they help us to make up our mind?

We suggest that argument maps are first and foremost a tool for determining positions proponents (including oneself) may adopt, and for checking whether these positions satisfy minimal standards of rationality, i.e. are "dialectically coherent." While arguments constrain the set of positions proponents can reasonably adopt, there will in practice always be a plurality of different, opposing positions which remain permissible.[25]

Such positions can be conceptualized and articulated on different levels of detail.

---

[24] The proper analysis and evaluation of non-deductive reasoning poses serious theoretical problems and goes beyond the scope of this chapter. For a comprehensive state-of-the-art presentation compare Spohn (2012).

[25] A prominent rival approach to the one presented in this chapter are Dung-style evaluation methods for complex argumentation, which have been developed in Artificial Intelligence over the last two decades (see Bench-Capon and Dunne 2007; Dung 1995). Dung-style evaluation methods impose far-reaching rationality constraints; e.g. non-attacked arguments *must* be accepted, and undefended arguments *must not* be accepted. According to the approach championed in this chapter, in contrast, any argument can be reasonably accepted, as long as the proponent is willing to give up sufficiently many beliefs (and other arguments).

- On the macro level, a complete (partial) position specifies for all (some) arguments in the debate whether it is accepted or refuted. To accept an argument means to consider all its premises as true. To refute an argument implies that at least one of its premises is denied (whereas such a coarse-grained position does not specify which premise).
- On the micro level, a complete (partial) position consists in a truth-value assignment to all (some) statements (i.e. premises and conclusions) that occur in the debate's arguments.

There is no one-to-one mapping between coarse- and fine-grained positions. Different fine-grained formulations may yield one and the same coarse-grained articulation of a proponent's position. Fine-grained positions are more informative than coarse-grained ones.

These two types of articulating a position come along with coherence standards, i.e. minimal requirements a reasonably adoptable position must satisfy. The basic rationality criterion for a complete macro position is:

- [No accepted attack] If an argument or thesis A is accepted, then no argument or thesis which attacks A is accepted.

A partial macro position is dialectically coherent if it can be extended to a complete position which satisfies the above criterion.

Consider for illustrative purposes the two macro positions (articulated on the background of the nuclear energy debate) which are shown in Fig. 3.5. The left-hand position is complete in the sense that it assigns a status to every argument in the map. Moreover, that position satisfies the basic rationality criterion. There is no attack relation such that both the attacking and the attacked item are accepted. The right-hand figure displays a partial macro position, which leaves some arguments without status assignment. That position violates constraint [No accepted attack] twice, as indicated through a flash of lightning.

Complete micro positions must live up to a rationality criterion which is articulated in view of the inferential relations between statements (rather than the dialectic relations between arguments).

- [No contradictions] Contradictory statements are assigned complementary truth-values.
- [Deductive constraints] There is no argument such that, according to the position, its premises are considered true while its conclusion is considered false.

A partial micro position is dialectically coherent if it can be extended to a complete position which satisfies the above criteria.

Consider for illustrative purposes the two arguments [Pro1.1] and [Con2.1] we have reconstructed formerly. A position which takes all premises of [Pro1.1] to be true but denies its conclusion, or which assents to the conclusions of both [Pro1.1] and [Con2.1] is obviously not dialectically coherent; it directly violates one of the above constraints. A partial position according to which all premises of [Pro1.1] and [Con2.1] are true is not dialectically coherent, either, because truth-values of

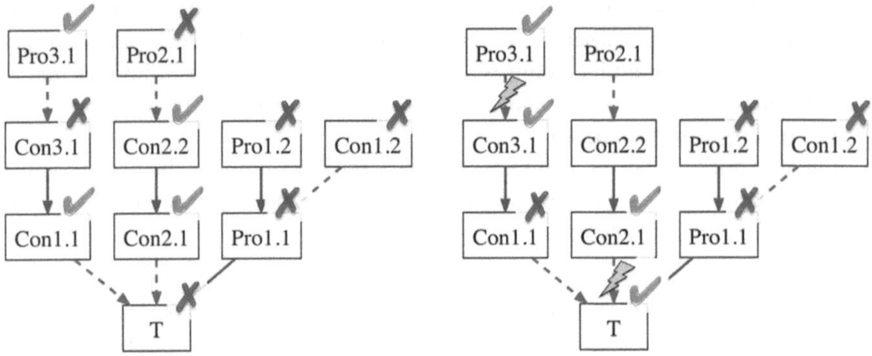

Fig. 3.5 Two macro positions, visualized against the background of the nuclear energy debate's argument map. "Checked" arguments are accepted, "crossed" arguments are refuted, "flashes" indicate local violations of rationality criteria (see also text)

the remaining statements (i.e. conclusions) cannot be fixed without violating one of the above constraints.

A micro or macro position which is not dialectically coherent violates basic logical/inferential constraints that have been discovered and articulated in the debate. (Note that this standard of coherence is even weaker than the notion of logical consistency.) If a proponent's position is not dialectically coherent, the proponent has not fully taken into account all the considerations that have been put forward so far. Either she has ignored some arguments, or she has not correctly adapted her position in regard of some arguments. As new arguments are introduced into a debate, previously coherent positions may become incoherent and in need of revision.

Argument maps and the articulation of positions in view of such maps may hence help proponents to arrive at well-considered, reflective positions that do justice to all the considerations set forth in a deliberation. Suppose, for example, a stakeholder newly realizes that her position is attacked by an argument she considers *prima facie* plausible. That discovery may – indeed: should – lead her to modify her stance. But there are different, equally reasonable ways to revise her position: she may decide to refute the previously ignored argument despite its *prima facie* plausibility, or she concedes the criticism and gives up the argument that is attacked.

Coherence checking is hence a proper way for balancing and aggregating conflicting normative arguments. Let us suppose that all *descriptive* premises in the arguments pro and con expanding nuclear energy were established and agreed upon. Whether a proponent assents to the central thesis [T] thus hinges only on her evaluation of the various *normative* premises, e.g. premise (5) in [Pro1.1] and [Con2.1], respectively. Typically, there will exist no dialectically coherent position according to which all ethical proscriptions, all decision principles, all evaluative statements and all claims to moral rights are simultaneously accepted. Only a subset of all normative statements that figure in a debate can be coherently adopted. And there are various such subsets. Coherence checking hence makes explicit the precise normative trade-offs involved when aggregating conflicting practical arguments.[26]

Over and above coherence checking, argument maps can be valuable tools for managing plurality and coping with conflicting positions. In terms of argument mapping, actual dissent between opponents can stem from two causes: (i) the proponents have overlooked arguments put forward by their respective opponent; (ii) some arguments and theses are evaluated differently. Re (i): If dissent arises, among other things, because one opponent has missed certain arguments, the opponents should first of all come to agree on, and possibly expand, the argument map, whereupon the positions held by the opponents will be re-evaluated. At best, dissent is dissolved right after that. Re (ii): If there is dissent in spite of agreement on the set of relevant arguments, one may proceed as follows. One firstly identifies the theses and arguments mutually agreed on by the opponents. Based on this common ground, one then tries to determine or develop consensual policies. For policy deliberations, this translates as follows: the argument maps can be used for developing robust policy proposals, i.e. policy measures that are compatible with many different positions and sets of basic moral assumptions.

Plurality management may also allow one to identify promising argumentation strategies for reducing disagreement. The reconstruction may for instance reveal that there is a central argument which is simply not agreed upon because its empirical assumptions are still controversial. Consensus on the central normative thesis might then be reached by arguing about and clarifying the empirical assumption (which is sometimes easier than agreeing on basic normative evaluations). In addition, formal models of debate dynamics suggest, quite generally, that one should argue in an opponent-sensitive way (i.e. on the basis of one's opponents' assumptions) in order to reduce mutual disagreement (see Betz 2013:12). The detailed analysis of a debate is certainly helpful in identifying such argumentative moves.

The very basic point of plurality management is illustrated by Fig. 3.6. It shows two macro positions that disagree on most of the relevant issues (arguments in the debate) but agree on some core points: the central thesis should be refuted; it is attacked by an argument that should be accepted; and the sole justification of the central thesis should be rejected. This core agreement may suffice to agree on

[26] Sometimes one and the same ("prima facie") normative principle, when applied to a complex decision situation, gives rise to conflicting implications. This is paradigmatically the case in dilemmatic situations, where one violates a given norm no matter what one does. In argument-mapping terms: given all descriptive premises are accepted, there is no coherent position according to which the "prima facie" principle is true. In regard of such cases, we suggest to systematize the aggregation and balancing process through *specifying* the normative principle in question such that the differences between alternative choices are made explicit. E.g. rather than arguing with the principle "You must not lie" in a situation where one inevitably either lies to a stranger or to one's grandma, one should attempt to analyze the reasoning by means of the two principles "You must not lie to relatives" and "You must not lie to strangers", which can then be balanced against each other.

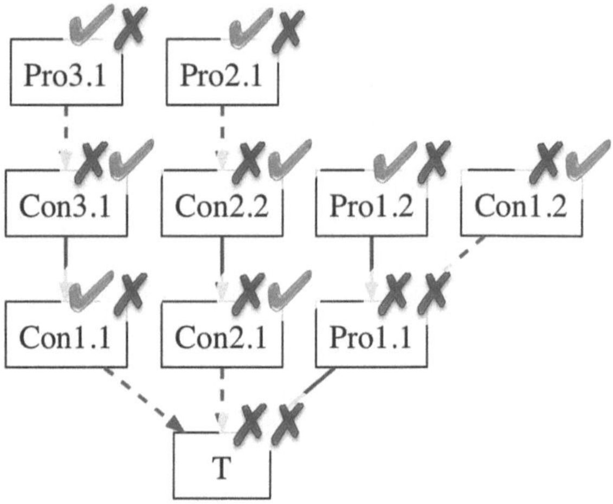

Fig. 3.6 Two macro positions, visualized against the background of the illustrative argument map

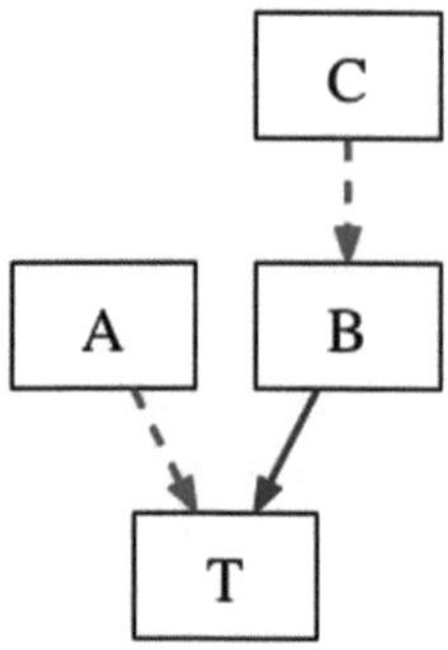

Fig. 3.7 A simple, abstract argument map

policy questions, further dissent concerning other arguments is then irrelevant (regarding *policy* consensus formation).

Let us briefly return to our third criticism of pro/con lists: improper aggregation methods. It should be clear by now that numbers do not count. We should not simply add up accepted *pros* and *cons*. A single *pro* argument may override a dozen *con* arguments. The left-hand macro position in Fig. 3.6, which is dialectically coherent, accepts 3 out of 4 *pro* arguments and only 1 out of 5 *con* arguments, but denies the central thesis nonetheless.

The process of specifying a dialectically coherent (macro or micro) position in view of an argument map can be modelled by means of a decision tree. To illustrate this process we shall consider a simplified dialectical structure that consists of three arguments A, B, C and a thesis T. We assume that A attacks T, B supports T, and C attacks B (Fig. 3.7).

Each argument has but one premise whose truth-value is not fixed through background knowledge, labelled a, b, c respectively. In order to find a dialectically

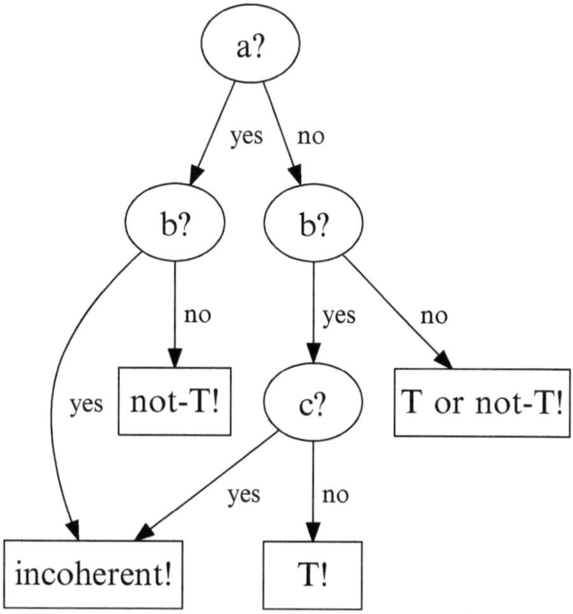

Fig. 3.8 Decision tree for determining whether to accept the central thesis in the argument map depicted in Fig. 3.7

coherent micro position on this map and to determine whether one should accept the central thesis, one may execute the decision tree shown in Fig. 3.8.[27]

We have started this section with the issue of aggregating conflicting reasons. Argument maps per se do not resolve this problem, they do not provide an algorithm for weighing conflicting reasons. They provide a detailed conceptual framework in which this task can be carried out. The resolution of normative conflicts will essentially depend on the acceptance/refutation of key premises in the arguments. These premises will also include conflicting decision principles. The map does not tell you how to do it, it only shows between which (sets of) normative statements one has to choose.

4.3 An Illustrative Case Study

This section illustrates the above methods by reporting how argument maps have been used as reasoning tools in climate policy advice.[28] Climate engineering (CE) refers to large-scale technical interventions into the earth system that seek

[27] "Yes" stands for statement accepted; "no" for statement not accepted. For the sake of simplicity, we do not distinguish between denying a statement and suspending judgement.

[28] This section is adapted from http://www.argunet.org/2013/05/13/mapping-the-climate-engineering-controversy-a-case-of-argument-analysis-driven-policy-advice/ [last accessed 16.03.2015].

to offset the effects of anthropogenic GHG emissions. CE includes methods which shield the earth from incoming solar radiation (solar radiation management) and methods which take carbon out of the atmosphere (carbon dioxide removal).[29]

In 2010, the German Ministry of Education and Research (BMBF) commissioned six individual scoping studies on different aspects of CE. Eventually, these individual studies were to be integrated into a single, interdisciplinary assessment. Betz and Cacean compiled a report on ethical aspects (eventually translated and published as Betz and Cacean 2012).

The overall aim in writing the study was to provide neutral policy advice on ethical issues of CE. To achieve this goal, Betz and Cacean (2012) decided to carry out an analysis of the various (ethical) arguments pro and con climate engineering methods. Splitting up the analysis into consecutive sub-tasks and including feedback rounds, they

- compiled a comprehensive commented bibliography of the CE discourse with a focus on ethical arguments (including scientific articles, policy statements, media reports, popular science books, etc.),
- sketched the overall dialectical structure and the individual arguments, which provided a first argument map,
- presented the preliminary argument map at project workshops to get feedback,
- and, finally, revised their interpretation of the debate and reconstructed the arguments in detail (as premise-conclusion structures).

The immediate result of this procedure was a comprehensive argument map, which was then used in the BMBF project in order

1. to compile the report "Ethical Aspects";
2. to assist policy makers in acquiring a coherent position (by evaluating alternative core positions proponents and policy makers may adopt);
3. to merge the various disciplinary studies in a final assessment report.

Re (1): The scoping study on ethical aspects of climate engineering contains a macro map of the debate that structures the entire report. Each chapter is devoted to a sub-debate of the controversy. The chapters in turn feature micro maps that display the internal structure of the sub-debates and visualize the individual arguments plus their dialectic relations. The arguments are then discussed in detail in the chapter texts. Central arguments are reconstructed as premise-conclusion structures.

Re (2): Betz and Cacean also used the argument map to assist stakeholders in acquiring a coherent position.

Thus, they have identified alternative core positions the ministry, or another stakeholder, may adopt. Such a core position might, for example, consist in saying that CE should be researched into so as to have these methods ready for deployment

---

[29] On the ethics of climate engineering and the benefits of argumentative analysis in this field compare Elliott (2016).

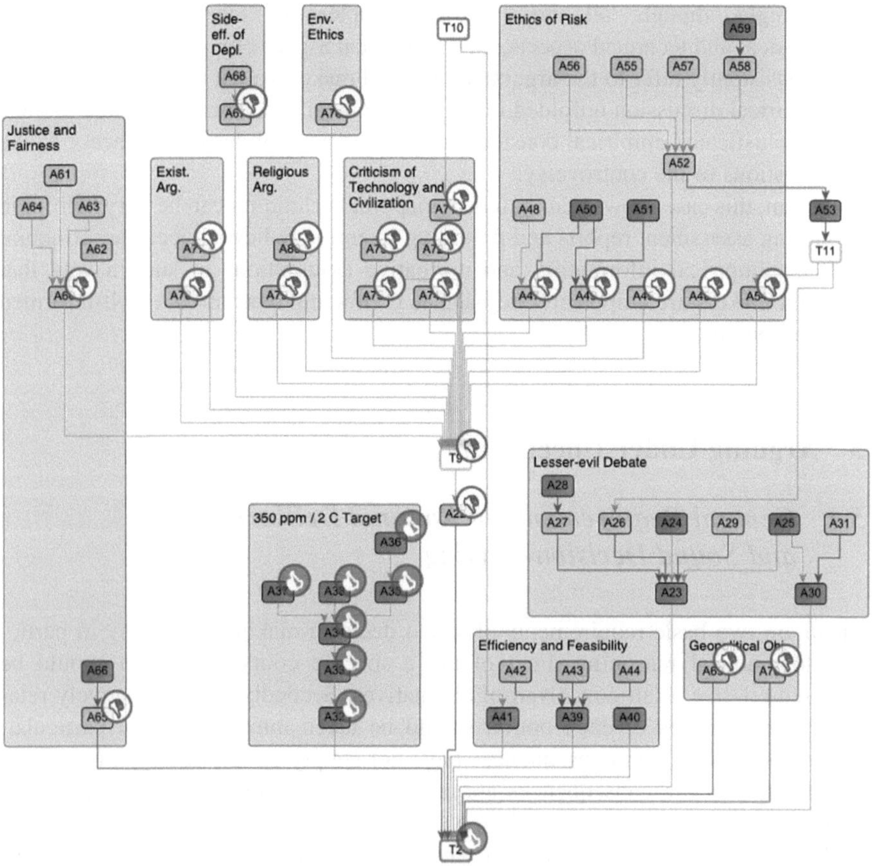

Fig. 3.9 Illustrative core position (here: *thumbs up*) and its logico-argumentative implications (here: *thumbs down*) in a detailed reconstruction of the moral controversy about so-called climate engineering (Source: Betz and Cacean 2012:87)

in time. They have then visualized the core position in the argument map and calculated the logico-argumentative implications of the corresponding stance (cf. Fig. 3.9). The enhanced map shows, accordingly, which arguments one is required to refute and which theses one is compelled to accept if one adopts the corresponding core position. For example, proponents who think that ambitious climate targets will make some sort of climate engineering inescapable are required to deny religious objections against CE deployment. By spelling out such implications, Betz and Cacean tried to enable stakeholders to take all arguments into account and to develop a well-considered position.

Re (3): The argument map also proved helpful in integrating the various discipline-specific studies into a single, interdisciplinary assessment report (Rickels et al. 2011). So, the assessment report, too, starts with a macro map, which depicts the overall structure of the discourse, and lists the pivotal arguments. Most interestingly, though, all the empirical chapters of the assessment report (on physical and technical aspects, on sociological aspects, on governance aspects, etc.) consistently refer to the argument map and make explicit to which arguments the empirical discussion unfolded in the chapter is related. This allows one to trace back sophisticated empirical considerations to the general debate and hence to the key questions of the controversy.

In sum, this case shows that argument mapping techniques can be very helpful in compiling assessment reports and providing scientific policy advice: they structure relevant empirical information and normative assumptions in such a way that decision makers are empowered to balance conflicting reasons in a well-informed and transparent way.

5 Arguing Under Uncertainty

5.1 General Requirements of Rational Deliberation and Sound Decision-Making

There are two basic requirements of sound decision-making that apply in particular to practical reasoning. First of all, a specific course of action should be assessed relative to all conceived-of alternatives. Secondly, all (normatively relevant) consequences of each option should be taken into account; in particular, uncertainty about such consequences must not simply be ignored (e.g. by falsely pretending that the consequences are certain or by ignoring some consequences altogether).[30]

There are two different ways in which these requirements can be applied to the *argumentative turn*, the argumentation-theoretic paradigm of practical reasoning. We have seen that every practical argument relies on a (frequently implicit) premise which states a more or less general decision principle (cf. Sect. 3.4). A decision principle licenses the inference from descriptive and normative statements to a normative conclusion. Now, the strong interpretation of the requirements demands that every individual decision principle (i.e. every individual practical argument) reasons for or against an action *in view of* all alternatives and all plausible outcomes. Arguments that fail to do so can accordingly be dismissed as defect. The alternative, weak interpretation of the requirements merely demands that all alternative options and all their plausible outcomes be considered in the entire debate, but not necessarily in each individual argument.

[30] Steele (2006) interprets the precautionary principle as a meta-principle for good decision-making which articulates essentially these two requirements.

This choice boils down to the following question: should we allow for decision principles which individually do not satisfy standards of good decision-making? – Yes, we think so. The following simplified example is a case in point:

Argument A

(1) The 2-degree-target will only be reached if some CE technology is deployed.
(2) The 2-degree-target should be reached.
(3) *Practical-Syllogism*-Principle (see below).
(4) **Thus**: Some CE technology should be deployed.

Argument B

(1) CE technologies are risk technologies without a safe exit option.
(2) Risk technologies without a safe exit option must not be deployed.
(3) **Thus**: No CE technology may be deployed [contrary to A.3 above].

None of these arguments considers explicitly all options and all potential outcomes. (This is because the antecedent conditions of their decision principles, A.3 and B.2, do not do so.) In combination, however, these two arguments allow for a nuanced trade-off between conflicting normative considerations. Risk-averse proponents may stick to argument B and hence give up the 2-degree-target (premise A.1) in order to reach a dialectically coherent position; others may prioritize the 2-degree-target and accept potential negative side-effects, in particular through denying that these side-effects are a sufficient reason for refraining from CE (i.e. they deny premise B.2). In sum, practical reasoning and, in particular, coherence checking is performed against the entire argument map; as long as all normatively relevant aspects are adequately represented somewhere in the map, practical reasoning seems to satisfy the general requirement of sound-decision making. There is thus no need for explicitly considering all options and all potential outcomes in each and every single argument.

5.2 Decision Principles for Reasoning Under Great Uncertainty

In the remainder of this chapter, we will present some argument schemes (in the form of decision principles that can be added as a premise to an argument reconstruction) which may allow argument analysts to reconstruct very different types of normative arguments. Such argument schemes can facilitate the reconstruction process and are mainly of *heuristic value*. There are certainly good reconstructions which do not correspond to any of these schemes. And schemes might have to be adapted in order to take the original text or plausibility etc. into account. That is, schemes are rather prototypes that will frequently provide a first version of an argument reconstruction, which will be further improved in the reconstruction process.

It is characteristic for practical arguments under uncertainty that their descriptive premises make explicit the uncertainty one faces. One way to arrive at (more or less) plausible decision principles for reasoning under uncertainty is hence to weaken their descriptive premises by introducing modal qualifications. The first six decision principles offer alternative qualifications of the descriptive premises (corresponding to apodictic, probabilistic and possibilistic versions). In general, the more far-reaching the qualification and the weaker the descriptive premises, the stronger and hence more questionable the corresponding decision principle.

Just to be clear: we are not advocating any of these decision principles. Following the idea that argument maps are tools which support agents in balancing conflicting normative reasons, the principles stated below will figure as premises in different arguments and will have to be weighed against each other on a case-specific basis.

The first principle states that any measure which is required to reach a goal should be taken – provided the goal should be attained.

[Practical Syllogism Principle]
If

(1) It ought to be the case that S.
(2) S [**will not/is unlikely to/might not**] be the case unless agent A does X.

then

(3) Agent A ought to do X.

While the apodictic version of this principle is analytic, the possibilistic version is arguably *very* weak, we have merely mentioned it for reasons of systematic completeness. This observation implies the following for the aggregation of conflicting arguments: when coherence checking reveals that we face a choice, we are rather prepared to give up the possibilistic principle than the probabilistic or the apodictic version. Similar remarks apply to the principles below.

Practical arguments frequently justify options not because they are necessary for attaining some goal but because they are optimal. Such arguments could be reconstructed with the following principle:

[Optimal Choice Principle]
If

(1) It *prima facie* [i.e. without considering negative side-effects that are inevitable when bringing about S] ought to be the case that S.
(2) S [**will/is likely to/might**] be the case if agent A does X.
(3) There is no alternative to X for agent A that [**will/is likely to/might**] bring about S and is more suitable than X.

(4) The certain, likely and possible side-effects of agent A doing X are collectively negligible as compared to the [**certain/likely/possible**] realization of S.

then

(5) **Thus**: Agent A ought to do X

The underlying idea is that conditions (1) and (4) collectively guarantee that S ought to be the case *all things considered* and that (2) and (3) imply that X is [likely/possibly] the optimal means to reach S.

Deontological reasons may be analysed along the following lines.

[Prohibition Principle]
If

(1) Acts of type T are categorically impermissible.
(2) Agent A doing X is [**certainly/likely/possibly**] an act of type T.

then

(3) Agent A must not do X.

The apodictic version of this principle is, as in the case of the Practical Syllogism, analytic. As an alternative to modal qualifications, uncertainties may be made explicit in the characterization T of an act; e.g.: "an attempted murder", that is an act (of a certain kind) that leads with some probability to some consequence. In such a case, premise (2) need not be qualified.

Rights-based considerations pose no principle problems for argument analysis, either.

[Principle of Absolute Rights Violation]
If

(1) Persons P possess the absolute right to be in state R.
(2) Agent A doing X [**certainly/likely/possibly**] prevents persons P from being in or achieving state R.

then

(3) Agent A must not do X.

The following principle speaks against some action based on the fact that the act violates *prima facie* rights that are not overridden (compare for example argument B in Betz (2016)).

[Principle of Prima Facie Rights Violation]
If

(1) Persons P possess the *prima facie* right to be in state R.
(2) Agent A doing X [**certainly/likely/possibly**] prevents persons P from being in or achieving state R.
(3) There exist no collectively weightier rights (than someone being in state R) whose realization is [**certainly/likely/possibly**] jeopardized when not doing X.

then

(4) Agent A must not do X.

Standard approaches in formal decision theory can be re-interpreted as decision principles, which in turn correspond to specific types of arguments (see also Betz (2016): Sect. 3). We illustrate this fact by means of two prominent examples. The following decision principle represents the criterion of expected utility maximization (e.g. Savage 1954).

[Principle of Expected Utility Maximization]
If

(1) The option o^+ has an expected utility of EU^+, according to probabilistic forecasts P and utility function U.
(2) There is no alternative option to o^+ which has an expected utility equal to or greater than EU^+, according to probabilistic forecasts P and utility function U.
(3) The probabilistic forecasts P are reliable.
(4) Utility function U adequately combines all normative dimensions that are relevant for the assessment of o^+ (and its alternatives).

Then

(5) Option o^+ ought to be carried out.

Finally, consider a principle that grasps maximin reasoning under great uncertainty (see Gardiner 2006).

[Worst Case Principle]
If

(1) Some available options may have catastrophic consequences.
(2) There are no options whose potential gains would outweigh, if realized, the worst possible consequences that may come up. [Counterfactual comparison of potential best and worst case]
(3) There are no reliable probabilistic forecasts of the available options' consequences, especially not of their worst possible consequences.
(4) There is no other available option whose worst possible consequence is (weakly) preferable to the worst possible consequence of option o^+.

then

(5) Option o^+ ought to be carried out.

For various examples of worst case arguments compare Betz (2016:Sect. 3.1).

6 Outlook

In this chapter we surveyed methods of argumentation analysis, with a special focus on justifying and criticising decisions under great uncertainty. Our approach starts with a systematic account of the aims of argument analysis, including the various dimensions in which an argumentation may be evaluated and the various standards that guide the reconstruction of arguments. On this basis, we introduced and exemplified the basic procedures for identifying, reconstructing and evaluating individual arguments as well as complex argumentation and debates. We then explained how such reconstructions of complex controversies may serve as reasoning tools. Finally, we discussed a range of decision principles that figure prominently in practical arguments under great uncertainty.

These methods have been developed as tools for clarifying and evaluating existing arguments and debates. The argumentative approach, however, has far greater potential. Concepts and techniques of argumentation analysis may be used to effectively *improve* practical reasoning in a variety of contexts. An argumentative approach enables experts and policy advisors to design scientific assessments and to provide decision-relevant scientific insights without being policy-prescriptive; it helps citizens and stakeholders to articulate their standpoints and to meaningfully contribute to intricate debates; it assists moderators in steering a controversy and managing a plurality of opinions; and it supports decision makers in balancing conflicting reasons in a transparent and well-informed way. We are convinced that a focus on argumentation will improve the deliberative quality of policy debates. Argumentation and argument analysis ultimately serve an emancipatory agenda. All too often, citizens and stakeholders are intellectual captives of unchallenged assumptions. Argumentation analysis frees people who are lost in the communicative labyrinth of reasons – it empowers them to speak up, to argue their views, and to scrutinize positions, held by themselves or others.

Resources Supporting Argument Analysis

Bowell, Tracy, and Gary Kemp. 2015. *Critical Thinking. A Concise Guide. 4th ed.* London: Routledge.

Chapter 5 gives a very accessible yet reliable introduction to techniques of argument reconstruction focusing on the analysis of individual arguments and complex argumentation.

Two online tutorials focusing on analysing complex argumentation are:

- Course "Argument Diagramming" at Carnegie Mellon University: http://oli.cmu.edu/courses/free-open/argument-diagramming-course-details/.
- *Critical Thinking Web*: http://philosophy.hku.hk/think/.

A more extensive treatment of fallacies can be found in the *Internet Encyclopedia of Philosophy*: http://www.iep.utm.edu/fallacy/.

Argunet is an argument mapping software designed to support the reconstruction of complex argumentation and debates: http://www.argunet.org/.

Links were correct on 22.07.2015.

References

Bench-Capon, T. J. M., & Dunne, P. E. (2007). Argumentation in artificial intelligence. *Artificial Intelligence, 171*, 619–641.

Betz, G. (2010). *Theorie dialektischer Strukturen*. Frankfurt am Main: Klostermann.

Betz, G. (2013). *Debate dynamics: How controversy improves our beliefs*. Dordrecht: Springer.

Betz, G. (2016). Accounting for possibilities in decision making. In S. O. Hansson & G. Hirsch Hadorn (Eds.), *The argumentative turn in policy analysis. Reasoning about uncertainty* (pp. 135–169). Cham: Springer. doi:10.1007/978-3-319-30549-3_6.

Betz, G., & Cacean, S. (2012). *Ethical aspects of climate engineering*. Karlsruhe: KIT Scientific Publishing. doi:10.5445/KSP/1000028245.

Brun, G. (2014). Reconstructing arguments. Formalization and reflective equilibrium. *Logical Analysis and History of Philosophy, 17*, 94–129.

Brun, G., & Hirsch Hadorn, G. (2014). *Textanalyse in den Wissenschaften. Inhalte und Argumente analysieren und verstehen* (2nd ed.) Zürich: vdf.

Brun, G., & Rott, H. (2013). Interpreting enthymematic arguments using belief revision. *Synthese, 190*, 4041–4063.

Cheyne, C. (2012). The asymmetry of formal logic. In M. Peliš & V. Punčochář (Eds.), *The logica yearbook 2011* (pp. 49–62). London: College Publications.

Dung, P. M. (1995). On the acceptability of arguments and its fundamental role in nonmonotonic reasoning. Logic programming and n-person games. *Artificial Intelligence, 77*, 321–357.

Elliott, K. C. (2016). Climate geoengineering. In S. O. Hansson & G. Hirsch Hadorn (Eds.), *The argumentative turn in policy analysis. Reasoning about uncertainty* (pp. 305–324). Cham: Springer. doi:10.1007/978-3-319-30549-3_13.

Fischer, F., & Forester, J. (1993). *The argumentative turn in policy analysis and planning*. Durham: Duke University Press.

Fischer, F., & Gottweis, H. (2012). *The argumentative turn revisited. Public policy as communicative practice*. Durham: Duke University Press.

Gardiner, S. M. (2006). A core precautionary principle. *The Journal of Political Philosophy, 14*, 33–60.

Hansson, S. O. (2000). Formalization in philosophy. *The Bulletin of Symbolic Logic, 6*, 162–175.

Hansson, S. O. (2013). *The ethics of risk: Ethical analysis in an uncertain world*. New York: Palgrave Macmillan.

Hansson, S. O. (2016). Evaluating the uncertainties. In S. O. Hansson & G. Hirsch Hadorn (Eds.), *The argumentative turn in policy analysis. Reasoning about uncertainty* (pp. 79–104). Cham: Springer. doi: 10.1007/978-3-319-30549-3_4.

Hansson, S. O., & Hirsch Hadorn, G. (2016). Introducing the argumentative turn in policy analysis. In S. O. Hansson & G. Hirsch Hadorn (Eds.), *The argumentative turn in policy analysis. Reasoning about uncertainty* (pp. 11–35). Cham: Springer. doi:10.1007/978-3-319-30549-3_2.

Harsanyi, J. C. (1975). Can the maximin principle serve as a basis for morality? A critique of John Rawls' theory. *American Political Science Review, 69*, 594–606.

Jacquette, D. (1996). Charity and the reiteration problem for enthymemes. *Informal Logic, 18*, 1–15.

Lumer, C. (2011). Argument schemes. An epistemological approach. In F. Zenker (Ed.), *Argumentation. Cognition and community. Proceedings of the 9th international conference of the Ontario Society for the Study of Argumentation (OSSA), May 18–22, 2011*. Windsor: University of Windsor. http://scholar.uwindsor.ca/ossaarchive/OSSA9/papersandcommentaries/17/. Accessed 22.07.2015.

McNamara, P. (2010). *Deontic logic. Stanford Encyclopedia of Philosophy*. http://plato.stanford.edu/archives/fall2010/entries/logic-deontic/.

Morscher, E. (2009). *Kann denn Logik Sünde sein? Die Bedeutung der modernen Logik für Theorie und Praxis des Rechts*. Wien: Lit.

Morscher, E. (2013). How to treat naturalistic fallacies. In H. Ganthaler, C. R. Menzel, & E. Morscher (Eds.), *Aktuelle Probleme und Grundlagenfragen der medizinischen Ethik* (pp. 203–232). St. Augustine: Academia.

Paglieri, F., & Woods, J. (2011). Enthymematic parsimony. *Synthese, 178*, 461–501.

Pollock, J. L. (1987). Defeasible reasoning. *Cognitive Science, 11*, 481–518.

Rawls, John. 1999. *A theory of justice* (Rev. ed.). Cambridge, MA: Belknap Press.

Reinmuth, F. (2014). Hermeneutics, logic and reconstruction. *Logical Analysis and History of Philosophy, 17*, 152–190.

Rescher, N. (2001). *Philosophical reasoning*. Malden: Blackwell.

Rickels, W., et al. (2011). *Large-scale intentional interventions into the climate system? Assessing the climate engineering debate. Scoping report conducted on behalf of the German Federal Ministry of Education and Research (BMBF)*. Kiel: Kiel Earth Institute. http://www.kiel-earth-institute.de/scoping-report-climate-engineering.html?file=tl_files/media/downloads/scoping_reportCE.pdf. Accessed 22.07.2015.

Sather, T. (1999). *Pros and Cons. A debater's handbook* (18th ed.). London: Routledge.

Savage, L. J. (1954). *The foundation of statistics*. New York: Wiley.

Schefczyk, M. (2016). Financial markets: the stabilisation task. In S. O. Hansson & G. Hirsch Hadorn (Eds.), *The argumentative turn in policy analysis. Reasoning about uncertainty* (pp. 265–290). Cham: Springer. doi:10.1007/978-3-319-30549-3_11.

Singer, P. (1988). Ethical experts in a democracy. In D. M. Rosenthal & F. Shehadi (Eds.), *Applied ethics and ethical theory* (pp. 149–161). Salt Lake City: University of Utah Press.

Singer, P. (2002). *Animal liberation* (3rd ed.). New York: Harper Collins.

Skyrms, B. (2000). *Choice and chance. An introduction to inductive logic* (4th ed.). Belmont: Wadsworth.

Snoeck Henkemans, A. F. (2001). Argumentation structures. In F. H. van Eemeren (Ed.), *Crucial concepts in argumentation theory* (pp. 101–134). Amsterdam: Amsterdam University Press.

Spohn, W. (2012). *The laws of belief*. Oxford: Oxford University Press.

Steele, K. (2006). The precautionary principle: A new approach to public decision-making? *Law, Probability, and Risk, 5*, 19–31.

van Eemeren, F. H., & Grootendorst, R. (1992). *Argumentation, communication, and fallacies: A pragma-dialectical perspective*. Hillsdale: Lawrence Erlbaum.

van Eemeren, F. H., & Grootendorst, R. (2004). *A systematic theory of argumentation. The pragma-dialectical approach*. Cambridge: Cambridge University Press.

Walton, D. N. (1996). *Argument structure. A pragmatic theory*. Toronto: University of Toronto Press.

Walton, D. N., Reed, C. A., & Macagno, F. (2008). *Argumentation schemes*. Cambridge: Cambridge University Press.

Chapter 4
Evaluating the Uncertainties

Sven Ove Hansson

Abstract In almost any decision situation, there are so many uncertainties that we need to evaluate their importance and prioritize among them. This chapter begins with a series of warnings against improper ways to do this. Most of the fallacies described consist in programmatically disregarding certain types of decision-relevant information. The types of information that can be disregarded differ between different decisions, and therefore decision rules that exclude certain types of information should not be used. The chapter proceeds by introducing a collection of useful and legitimate rules for the evaluation and prioritization of uncertainties. These rules are divided into three major groups: rules extending the scope of what we consider, rules for evaluating each uncertainty, and rules for the comparative evaluation of uncertainties (in both moral and instrumental terms). These rules should be applied in an adaptable process that allows the introduction of new and unforeseen types of arguments.

Keywords Uncertainty • Decision rules • Argumentation • Fallacies • Scenarios • Epistemic defaults • Symmetry arguments • Expected utility • Hypothetical retrospection

1 Introduction

Uncertainty is one of the major complicating factors in many policy decisions. When we do not know what the effects will be of the options that we choose between, how can we then make a rationally defensible decision? As shown in Hansson and Hirsch Hadorn (2016), the term "uncertainty" has a wide meaning and covers more or less everything that we might wish to know, but yet do not know. Here the focus will be on lacking or incomplete factual knowledge. Uncertainty in this sense is often exacerbated by normative inconclusiveness that may result from incommensurability of decision-relevant values or other unresolved value issues (Möller 2016).

S.O. Hansson (✉)
Department of Philosophy and History, Royal Institute of Technology, Stockholm, Sweden
e-mail: soh@kth.se

S.O. Hansson, G. Hirsch Hadorn (eds.), *The Argumentative Turn in Policy Analysis*,
Logic, Argumentation & Reasoning 10, DOI 10.1007/978-3-319-30549-3_4

Perhaps unfortunately, the more closely you investigate a decision problem, the more uncertainties will turn up. The debate on nanotechnology provides an excellent example of this. A wide range of uncertainties have been brought up in the discussion of that technology. Some are quite down to earth, such as our lack of knowledge of the toxicity of new materials, but others have a more speculative flavour, such as the accidental creation of nano-robots that destroy the earth in the course of building more and more replicas of themselves. The latter scenario seems implausible, but if it takes place then that would be the end of humanity. So can we really afford not to take it into account?

In almost any decision situation, a large number of uncertainties can be pointed out. It can be argued that ideally, we should take all of them into account throughout the decision process. But in practice, doing so would in many cases make our decision-making extremely complex and time-consuming, thereby leading to delays and stalemates and in some cases possibly render us unable to make any decision at all. We therefore need means to evaluate uncertainties and prioritize among them. It is the purpose of the present chapter to provide argumentative methods that can be used for that purpose.

But before doing the constructive work I propose that we have a look at some ways to reason about uncertainties that tend to lead us wrong.

2 How Not to Argue

The notion of a fallacy is not entirely clear. The Oxford English Dictionary uses the phrase "deceptive or misleading argument" in defining it. This could be improved by observing that fallacies (in the philosophical sense) are argument patterns, rather than single arguments (Brun and Betz 2016). We can at least provisionally define a fallacy as a "deceptive or misleading argument pattern". In discussions of uncertainty and risk all kinds of fallacies known from other contexts, such as ad hominem, circular reasoning and the strawman, can be encountered. But there are also some types of fallacious reasoning that are specific for the subject-matter of uncertainties (Hansson 2004a). What follows is a list of some such uncertainty-specific fallacies. The first two of them concern categories of undesirable effects that are often dismissed for dubitable reasons.

2.1 The Fallacy of Disregarding Unquantifiable Effects

Risk assessment and cost-benefit analysis have a strong tradition of quantification. The aim is usually to produce a quantitative assessment, and therefore the focus is on quantifiable factors, such as the expected number of deaths and the expected economic gains or losses. Values that are difficult or impossible to quantify tend to fall outside of such comparisons. Examples of potential negative effects that tend to be neglected are cultural impoverishment, social isolation, and increased tensions between social strata. However, the lack of established quantitative measures for an effect does not necessarily imply that it is unimportant in policymaking. Therefore, it is a fallacy to neglect relevant uncertainties because they cannot be quantified. This fallacy can be avoided in at least three different ways: (i) making quantitative estimates in all cases even when it is difficult, (ii) providing a supplementary analysis of non-quantitative factors in addition to the traditional quantitative analysis, and (iii) replacing the quantitative analysis by a non-quantitative one. Obviously, the relative importance of quantitative vs. non-quantitative effects should be a key issue in the choice between these strategies.

2.2 *The Fallacy of Disregarding Indetectable Effects*

There may be strong reasons to believe that an effect exists even though we cannot discover it directly. This is particularly important for chemical exposures. We may for instance have strong experimental or mechanistic reasons to believe that a chemical substance has negative effects on human health or the environment, but these effects may still not be detectable. It is a little-known statistical fact that quite large effects can be indetectable in this sense. For a practical example, suppose that 1000 persons are exposed to a chemical substance that increases lifetime mortality in coronary heart disease from 10.0 to 10.5 %. Statistical calculations will show that this difference is in practice indistinguishable from random variations. If an epidemiological study is performed in which this group is compared to an unexposed group, then there is no possibility to discover the increased incidence of lethal heart disease. More generally speaking, epidemiological studies cannot (even under favourable conditions) reliably detect an increase in the relative risk unless this increase is greater than 10 %. For the more common types of lethal diseases, such as coronary disease and lung cancer, lifetime risks can be of the order of magnitude of about 10 %. Therefore, even in the most sensitive studies, an increase in lifetime risk of the size 10^{-2} (10 % of 10 %) or smaller may be indistinguishable from random variations (Hansson 1995, 1999b). However, effects of this size are usually considered to be of considerable concern from a public health point of view.

It is often claimed in public debates that if an exposure has taken place without any harmful effects being detected, then there is nothing to worry about. Most of these statements are made by laypersons, but sometimes they have been made by professed experts or by authorities with access to expertise. In 1950 Robert Stone, a radiation expert with the American military, proposed that humans be exposed experimentally to up to 150 roentgens (a dose that can give rise to acute radiation sickness) with the motivation that “it seems unlikely that any particular person would realize that any damage had been done on him by such exposure” (Moreno 2001:145). In 1996 the Health Physics Society proposed that “inability to detect any increased health detriment” should be used as a criterion of acceptability of radiation doses (Health Physics Society 1996. For details, see Hansson

2013:112–116). However, these arguments identify the problem of exposure-induced health effects in a way that is implausible and socially indefensible. For the public, the problem is not that there is a known link between exposure and disease. The problem is the preventable disease itself. The fallacious reasoning involved in these arguments can be called the "ostrich's fallacy", as a tribute the biological folklore that the ostrich buries its head in the sand, believing that what it cannot see is no problem (Hansson 2004a).

Obviously, in order to avoid this fallacy it is necessary to recognize and make use of other types of evidence than the most direct ones. For instance, protection against chemical hazards will have to be based in part on what is known about toxic mechanisms and about toxic effects in non-human species that are reasonably similar to us in terms of biochemistry and physiology (Rudén and Hansson 2008).

2.3 *The Fallacy of Disregarding Benefits*

Since risks are by definition possibilities of undesired events, risk-taking cannot be justified for its own sake. It will have to be justified by the benefits that it gives rise to (Hansson 2013:117–119). Therefore a systematic discussion of what risks to take will have to be based on (i) a characterization of the risks, (ii) a characterization of the associated benefits, and (iii) an argumentation on whether the risks are worth taking in order to obtain these benefits. However, some discussants wish to do without the second of these components. This is the fallacy of disregarding benefits. It comes in three variants.

The first variant consists in trying to determine a level of "acceptable risk" (also called "de minimis risk"). The idea is that when a risk is below a certain level, it is acceptable even if no benefits come with it. This was a popular approach in the 1960s and 1970s, when it was often developed in the form of an "acceptable" lethal risk of 1 in 100,000 or 1 in 1000,000 (Fiksel 1985). But obviously, there is no reason to accept frivolous risk-taking, whatever the probability. Although many attempts were made to determine a level of "acceptable risk", it soon became clear that a general-purpose level of acceptable risk cannot be determined for the simple reason that the acceptability of a risk-generating activity depends not only on the risk but also on the associated benefits (Bicevskis 1982).

The second variant of the fallacy consists in disregarding substantial benefits when assessing a risk. One example of this is the argumentation used by Greenpeace against the introduction of genetically modified cultivars in agriculture. In the 1970s when genetic modification was a new technology, scientists implemented a voluntary moratorium until they had evaluated the hazards and found them manageable (Berg et al. 1974). The initial uncertainties connected with the technology per se have now since long been resolved (Berg and Singer 1995), and particular uses of the technology can be scientifically evaluated in terms of their specific positive and/or negative effects. The technology has important medical applications, and it also has an (unfortunately largely unused) potential to decrease the environmental damage caused by agriculture (Hansson and Joelsson 2013). However, the potential benefits of the technology are denied by Greenpeace and a couple of other organizations, even in the case of Golden Rice that has proven life-saving effects for children with vitamin A deficiency (Revkin 2011, 2013).

The third variant of the fallacy consists in using the benefits a certain risk provides in one context as an argument for accepting the same risk in contexts where these benefits do not arise. An unusually clear example of this fallacy is the claim repeatedly made by Wade Allison, a professor of physics at Oxford University, that if a radiation dose is acceptable in radiotherapy, then that same dose is also an acceptable exposure in the nuclear industry. "Nuclear technology cures countless cancer patients everyday – and a radiation dose given for radiotherapy is no different in principle to a similar dose received in the environment." (Allison 2011:193) This is a serious fallacy, since in oncology, the only chance to save the patient's life may sometimes be a therapy including high doses of ionizing radiation that significantly increase the patient's risk of contracting a new cancer at a later point in time. Extensive epidemiological studies show that high dose radiotherapy leads to significant risks of new, radiation-induced tumours (Hansson 2011).

The only way to avoid this fallacy, in all its variants, it to always look for, and take into account, both the positive and negative effects of a potential action or activity. In the end, the positive and negative uncertainties will have to be weighed against each other – but of course this weighing need not be quantitative.

2.4 *The Fallacy of Cherry-Picking Uncertainties*

One of the major problems with uncertainties is that there are so many of them. It is possible to construct chains of events leading from almost any human activity to a disaster. Obviously, a biased or unsystematic selection of uncertainties can lead us severely wrong. Many forms of pseudoscience are characterized by cherry-picking uncertainties that support a particular claim. For instance, anti-vaccination activists tend to focus on various potential side-effects that vaccines might have (Betsch and Sachse 2013; Kata 2010). Although some of these proposed side effects are rather far-fetched, absolute certainty that they cannot occur may not be obtainable. However, what is lacking on the anti-vax webpages is a discussion of all the uncertainties that will emerge if we refrain from vaccination, thereby relinquishing our protection against devastating epidemics. Other examples of the same nature can be found in climate science denialism. Activists renouncing the evidence of anthropogenic climate change put much emphasis on uncertainties that refer to possible overestimates of the anthropogenic effects on the climate, while entirely disregarding uncertainties referring to the possibility that those effects might be more severe than what is assumed in the standard models (Goldblatt and Watson 2012). In many areas, a biased selection of uncertainties can be used to argue in favour of almost any policy option.

In order to avoid this fallacy, a non-biased assessment has to be made of the uncertainties at hand in order to determine which of them have a potential impact on the decision. In this process, uncertainties supporting different decision alternatives must be taken into account. How this can be done in practice will be discussed in Sects. 3, 4, 5, and 6.

2.5 The Fallacy of Disregarding Scientific Knowledge

We are probably all more inclined to believe in the scientific results that we like than in those that we dislike. If uncurbed, this tendency can lead to science denialism that impairs our ability to evaluate uncertainties. A major example is the tobacco industry's denial of scientific evidence showing the fatal effects of their products. This is an extreme example since the perpetrators knew that their product was killing customers and that their campaigns against medical science would have fatal consequences (Proctor 2004). More typically, science denialism is advanced by people who seriously believe what they are saying. However, the practical effect can nevertheless be the same: decisions that go wrong because important scientific information is not taken into account. (More will be said about this in Sect. 5.)

2.6 The Fallacy of Disregarding Scientific Uncertainty

It is important to make use of solid scientific evidence whenever it is available. It is equally important to make a realistic estimate of scientific uncertainty whenever it is present. The reason for this is that there are cases when scientific uncertainty can have impact on a decision. For a simple example, consider the following example:

> New scientific evidence indicates that a common preservative agent in baby food may have a small negative effect on the child's brain development. According to the best available scientific expertise, the question is far from settled but the evidence weighs somewhat in the direction of there being an effect. A committee of respected scientists unanimously concluded that although the evidence is not conclusive it is more probable that the effect exists than that it does not. What should the food safety agency do?

I believe that most of us – in particular most parents – would recommend the agency to prohibit the substance, but lift the ban later if the suspected effect is shown not to exist. However, this contradicts a view with highly vociferous proponents, namely the view that only well-established scientific fact should be used in decision-making. This view is usually described as the application of "sound science". It means, in practice, that if there is scientific uncertainty about the existence of some possible danger, then that danger as treated in the same way as if its probability was known to be zero. "Sound science" is strongly associated with corporate proponents of pseudoscience, in particular the tobacco industry who have used it to delay action against passive smoking (Oreskes and Conway 2010;

Mooney 2005; Ong and Glantz 2001). However, there can be no doubt that the doctrine of "sound science" is a fallacy. Practical rationality demands that we take all the relevant evidence into account, and therefore it is irrational to disregard well-grounded evidence of danger when it is not strong enough to dispel all doubts. We would not have survived as a species if our forefathers on the savannah did not hurry up into the trees until there was no shadow of a doubt that the lions were after them.

2.7 *The Fallacy of Treating Uncertain Probability Estimates as Certain*

Conventionally, a distinction is made between decision-making under risk and under uncertainty (Hansson and Hirsch Hadorn 2016). By risk in this context is meant known probabilities. Since most probabilities are uncertain, clear-cut examples of decision-making under risk are not easy to find. Perhaps the gambler's decisions at the roulette table are as close as we can get to decision-making under risk. Given that the wheel is fair, the probabilities of various outcomes – gains and losses – are easily calculable, and thus knowable, although the gambler may not take them into account.

There is, however, a strong tendency in decision-supporting disciplines to proceed as if reasonably reliable probability estimates were available for all possible outcomes. Once a probability estimate has been produced, it is treated as a "true" and fully certain probability. In this way all decisions are dealt with as if they took place under epistemic conditions analogous to gambling at the roulette table. In honour of the dress code at some casinos, I have proposed to call this the *tuxedo fallacy* (Hansson 2009).

One interesting historical example is the strong belief that was common in the 1970s in experts' estimates of the probability a core damage in nuclear reactors. Although these estimates were based on extensive and competent technical analysis, they were fraught with uncertainties, in particular uncertainties concerning unknown accident mechanisms and probabilistic dependences between mutually aggravating faults. However, in the public debate they were often treated as known with certainty. Today we have experienced quite a few accidents with core damage and therefore we know that these early estimates were much too low.[1]

[1] The highly influential WASH-1400 report in 1975 predicted that the frequency of core damages (meltdowns) would be 1 in 20,000 reactor years. We now have experience from about 15,000 reactor years, and there have been ten accidents with core damages (meltdowns), i.e. about 1 in 1500 reactor years. (There have been four reactor explosions, namely one in Chernobyl and three in Fukushima Dai-ichi, adding up to a frequency of 1 in 3750 reactor years) (Escobar Rangel and Lévêque 2014; Ha-Duong and Journé 2014; Cochran 2011).

2.8 *The Fallacy of Monetizing All Values*

In cost-benefit analysis (CBA), options for a (usually public) decision are compared to each other by means of a careful calculation of their respective consequences. These consequences can be different in nature, e.g. economic costs, risks of disease and death, environmental damage etc. In the final analysis, all such consequences are assigned a monetary value, and the option with the highest value of benefits minus costs is recommended or chosen. The assumption is that in order to compare different consequences their values have to be expressed in the same unit – how could they else be compared? This has led to controversial practices such as putting a price on human lives that have been subject to extensive criticism (Anderson 1988; Sagoff 1988).[2]

It does not take much reflection to realize that we do not need to express values in the same unit – monetary or not – in order to be able to compare them. Most of the value comparisons that we make in our everyday lives are performed with non-numerical values. For instance, I assign higher value to some pieces of music than to others, but I am not able to specify these assessments in numerical terms. Perhaps more to the point, most of the difficult decisions taken by political leaders and the leaders of companies and organizations do not take the form of reducing all value dimensions to one in order to attribute a numerical value to each aspect, indicating the performance on the corresponding value dimension. Instead, the pros and cons of different options are weighed against each other by means of deliberations and comparisons that refer directly to the different dimensions of the problem, rather than trying to reduce all of them to one dimension. Therefore the claim that we have to assign comparable numbers to options (for instance by monetizing them) in order to compare them is a fallacy.

This fallacy has led to misguided attempts to achieve "consistency" across policy contexts. For instance, it has often been claimed that the "life value", i.e. the value of saving a life, expressed as a sum of money, should be the same in all contexts. However, we may have good reasons to pay more for saving a life against one danger than against another. For instance, we may choose to pay more per life saved in a law enforcement programme that reduces the frequency of manslaughter than what we pay for most other life-saving activities. One reason for this is the disruptive effects that violent crime has on both individual and social life. There are also good reasons why we are willing to pay more for saving a trapped miner's life than what we would pay for a measure in preventive medicine that has the expected effect of saving one (unidentified) person's life. The miner is an individual to whom others have person-related obligations, and we may also consider the general social effects of a decision to let people die who could have been saved.

---

[2] The same problem arises when the outcome of some other tool for multicriteria decision-making, for instance sustainability analysis, is reduced to a single aggregate value.

2.9 The Fallacy of Naturalness

It is sometimes assumed that we can find out whether something is good or bad for us by considering whether it is natural. In particular, naturalness is often seen as a guarantee against danger. This argument comes in two major forms.

First version:
X occurs naturally.
Therefore: X, when produced naturally, is better than artificially produced X.

Second version:
X occurs naturally.
Therefore: X, whether produced naturally or artificially, is good.

The first version of the fallacy can be called the "health food store variant" since it is particularly frequent in health food shops where synthetic chemicals are commonly claimed to be in some way inferior to naturally occurring instances of the same molecules. For instance, vitamin C from plants is considered healthy whereas synthetically produced L-ascorbic acid is considered unhealthy. Naturalness also plays an important role in some forms of non-scientific medicine, in particular "herbal medicine".

The second variant is common among proponents of nuclear technologies who claim that radiation doses at the same level as background radiation cannot be dangerous (See for instance Allison 2011; Jaworowski 1999).

In both its forms, the naturalness argument is a fallacy. The fact that something occurs naturally does not prove that it is harmless, and neither does it prove that it is safe to increase our exposure to it. Nature is full with dangers and it is simply wrong to conclude that since something is natural, it is harmless. Many plants are poisonous and the vast majority of them have no therapeutic potential. Therefore, that a drug is herbal does not make it efficient or for that matter harmless. To the contrary, serious side effects have followed from the use of such drugs (Levine et al. 2013; Saper et al. 2004; Lietman 2012; Shaw et al. 2012). Equally obviously, the presence of ionizing radiation in nature does not prove its harmlessness. The fallacy of taking naturally occurring products and exposures to be harmless is a variant of the somewhat more general fallacy argumentum ad naturam (appeal to nature) (Baggini 2002).

3 How to Argue

Most of the fallacies mentioned above have in common that they induce us to programmatically disregard certain types of decision-relevant information. (The only exception is the fallacy of naturalness, that does not follow this pattern.[3])

[3] However, as pointed out to me by Gertrude Hirsch Hadorn, the fallacy of naturalness usually tends to involve neglect of scientific information, and it can then be subsumed under the general category of neglect of decision-relevant information.

They can therefore be subsumed under a joint larger category, *fallacies of excluding decision-relevant information*. Even when decisions are complex, a rational decision maker should not follow decision rules that require the exclusion of certain types of relevant information. Obviously, in each particular decision the decision maker should focus on the most important information, but the types of information that can in practice be only cursorily attended to will differ between different decisions. In some decisions, non-quantifiable effects may be small and unimportant; but in other decisions they may be critical. In some decisions, scientific uncertainties may be negligible, whereas in other decisions they may be the predominant problem that has to be dealt with, etc. Therefore, consistent exclusion of certain types of information is not a good strategy for dealing with uncertainty or for decision-making in general. Instead I propose that three other general argumentative strategies should be pursued. The first of them is the very opposite of the one we just put aside, namely to *search for pertinent uncertainties and other decision-relevant circumstances that we have not yet observed*, in order to ensure that none is inadvertently left out. The second is to *evaluate each of these uncertainties and circumstances* in order to find out which of them have, in the particular context, considerable impact on the decision. The third is a *comparative evaluation* in which the arguments that point in different directions are weighed against each other. These strategies are developed in the three sections that follow.

4 Extending the Scope of What We Consider

Two major methods are proposed to decrease the risk that we miss something important in the evaluation of uncertainties. One is to search directly for uncertainties that we have not yet identified. The other, more elaborate one, is to develop scenarios in which new uncertainties may crop up.

4.1 Uncertainty Inventories

In many areas of decision-making there are lobbyists and others who promote the implementation and use of new technologies, and in some areas there are also opponents who argue in the opposite direction. For instance, in many environmental decisions there are activists arguing for strict regulations, and industry representatives arguing in the opposite direction. The situation is similar in many other issues. But there are also issues in which stakeholders have only been mobilized on one side of the issue (Cowles 1995). In particular in the latter cases active measures are required to ensure that decisions are based on a non-partisan selection of decision alternatives and a non-partisan collection and description of their possible consequences. Therefore it is not sufficient to base our deliberations on the arguments that have been put forward spontaneously.

Depending on the circumstances, there are many ways to search for decision-relevant arguments. Sometimes a very general search method such as brainstorming can be useful. Often it is useful to bring in experts or interest groups representing aspects that seem to have been underrepresented in the process thus far. As one example of this, social aspects are sometimes marginalized in discussions on urban and infrastructure planning. This can at least in part be remedied by engaging expertise on the potential social impacts of urban design. In many cases, the aspects relevant for environmental protection, public health or consumer interests will not be covered "spontaneously" but have to be introduced. Of course, even with such methods we cannot expect to achieve a complete list of the decision-relevant issues. What we can do, however, is to reduce any initial bias in the selection of aspects to which attention will be paid. We can also make sure that uncertainties, and not only that for which we have sufficient evidence, are taken into account.

4.2 Scenario Development

A scenario, in the sense in which the word is used here, is "a sketch, outline, or description of an imagined situation or sequence of events" (OED). The term has been used in the decision sciences since the 1960s for a narrative summarizing either a possible future development that leads up to a point where a decision will be made, or a possible development after a decision has been made. Scenario planning methodology was developed in post World War II defense planning in the U.S., and significantly enhanced in the 1970s, in particular by employees of Royal Dutch Shell company (Börjeson et al. 2006; Wack 1985a, b). Today, scenarios are used in a wide range of applications, including military planning, technology assessment, evaluation of financial institutions (stress testing), and climate science. The climate change scenarios developed by the IPCC have a central role in the integration of science from different fields that provides the background knowledge necessary both for international negotiations on emission limitation and in national policies for climate mitigation and adaptation.

In all these applications, the use of multiple scenarios is essential. It was noted already in 1967 by Herman Kahn and Anthony J. Wiener, two of the pioneers in future studies, that the use of multiple scenarios is necessary since decision-makers should not only consider the development believed to be most likely but also take less likely possibilities into account, in particular such that would "present important problems, dangers or opportunities if they materialized" (Kahn and Wiener 1967:3).

Such an approach conforms to how future technologies are often discussed in modern societies. In public discussions on contested technologies such as biotechnology and nanotechnology a multitude of possible (or at least allegedly possible) future scenarios have been put forward. There is no way to determine a single "correct" scenario on which to base our deliberations. We have to be able to base our decisions on considerations of several of them. Another way of saying this is that scenarios help us to deal with uncertainties. Each of the major possibilities that we are uncertain between can be developed into a scenario so that it can be studied and evaluated in detail.

5 Evaluating Each Uncertainty

In many cases, science provides us with efficient means to evaluate uncertainties and classify them as more or less plausible. There are also other types of arguments, in addition to the scientific ones, that can be used to evaluate uncertainties in terms of their plausibility. They will be referred to in what follows as arguments pertaining to epistemic defaults and to effect size defaults. But let us first consider how science can be used to evaluate uncertainties.

5.1 Scientific Evaluation of Arguments

Many uncertainties refer to "what science does not know", but in some cases (such as the claims of climate science denialists) inaccurate descriptions of scientific uncertainty are actively promoted. It is important to clarify in each individual case whether a purported uncertainty refers to issues that science can or cannot settle. The answer to this question is not always a simple "yes" or "no". In some cases the answer will depend on the burden of evidence that one wishes to apply. For example, suppose that someone brings up the supposition that a particular drug causes glaucoma. Such a statement can never be disproved. For statistical reasons, a very low increase in the frequency of glaucoma among patients using the drug will be impossible to detect. Science can, however do two things in a case like this, two things that are important enough. First, it can answer the question whether or not the effect occurs with a frequency above the detection limit (Hansson 1995). Secondly, it can answer the question whether there are any valid reasons to suspect this drug, rather than any other drug, of the effect in question. If the answer to the first question is that no effect can be detected, and the answer to the second question is that there are no valid reasons to suspect this drug rather than any other drug of the effect, then that is sufficient reason to strike this uncertainty from the agenda – even though science cannot provide a proof that the drug does not at all have the effect in question.

We can apply this to the supposition that MMR vaccine causes autism. This claim was put forward by Andrew Wakefield in 1998, but the study purported to show the connection has been proven to be fraudulent (Deer 2011). In spite of this, anti-vaccination activists still make the connection, claiming that there is remaining scientific uncertainty in the issue. However, extensive scientific studies have shown (1) that there is no detectable increase in the frequency of autism among children receiving the vaccine (Maglione et al. 2014), and (2) that there is no credible reason, such as a plausible mechanism, to assign this effect to the vaccine. Of course, science has not disproved the supposed connection, but only in the same sense that science has not disproved that the frequency of autism is increased by any other factor in a child's life that you can think of, such as riding the merry-go-round, eating strawberries, or drinking carbonated drinks. Therefore the uncertainty about a vaccine-autism connection should be struck from the agenda.

The vaccine example also shows the practical importance of evaluating uncertainties scientifically. The decreased vaccination rate that followed from the Wakefield scam has led to measle epidemics in which several children have died and others have been permanently injured (Asaria and MacMahon 2006; McBrien et al. 2003). This could have been avoided if proper use had been made of science. In this case the purported uncertainty can for all practical purposes be dispelled with the help of solid scientific information. When science can answer a question we had better use that answer.

5.2 *Epistemic Defaults: Novelty and Complexity*

Unfortunately, there are many questions that science cannot answer, and often we have to make decisions in spite of scientific uncertainty in key issues. Fortunately, in many of these cases there are other types of valid arguments that can help us. To begin with there are two epistemic defaults that can often help us evaluate uncertainties that science cannot resolve.

The first of these is the *novelty default*: We typically know less about new phenomena than about old ones. This can be a good reason to pay more attention to uncertainties that refer to new risk factors or new technologies. Hence, it would seem reasonable to pay more attention to uncertainties relating to fusion energy (from which we have no experience) than to uncertainties about any of the energy sources currently in use.

The novelty default has an interesting application in particle physics. Before new and more powerful particle accelerators were built, physicists have sometimes feared that the new levels of energy might generate a new phase of matter that accretes every atom of the earth. On some occasions, in particular before the start of the Large Hadron Collider at CERN, concerns have also spread among the public. The decisions to regard these fears as groundless have largely been based on observations showing that the energy levels in question are no genuine novelties since the earth is already under constant bombardment from outer space of particles with the same or higher energies (Ball 2008; Ellis et al. 2008; Overbye 2008; Ruthen 1993).

In other cases, proposed activities are really novel and the worries that this gives rise to cannot be so easily dispelled. For instance, consider the proposals that have been put forward to reduce the greenhouse effect by injecting substances into the stratosphere that will deflect incoming sunlight (Elliott 2016). Critics have produced long lists of possible negative effects of this technology: it may change cloud formation, the chemical composition of the stratosphere can be affected in undesired ways, down-falling particles may disturb ecosystems, etc. Perhaps most importantly, some negative effect may follow that we have not been able to think of. All these fears have to be taken seriously since the technology is genuinely new.[4] If a new technology is introduced, the uncertainties will be gradually reduced as we gain experience from it.

The other epistemic default is the *complexity default*. Uncertainty is usually larger in more complex systems. Systems such as ecosystems and the atmospheric system are known to have reached some type of balance that may be impossible to restore after a major disturbance. In fact, experience shows that uncontrolled interference with such systems may have irreversible consequences. One example of this is the introduction of invasive species into a new environment. The introduction can be small-scale and just consist in the release of a small number of plants or animals, but the effects on the ecosystem can be large and include the loss of original species (Clavero and García-Berthou 2005; Molnar et al. 2008; McKinney and Lockwood 1999). This is a good reason to take uncertainties about effects on ecosystems seriously.

Essentially the same can be said about uncontrolled interference with social and economic systems. Although politically controversial, this is a valid argument for piecemeal rather than wholesale economic reforms.

It might be argued that we do not know that these systems can resist even minor perturbations. If causation is chaotic, then for all that we know, a minor modification in the liturgy of the Church of England may trigger a major ecological disaster in Africa. If we assume that all causal connections between events are chaotic, then the very idea of planning and taking precautions seems to lose its meaning. Such a world-view would leave us entirely without a guidance, even in situations when we now tend to consider ourselves well-informed. Fortunately, experience does not bear out this grim epistemology. Accumulated empirical experience and the outcomes of theoretical modelling strongly indicate that certain types of influences on ecological systems can be withstood, whereas others cannot, and the same applies to social and economic systems. It is at least in many cases a feasible strategy to reduce the risk of inadvertent irreversible changes by making alterations in complex systems in a step-by-step fashion (excepting of course the cases when we have good knowledge about how the system will respond to large changes) (Hirsch Hadorn 2016).

---

[4] Experiences from volcanic emissions can be used to some extent, but there are important differences in chemical composition and atmospheric distribution.

5.3 Effect Size Defaults: Spatiotemporal Limitations

Another factor in judging the seriousness of uncertainties is the potential size of the effects that we are uncertain of. Spatial limitations are an important factor in this respect. In some cases, we know that the effect will only be local. In other cases we cannot exclude widespread, perhaps global effects. Uncertainties referring to effects of the latter type should, other things being equal, be given higher priority. In addition we also have to consider temporal limitations. An uncertainty is more serious if it refers to effects that may be long-lived or even permanent than if only short-lived effects can be expected.

Ecotoxicological risk assessment provides an excellent example of this. A substance can be toxic to a biotope by having a deleterious effect on any of its species, and most biotopes have a large number of species. It is in practice not feasible to investigate the effects of a substance other than on a small number of indicator species. Therefore, even if tests have been performed on a substance and no ecotoxic effects were discovered, there is a remaining uncertainty about its effects on the environment. However, the fate in the environment of a chemical substance is often much easier to determine than its toxicity. Some substances degrade readily in relatively short time. Others are persistent, i.e. they disintegrate very slowly or practically speaking, not at all. Some of the persistent substances are also bioaccumulating, which means that their concentration tends to increase in organisms (due to low excretion rates). Persistent and bioaccumulating substances spread at surprisingly high speed to ecosystems all over the world. For instance, polar bears in the Arctic have increasing concentrations of mercury, DDT, PCB, and other toxic pollutants that have reached them through winds and water and through bioaccumulation up the food chain (Dybas 2012). In addition to these known toxicants, the bodies of polar bears also contain many other persistent and bioaccumulating substances whose effects are unknown (McKinney et al. 2011). If any of these substances should turn out to have serious toxic effects in the long run – on polar bears or on any of the many other organisms in which they are accumulated – the consequences can be both serious and very long-lasting. This is a reason to be more worried about the release into the environment of these substances than of other substances that also have unknown toxicity but are known not to be persistent or bioaccumulating. From a general decision-theoretical point of view this means that we apply a criterion of spatio-temporal limitedness: lack of such limits justifies higher priority to uncertain hazards.

Environmental policies offer many other examples of the same principle. Long-range transport of pollutants is recognized as an important factor in assessing polluting activities. For instance, the discovery in the 1960s that long-range transport of sulphur oxides and nitrogen oxides gives rise to acid rain far away from the sources of pollution was crucial for the development of international measures against these emissions (Fraenkel 1989; Likens et al. 1972). And of course, today the fact that the climate effects of greenhouse gas emissions are global is an essential part of the reason why concerted international action is needed to mitigate the problem.

6 Comparing and Weighing

After we have identified and assessed the various (positive and negative) effects of decision options, it remains to weigh them against each other. Contrary to what is sometimes claimed by advocates of quantitative methods for decision support, such weighing does not require comparisons in quantitative terms. This was made very clear in a famous letter by Benjamin Franklin in 1772 to the chemist Joseph Priestley:

> When these difficult Cases occur... my Way is, to divide half a Sheet of Paper by a Line into two Columns, writing over the one Pro, and over the other Con. Then during three or four Days Consideration I put down under the different Heads short Hints of the different Motives that at different Times occur to me for or against the Measure. When I have thus got them all together in one View, I endeavour to estimate their respective Weights; and where I find two, one on each side, that seem equal, I strike them both out: If I find a Reason pro equal to some two Reasons con, I strike out the three... and if after a Day or two of farther Consideration nothing new that is of Importance occurs on either side, I come to a Determination accordingly. (Franklin 1970:437–438)

Obviously, when appropriate and comparable numbers can be assigned for all the pros and cons, then we can quantify this procedure by assigning a number to each item, representing its weight, and adding up these numbers in each column. This is the moral decision procedure proposed by Jeremy Bentham a few years later (Bentham 1780:27–28). However, in the cases when appropriate numbers are not available – and these are the cases that concern us here – we can stick to Franklin's non-quantitative method. The next subsection is devoted to symmetry arguments about uncertainties that can be used to strike out outbalancing items in the way proposed by Franklin.

6.1 Symmetry Arguments

In some decisions there are uncertainties that will be with us whatever option we choose. In other decisions, two uncertainties for one of the options cancel each other out. In both cases, we can – in the spirit of Franklin – reduce our list of uncertainties and thereby simplify the decision. For each of the two types of situations, a simple test is available (These tests were first proposed in Hansson 2004b).

For the first-mentioned situation we apply the *test of alternative causes*. It consists in investigating whether the uncertainty in question can be defeated by showing that we have at least as strong reasons to consider the possibility that either the same effect or some other effect that is at least as undesirable will come about if the action under consideration is *not* performed. If the same uncertainty (or equivalent uncertainties) can be found in both cases, then it is not decision-relevant.

For example, some opponents of nanotechnology claim that its development and implementation will give rise to a "nano divide", i.e. growing inequalities between those who have and those who lack access to nanotechnology (Moore 2002). However, this problem can easily be shown not to be specific for nanotechnology. An analogous argument can be made for any other new technology with wide application areas. We already have, on the global level, large "divides" in almost all areas of technology – including the most elementary ones such as sanitation (Bartram et al. 2005). Under the assumption that other technologies will be developed if we refrain from advancing nanotechnology, other "divides" will then emerge instead of the nano divide. If this is true, then the nano divide is a non-specific effect that does not pass the test of alternative causes, and therefore it does not have to be attended to in a decision whether to proceed with the development of nanotechnology.

For another example, consider a decision whether to build a nuclear plant or a coal plant under the (arguably dire) assumption that no other option is available.[5] An argument against the former option is that mistakes by operators can have unknown, undesirable effects. A potential counterargument is that operator mistakes are equally likely in a coal plant. However, the counterargument does not cancel out the corresponding argument against the nuclear plant since the worst potential consequences are smaller in a coal plant (and thus, operator mistakes are more undesirable in a nuclear plant). Therefore, the argument against the nuclear option that is based on mistakes by operators passes this application of the test of alternative causes.

In the other type of situation mentioned above, the *test of opposite effects* can be used. It consists in investigating whether an uncertainty can be outweighed by some other effect that (1) is opposite in value to the effect originally postulated (i.e. positive if the postulated effect is negative, and vice versa), and (2) has equal or larger moral weight than the postulated effect. Let us apply it to two examples.

In the first example, a breakthrough has been achieved in genetic engineering. Ways have been found to control and modify the metabolism of a species of microalgae with unprecedented ease. "Synthesizing a chemical with this technology is more like programming a computer than modifying an organism," said one of the researchers. A group of critics demand that the new technology be prohibited by international law. They point to its potential dangers, such as the spread of algae that produce highly toxic substances.

Here, we can apply the test of opposite effects. Expectedly we will then find that it is equally possible that this technology can be used to solve serious problems that confront mankind. Perhaps modified algae can make desalination cheap enough for

[5] This example was proposed to me by Gregor Betz.

large-scale irrigation. Perhaps such algae can be used to produce most of the energy that we need, without emitting greenhouse gases. Perhaps it can be used to produce much of the food that we need. Perhaps all pharmaceutical drugs can be produced at a price that will be affordable even in the poorest countries of the world. If any of this is true, then the prohibition rather than the use of this technology may have dire consequences. This means that the first argument has been defeated by equally strong arguments pointing in the opposite direction. Of course, the discussion does not stop there. It should be developed into a detailed discussion of more specified negative and positive effects – and in particular about what is required to realize the positive but not the negative ones.

In the other example, a company applies for an emission permit to discharge its chemical waste into an adjacent, previously unpolluted lake. The waste in question has no known ecotoxic effects. A local environmental group opposes the application, claiming that the substance may have unknown deleterious effects on organisms in the lake.

In this case as well we can apply the test of opposite effects. However, it does not seem possible to construct a positive scenario that can take precedence over this negative scenario. We know from experience that chemicals can harm life in a lake, but we have no correspondingly credible reasons to believe that a chemical can improve the ecological situation in a lake. (To the extent that this "can" happen, it does so in a much weaker sense of "can" than that of the original argument. This difference can be used in a specification that defeats the proposed counterexample.) Therefore, the environmental group's argument resists the test of opposite effects.

6.2 *Prudent Uses of Expected Utility*

Above I argued against the presumption that expected utility maximization, the standard method in risk analysis and cost-benefit analysis, is a "one size fits all" method for dealing with uncertainties. As we have seen, there are many decision situations in which important aspects cannot be captured with reasonable estimates of utilities (values) and probabilities, and the decision rule is also normatively assailable in some of its applications.

But obviously, this does not mean that the calculation of expected utility is always useless. In some decisions it may be a most valuable decision aid. The following is a case in point:

> A country is going to decide whether or not it will make the use of seat belts compulsory. The sole aim of this decision is to reduce the total number of traffic casualties. Calculations based on extensive experience from other countries show that the expected number of deaths in traffic accidents is 300 per year if safety belts are compulsory and 400 per year if they are optional.

Under the assumptions given there could not be much doubt that making seat belts mandatory would be the better decision. If the statistics is, as we suppose, reasonably reliable, then we can for practical purposes be sure that about 100 less people will die every year if seat belts are mandated than if they are not. Since this decision has as its sole purpose to reduce the number of victims of road death, this is about as close to an undefeatable argument as we can get.

We should observe, however, that two important conditions are satisfied in this example, and that if any of them fails then the argument loses its force.[6] The first of these conditions is that outcomes can be appraised in terms of a single number (in this case the number of persons killed) and that this number is all that counts. This assumption is usually made in discussions of road safety but it is by no means uncontroversial even in that context. For instance, a measure that is expected to save the lives of 125 drivers but at the same time cause 100 pedestrian casualties might not be as unanimously welcomed as one that just saves the lives of 25 drivers without any increased risks for anyone else.

The second condition is that a sufficient number of events is involved for the law of large numbers to apply. In our seat belt example it is the law of large numbers that makes us reasonably certain that about 100 more persons per year will be killed if seat belts are not compulsory than if they are not. The same type of argument cannot be used when this condition is not satisfied. In particular, it is not applicable when only a single or very few actions or decisions with uncertain outcomes are under review. The following example should make that clear:

> A trustee for a minor empties her bank accounts and buys shares for her in a promising company. He has good reasons to believe that with this investment the statistical expectation value of her fortune when she comes of age will be higher than if her money had remained on the bank accounts.
>
> Half a year later, the company runs into serious trouble and the shares lose most of their value within a few days. When the trusteeship ends, the beneficiary's fortune is worth less than a tenth of its original value.

The law of large numbers is not at play here. If the beneficiary had a multitude of fortunes, it would arguably be best for her to have them all managed according to the principle of maximizing expected utilities (provided of course that the risks connected with the different fortunes were statistically independent). But she had only one fortune. A decision criterion should have been chosen that protects better against large losses than what expected utility maximization does. Obviously, some decisions in global environmental issues have a similar structure. Just as the minor in our example had only one fortune, we have only one earth.

In summary, expected utility maximization cannot credibly be justified as a universal format for decision-making, but it can be justified if two criteria are satisfied, namely (1) that outcomes can be appraised in terms of a single number and that this number is all that counts, and (2) that one and the same type of action or decision is repeated sufficiently many times to make the law of large numbers applicable.

---

[6] For a more detailed discussion of this, see Hansson (2013:74–80).

6.3 Hypothetical Retrospection

As moral agents we need to go beyond the simple "me now" perspective. We need to see our own actions in other personal perspectives than "me" and other temporal perspectives than "now". This is what we teach our children when educating them to have empathy for others, i.e. see things from their perspective, and to plan and save for the future. Moral philosophers have devoted considerable efforts to developing and advocating one of these two extensions of the ethical perspective, namely the use of other person perspectives than "me". Much less effort has been devoted to the extension from "now" to the future, but for competent decision-making it may be equally important. It can be achieved with the method of hypothetical retrospection that I will now proceed to introduce. (It has previously been described in greater detail in Hansson 2007a, 2013:61–73).

In our everyday lives we often use a simple type of future-directed argument that can be called the "foresight argument". It consists in an attempt to see things the way that we will see them at some later point in time. Its simplest applications refer to situations that we treat as deterministic. For instance, some of the consequences of drinking excessively tonight can, for practical purposes, be regarded as foreseeable. Thinking in advance about these consequences may well be what deters a person from drunkenness.

When the foresight argument is applied to cases with risk or uncertainty, more than one future development has to be taken into account. An example: Betty considers whether she should sue her ex-husband for having taken several valuable objects with him that she sees as her private belongings. This is no easy decision to make since her case is difficult to prove and she wants to avoid a conflict that may harm the children. When contemplating this she has reasons to ponder how she would react to each of the major alternative outcomes of the legal process. She also needs to think through how she would later look back at having missed the chance of claiming her rights. Generally speaking, in cases of risk or uncertainty there are several alternative "branches" of future development. Each of these branches can be referred to in a valid argument about what one should do today. The foresight needed to deal with such cases must therefore be applied to more than one future development.

As a first approximation, we wish to ensure that whichever branch materializes, a posterior evaluation should not lead to the conclusion that what we did was wrong. We want our decisions to be morally acceptable (permissible) even if things do not go our way. This can also be expressed as a criterion of decision-stability: Our conviction that the decision was right should not be perturbed by information that reaches us after the decision. In order to achieve this, we have to consider, for each option in a decision, the major future developments that can follow if we choose that option.

Importantly, these deliberations should take into account the information that was available at the point in time of decision about other possible future developments than the one that actually took place. Suppose that Petra reflects (in actual retrospection) on her decision 5 years ago to sell her cherished childhood home in order to buy an apartment for herself and her husband. If she had known then what she knows today (namely that her husband would leave her 1 year later) then she would not have sold her childhood home. But when reconsidering the decision she has to see it in the light of what she had reasons to believe when she made it. Hypothetical retrospection is similar to actual retrospection in this respect. Suppose that Petra, 5 years ago, deliberated on whether or not to buy the apartment and that in doing so she performed hypothetical retrospection. Given that she had reasons to consider a divorce unlikely, she might then very well come to the conclusion that if she buys the apartment she will, 5 years later, consider the decision to have been right even in the improbable case of a divorce.

The aim of hypothetical retrospection is to make a decision such that whatever happens, the decision made will be acceptable from the perspective of actual retrospection. To achieve this, the decision has to be acceptable from each viewpoint of hypothetical retrospection. There may be cases in which this cannot be achieved, i.e., cases in which there is no decision alternative that appears to be acceptable come whatever may. Such situations are similar to moral dilemmas, and just as in moral dilemmas we will have to choose one of the (unacceptable) alternative that comes closest to being acceptable (Hansson 1999a). If no available alternative is acceptable from every future viewpoint, then we should determine the lowest level of unacceptability that some alternative does not exceed in any branch, and choose one of the alternatives that does not exceed it.

6.4 Moral Argumentation

Many of the difficult issues when evaluating uncertainties are interindividual, i.e. they refer to the distribution of potential advantages and disadvantages between different persons. One of the central problems in moral philosophy is to determine when it is allowable to subject another person to a disadvantage, typically in order to obtain an advantage for oneself. This problem has turned out to be particularly difficult to deal with when the extent of the disadvantage is uncertain, for instance when it has the form of exposure to a risk.

Leaving out the fine details, there are two major styles of argumentation about what is allowed and what is not. One consists in weighing the advantages of an option against its disadvantages. The other consists in setting limits for what one may and may not do. In our everyday discussions on moral issues, we tend to shift freely between the weighing and the limit-setting ways of arguing. In academic moral philosophy it is more common to develop one of them into an all-encompassing moral theory that excludes the other. Utilitarianism is based on the exclusive use of weighing, whereas deontological and rights-based ethics are entirely based on the limit-setting mode of argumentation. Both these purified approaches have the advantage of being more consistent than quotidian moral argumentation, but they also have the disadvantage of sometimes leading to implausible conclusions. In particular, they both have prominent problems in situations involving risk and uncertainty. We tend to accept much larger risks when they are associated with important benefits than when they are not. Therefore, some form of weighing has to take place, which means that the limit-setting mode of thinking is insufficient. But on the other hand, we also tend to regard large risks, in particular large risks to human health and human lives, as unacceptable whatever the benefits. The weighing of fatality risks against monetary advantages is commonly perceived as morally awkward, even by those who consider it to be an unavoidable component of rational decision-making (Hansson 2007b). This means that the weighing approach is also incapable of solving the problem on its own (Hansson 2013:21–43).

A plausible response to this conundrum is to abandon the argumentative limitations of the purified theories. We can then revert to the common approach in most non-regimented moral discussions, namely to allow both weighing and limit-setting moral arguments. One plausible option is to assume that each of us has a defeasible right not to have risks imposed on oneself by others. By "defeasible" is meant that this right can be overruled. However, it cannot be overruled just by identifying some benefit that is larger than the risk. A risk to you may be outweighed by a larger benefit if that benefit accrues to you, but not if it accrues to someone else. This gives us reason to consider the risk-benefit balance for each person, not just the aggregate balance that sums up all risks and benefits irrespective of who receives them.

This may seem to result in a too demanding criterion for risk acceptance. To make it socially tenable we will have to introduce the notion of mutually beneficial risk exposures. For instance, if you drive a car in my hometown you impose a (hopefully small) risk on me of being a victim in a traffic accident. Similarly, if I drive a car where you live I expose you to a similar risk. Provided that we both have much to gain from being allowed to drive a car, we would both gain from allowing each other to do so (under appropriate restrictions specified in traffic rules).

We can generalize this mode of thinking and allow for a wider range of "risk exchanges", thus accepting *risks that are parts of a social system of reciprocal risk exposures that are beneficial to all members of society* (Hansson 2013:97–110). This is a stricter criterion than the traditional utilitarian one. In a standard utilitarian risk calculus, exposing you to a risk can be justified by benefits to other persons. In the reciprocal approach, such an argument is not accepted. There has to be a positive benefit-risk balance for each person.

7 Pulling It All Together

How should the argumentative methods introduced in Sects. 4, 5, and 6 be combined? Basically we need a flexible and iterable process where each instrument for analysis can be used more than once (Brun and Betz 2016). Since we need to know the arguments before evaluating them, inventorying and scenario development (as described in Sect. 4) should normally take place first. It also makes sense to perform the evaluations of each individual option (as described in Sect. 5) before the comparative evaluation (as described in Sect. 6). Hypothetical retrospection and moral argumentation operate on an overarching level and are therefore suitable in the final stage of the process. However, it should be no surprise if new arguments or new options for decision-making come up at a late stage in the process. An argumentative process must be open in the sense of allowing for new inputs and for unforeseen types of arguments. This openness is one of its major advantages over traditional, more strictly rule-bound forms of uncertainty management. Therefore, tools and structures such as those introduced in this chapter have to be applied in an adaptable and creative way that recognizes the widely different conditions under which decisions are made.

Recommended Readings

Halpern, J. (2003). *Reasoning about uncertainty*. Cambridge, MA: MIT Press.

Hansson, S. O. (2007). Philosophical problems in cost-benefit analysis. *Economics and Philosophy, 23*, 163–183.

Hansson, S. O. (2013). *The ethics of risk. Ethical analysis in an uncertain world*. New York: Palgrave Macmillan.

Roeser, S., et al. (2012). *Handbook of risk theory*. Dordrecht: Springer.

References

Allison, W. (2011). We should stop running away from radiation. *Philosophy and Technology, 24*, 193–195.

Anderson, E. (1988). Values, risks and market norms. *Philosophy and Public Affairs, 17*, 54–65.

Asaria, P., & MacMahon, E. (2006). Measles in the United Kingdom: Can we eradicate it by 2010? *BMJ, 333*, 890–895.

Baggini, J. (2002). *Making sense: Philosophy behind the headlines*. Oxford: Oxford University Press.

Ball, P. (2008, May 2). Of myths and men. *Nature News*. http://www.nature.com/news/2008/080502/full/news.2008.797.html. Accessed Jan 2013.

Bartram, J., Lewis, K., Lenton, R., & Wright, A. (2005). Focusing on improved water and sanitation for health. *Lancet, 365*, 810–812.

Bentham, J. (1780). *An introduction to the principles of morals and legislation*. London: T. Payne. http://gallica.bnf.fr/ark:/12148/bpt6k93974k/f2.image.r=.langEN

Berg, P., & Singer, M. F. (1995). The recombinant DNA controversy: Twenty years later. *Proceedings of the National Academy of Sciences, 92*, 9011–9013.

Berg, P., Baltimore, D., Boyer, H. W., Cohen, S. N., Davis, R. W., Hogness, D. S., Nathans, D., et al. (1974). Potential biohazards of recombinant DNA molecules. *Science, 185*, 303.

Betsch, C., & Sachse, K. (2013). Debunking vaccination myths: Strong risk negations can increase perceived vaccination risks. *Health Psychology, 32*, 146.

Bicevskis, A. (1982). Unacceptability of acceptable risk. *Search, 13*, 31–34.

Börjeson, L., Höjer, M., Dreborg, K.-H., Ekvall, T., & Finnveden, G. (2006). Scenario types and techniques: Towards a user's guide. *Futures, 38*, 723–739.

Brun, G., & Betz, G. (2016). Analysing practical argumentation. In S. O. Hansson & G. Hirsch Hadorn (Eds.), *The argumentative turn in policy analysis. Reasoning about uncertainty* (pp. 39–77). Cham: Springer. doi:10.1007/978-3-319-30549-3_3.

Clavero, M., & García-Berthou, E. (2005). Invasive species are a leading cause of animal extinctions. *Trends in Ecology and Evolution, 20*, 110–110.

Cochran, T. B. (2011, April 12). Statement on the Fukushima nuclear disaster and its implications for U.S. Nuclear Power Reactors. Joint Hearings of the Subcommittee on Clean Air and Nuclear Safety and the Committee on Environment and Public Works, United States Senate. http://www.nrdc.org/nuclear/files/tcochran_110412.pdf. Accessed 22 Mar 2015.

Cowles, M. G. (1995). Setting the agenda for a new Europe: The ERT and EC 1992. *JCMS: Journal of Common Market Studies, 33*, 501–526.

Deer, B. (2011). How the vaccine crisis was meant to make money. *BMJ, 342*, c5258.

Dybas, C. L. (2012). Polar bears are in trouble—And ice melt's not the half of it. *BioScience, 62*, 1014–1018.

Elliott, K. C. (2016). Climate geoengineering. In S. O. Hansson & G. Hirsch Hadorn (Eds.), *The argumentative turn in policy analysis. Reasoning about uncertainty* (pp. 305–324). Cham: Springer. doi:10.1007/978-3-319-30549-3_13.

Ellis, J., Giudice, G., Mangano, M., Tkachev, I., & Wiedemann, U. (2008). Review of the safety of LHC collisions. *Journal of Physics G: Nuclear and Particle Physics, 35*, 115004.

Escobar Rangel, L., & Lévêque, F. (2014). How Fukushima Dai-ichi core meltdown changed the probability of nuclear accidents? *Safety Science, 64*, 90–98.

Fiksel, J. (1985). Toward a De Minimis policy in risk regulation. *Risk Analysis, 5*, 257–259.

Fraenkel, A. A. (1989). The convention on long-range transboundary air pollution: Meeting the challenge of international cooperation. *Harvard International Law Journal, 30*, 447–476.

Franklin, B. (1970). *The writings of Benjamin Franklin* (Vol. V, pp. 1767–1772). New York: Haskell House.

Goldblatt, C., & Watson, A. J. (2012). The runaway greenhouse: Implications for future climate change, geoengineering and planetary atmospheres. *Philosophical Transactions of the Royal Society A: Mathematical, Physical and Engineering Sciences, 370*, 4197–4216.

Ha-Duong, M., & Journé, V. (2014). Calculating nuclear accident probabilities from empirical frequencies. *Environment Systems and Decisions, 34*, 249–258.

Hansson, S. O. (1995). The detection level. *Regulatory Toxicology and Pharmacology, 22*, 103–109.

Hansson, S. O. (1999a). But what should I do? *Philosophia, 27*, 433–440.

Hansson, S. O. (1999b). The moral significance of indetectable effects. *Risk, 10*, 101–108.

Hansson, S. O. (2004a). Fallacies of risk. *Journal of Risk Research, 7*, 353–360.

Hansson, S. O. (2004b). Great uncertainty about small things. *Techne, 8*, 26–35 [Reprinted in *Nanotechnology Challenges*: *Implications for Philosophy*, *Ethics and Society*, eds. Joachim Schummer, and Davis Baird, 315-325. Singapore: World Scientific Publishing, 2006.].

Hansson, S. O. (2007a). Hypothetical retrospection. *Ethical Theory and Moral Practice, 10*, 145–157.

Hansson, S. O. (2007b). Philosophical problems in cost-benefit analysis. *Economics and Philosophy, 23*, 163–183.

Hansson, S. O. (2009). From the casino to the jungle. Dealing with uncertainty in technological risk management. *Synthese, 168*, 423–432.

Hansson, S. O. (2011). Radiation protection – Sorting out the arguments. *Philosophy and Technology, 24*, 363–368.

Hansson, S. O. (2013). *The ethics of risk. Ethical analysis in an uncertain world.* New York: Palgrave Macmillan.

Hansson, S. O., & Hirsch Hadorn, G. (2016). Introducing the argumentative turn in policy analysis. In S. O. Hansson & G. Hirsch Hadorn (Eds.), *The argumentative turn in policy analysis. Reasoning about uncertainty* (pp. 11–35). Cham: Springer. doi:10.1007/978-3-319-30549-3_2.

Hansson, S. O., & Joelsson, K. (2013). Crop biotechnology for the environment? *Journal of Agricultural and Environmental Ethics, 26*, 759–770.
Health Physics Society. (1996). Radiation risk in perspective. Position statement of the Health Physics Society. https://www.hps.org/documents/radiationrisk.pdf. Accessed 28 May 2015.
Hirsch Hadorn, G. (2016). Temporal strategies for decision making. In S. O. Hansson & G. Hirsch Hadorn (Eds.), *The argumentative turn in policy analysis. Reasoning about uncertainty* (pp. 217–242). Cham: Springer. doi:10.1007/978-3-319-30549-3_9.
Jaworowski, Z. (1999). Radiation risk and ethics. *Physics Today, 52*, 24–29.
Kahn, H., & Wiener, A. J. (1967). *The year 2000: A framework for speculation on the next thirty-three years*. New York: Macmillan.
Kata, A. (2010). A postmodern Pandora's box: Anti-vaccination misinformation on the Internet. *Vaccine, 28*, 1709–1716.
Levine, M., Mihalic, J., Ruha, A.-M., French, R. N. E., & Brooks, D. E. (2013). Heavy metal contaminants in Yerberia shop products. *Journal of Medical Toxicology, 9*, 21–24.
Lietman, P. S. (2012). Herbal medicine development: A plea for a rigorous scientific foundation. *American Journal of Therapeutics, 19*, 351–356.
Likens, G. E., Herbert Bormann, F., & Johnson, N. M. (1972). Acid rain. *Environment: Science and Policy for Sustainable Development, 14*, 33–40.
Maglione, M. A., Das, L., Raaen, L., Smith, A., Chari, R., Newberry, S., Shanman, R., Perry, T., Goetz, M. B., & Gidengil, C. (2014). Safety of vaccines used for routine immunization of US children: A systematic review. *Pediatrics, 134*, 325–337.
McBrien, J., Murphy, J., Gill, D., Cronin, M., O'Donovan, C., & Cafferkey, M. T. (2003). Measles outbreak in Dublin, 2000. *The Pediatric Infectious Disease Journal, 22*, 580–584.
McKinney, M. L., & Lockwood, J. L. (1999). Biotic homogenization: A few winners replacing many losers in the next mass extinction. *Trends in Ecology and Evolution, 14*, 450–453.
McKinney, M. A., Letcher, R. J., Aars, J., Born, E. W., Branigan, M., Dietz, R., Evans, T. J., Gabrielsen, G. W., Peacock, E., & Sonne, C. (2011). Flame retardants and legacy contaminants in polar bears from Alaska, Canada, East Greenland and Svalbard, 2005–2008. *Environment International, 37*, 365–374.
Möller, N. (2016). Value uncertainty. In S. O. Hansson & G. Hirsch Hadorn (Eds.), *The argumentative turn in policy analysis. Reasoning about uncertainty* (pp. 105–133). Cham: Springer. doi:10.1007/978-3-319-30549-3_5.
Molnar, J. L., Gamboa, R. L., Revenga, C., & Spalding, M. D. (2008). Assessing the global threat of invasive species to marine biodiversity. *Frontiers in Ecology and the Environment, 6*, 485–492.
Mooney, C. (2005). *The Republican war on science*. New York: Basic Books.
Moore, F. N. (2002). Implications of nanotechnology applications: using genetics as a lesson. *Health Law Review, 10*, 9–15.
Moreno, J. D. (2001). *Undue risk. Secret state experiments on humans*. New York: Routledge.
Ong, E. K., & Glantz, S. A. (2001). Constructing 'Sound Science' and 'Good Epidemiology': Tobacco, lawyers, and public relations firms. *American Journal of Public Health, 91*, 1749–1757.
Oreskes, N., & Conway, E. M. (2010). *Merchants of doubt: How a handful of scientists obscured the truth on issues from tobacco smoke to global warming*. New York: Bloomsbury Press.
Overbye, D. (2008, April 15). Gauging a Collider's odds of creating a black hole. New York Times. http://www.nytimes.com/2008/04/15/science/15risk.html. Accessed 24 Aug 2015.
Oxford English Dictionary (OED). (2015). *fallacy, n.* Oxford University Press. http://www.oed.com. Accessed 28 May 2015.
Proctor, R. N. (2004). The global smoking epidemic: A history and status report. *Clinical Lung Cancer, 5*, 371–376.
Revkin, A. C. (2011, July 22). On green dread and agricultural technology. *New York Times*. http://dotearth.blogs.nytimes.com/2011/07/22/on-green-dread-and-agricultural-technology/. Accessed 28 May 2015.

Revkin, A. C. (2013, August 27). From Lynas to Pollan, agreement that golden rice trials should proceed. *New York Times*. http://dotearth.blogs.nytimes.com/2013/08/27/from-mark-lynas-to-michael-pollan-agreement-that-golden-rice-trials-should-proceed/. Accessed 28 May 2015.

Rudén, C., & Hansson, S. O. (2008). Evidence based toxicology – 'Sound science' in new disguise. *International Journal of Occupational and Environmental Health, 14*, 299–306.

Ruthen, R. (1993). Strange matter. *Scientific American, 269*, 17.

Sagoff, M. (1988). Some problems with environmental economics. *Environmental Ethics, 10*, 55–74.

Saper, R. B., Kales, S. N., Paquin, J., Burns, M. J., Eisenberg, D. M., Davis, R. B., & Phillips, R. S. (2004). Heavy metal content of ayurvedic herbal medicine products. *Jama, 292*, 2868–2873.

Shaw, D., Graeme, L., Pierre, D., Elizabeth, W., & Kelvin, C. (2012). Pharmacovigilance of herbal medicine. *Journal of Ethnopharmacology, 140*, 513–518.

Wack, P. (1985a). Scenarios, uncharted waters ahead. *Harvard Business Review, 63*, 73–89.

Wack, P. (1985b). Scenarios, shooting the rapids. *Harvard Business Review, 63*, 139–150.

Chapter 5
Value Uncertainty

Niklas Möller

Abstract In many decision-situations, we are uncertain not only about the facts but also about our own values that we intend to apply to the problem. Which values are at stake, and whether and how those values compare may not always be clear to us. This chapter introduces the issue and discusses some ways to deal with value uncertainty in practical decision-making. In particular, four types of uncertainty of values are introduced: uncertainty about which values we endorse, uncertainty about the specific content of the values we do endorse, uncertainty about which among our values apply to the problem at hand, and the relative weight among different values we endorse. Various ways of contributing to solving value uncertainty are then discussed: contextualization, hierarchy of values, assigning strength to values, embedding and transforming the problem. Furthermore, two methods of dealing with value uncertainty remaining even after these methods have been applied are treated.

Keywords Value uncertainty • Fact-value distinction • Contextualization • Normative theorizing • Argumentation • Reflective equilibrium

1 Introduction

When we talk about 'uncertainty' in decision problems, we typically think of factual uncertainty, that is, uncertainty about how the facts stand. What are the potential outcomes of an action or policy, and what are the probabilities for those different outcomes? These are two typical questions we ask ourselves, often without finding a satisfactory answer. We are then uncertain about the facts. Our knowledge about the values themselves, on the other hand, is often taken for granted: we do not know how the world is, but we do know how we want it to be. We want to save lives, ensure freedom, welfare and security. More often than

N. Möller (✉)
Department of Philosophy and the History of Technology, Royal Institute of Technology (KTH), Stockholm, Sweden
e-mail: nmoller@kth.se

S.O. Hansson, G. Hirsch Hadorn (eds.), *The Argumentative Turn in Policy Analysis*, Logic, Argumentation & Reasoning 10, DOI 10.1007/978-3-319-30549-3_5

we perhaps want to admit, however, we are unsure how to evaluate the potential outcomes. We are then uncertain about our values.

Value uncertainty is far more common than the (typically absent) discussion of the phenomenon would suggest. In many decisions, we are uncertain not only about the facts of the matter but also about which values we intend to apply to the problem. This chapter introduces the issue and discusses some ways to deal with value uncertainty in practical decision-making.

I will proceed as follows. In the next section, I will introduce the topic by discussing some central distinctions for value uncertainty: in particular that between facts and values, and between the subjective and the objective. My stance towards the controversies about the fact-value distinction is that rather than undermining the distinction, they motivate awareness about the distinction being one of degree rather than kind; there are still good pragmatic reasons to use it. As to the complex question about the status of values, whether they are subjective or in some sense transcend the individual or interpersonal evaluation, what matter for our decision-making are the *actual* commitments we have, and so our subjective values are central for the current chapter.

In Sect. 3, I will distinguish several important aspects of value uncertainty. I will argue that most of us are uncertain about our values in the sense that there are *hypothetical* situations in which we would not be certain about what we prefer. What mainly matters for decision-making, however, is the actual decision situation we confront, and it is value uncertainty in this more local sense which we will be focusing on in the current chapter. Other distinctions I introduce are whether we have full or only partial information, and different kinds of strength of preferences. Moreover, I will distinguish between four types of uncertainty of values: uncertainty about *which* values we endorse, uncertainty about the specific *content* of the values we do endorse, uncertainty about *which* among our values *apply* to the problem at hand, and *the relative weight* among different values we do endorse. Lastly, I introduce uncertainty about moral theories, a form of value uncertainty sometimes discussed in moral philosophy.

In Sect. 4, I will introduce some methods contributing to solving value uncertainty by *specifying* the problem. The aim here is to clarify what the salient factors may be, as such clarification often lessens the uncertainty. One central method here is *contextualization*, making explicit the relevant context in which the value will be applied. I will also discuss the importance of clarifying the hierarchy among our values as well as how much weight the values carry, especially for situations where there are conflicting values at place. Two further methods introduced are modifying the embedding (framing) of the problem, and transforming the problem, for example by postponing our original decision or make the overall problem into several, smaller, decisions.

In Sect. 5, I will discuss methods for what to do when clarifying is not enough. While more clearly specifying the problem often may lessen or even solve the problem, it may of course remain even in the most detailed and thought-through characterization of what is at stake. Two approaches will be introduced. The first comes from the debate in philosophy about moral uncertainty, where it is argued that there are rational decision methods for what to do even when we remain uncertain about which moral theory is the correct one. Some theorists argue that we should then compare the recommendations given by all of the theories we put some credence in, and, for example, choose the alternative that would maximize the expected moral value. Other theorists argue that we should instead pick the one moral theory we put *most* faith in and stick to that, no matter our moral uncertainty.

This first approach is limited to uncertainty about moral *theories*, but I will also raise some skeptical points against its viability in that area. The second approach, however, I take to be a more promising way forward. In fact, it amounts to the overall theme of the present anthology (Hansson and Hirsch Hadorn 2016), pointing to *argumentation* as the solution to uncertainty. Here, I will in particular introduce the method of reflective equilibrium, a central method in current normative philosophy; but in more general terms, the entire anthology exemplifies ways in which the argumentative process always offers a potential way forward where there is uncertainty.

2 What Is Value Uncertainty?

Let us start by characterizing the phenomenon of value uncertainty in more detail, starting with the case of Eve, which will follow us throughout the chapter. Eve hesitates whether or not she should give money to the woman who often sits begging outside her supermarket. A lot of her indecision is due to factual circumstances: although she is pretty convinced the woman would not sit there were she not poor, she does not know *how* poor the woman in fact is. And she does not know whether giving the woman money, while helping her out in the short perspective, contributes to retaining her poverty in the long run. Some of these facts are comparably easy to gain access to, while some are much harder. But we typically take there to be a fact of the matter in relation to questions such as these. Some of her uncertainty, however, has to do not with facts but with what should guide Eve's decision, even *given* a certain set of facts. She may wonder whether she should show kindness to the woman and give her money, or whether she should promote the autonomy of the woman by refraining from doing so. These latter questions are questions of value rather than of fact.

In this vein, value uncertainty may be characterized in relation to the factual: value uncertainty is uncertainty beyond factual uncertainty (see Hansson 2016). As the term suggests, value uncertainty is uncertainty about what we *value*. In this chapter, this will be interpreted broadly, pertaining not only to uncertainty explicitly expressed in terms of values, but also about uncertainty expressed in terms of preferences, norms, principles or (moral or political) theories. Moreover, the uncertainty may be both about *what* we value – e.g. freedom, security, a morning cup of coffee – and about *how much* value we assign to that which we value. Consequently, uncertainty expressed as 'is gender equality an important value to me?', 'is less expensive energy preferable to more expensive but a more sustainable means of energy production?', or 'should I follow the principle to harm someone only in self-defense?' are, to the extent the question relates to non-factual issues, all examples of value uncertainty. Consequently, when expressions such as 'uncertainty about our values' etc. are used in the chapter, it should be understood in the broad sense. I will, however, sometimes explicitly mention norms, principles etc. in order to remind the reader of the broad notion of value used, or when focus is directed specifically at these aspects of the notion.

2.1 Facts Versus Values

Before we engage further with value uncertainty, it should be mentioned that the distinction between factual uncertainty and value uncertainty makes sense only to the extent that facts and values are distinguishable. A contemporary theme in philosophy has been to critically evaluate the extent to which they are (Putnam 1990, 2002; Quine 1953). The perhaps most influential thought here is that the class of propositions we take to correspond to facts, on a closer look turns out to be essentially dependent on values. Even science, the paradigm of fact-investigating endeavor, contains values, for the simple reason that there is no theory-neutral description of the world. What we take to be a fact depends on the theory choices we make, and we cannot choose among competing theories without values. These so-called epistemic values – *coherence*, *simplicity*, *reasonableness* etc. – are integral to the entire process of assessment in science. Hence, our fundamental knowledge of the world is value-dependent (McMullin 1982; Lakatos and Musgrave 1970; Kuhn 1962).

The standard retort in view of these concerns is that the *epistemic values* of science and other fact-stating enterprises are different from the *action-guiding* values we are talking about here; practical values are guiding us in knowing what to *do* rather than what to *believe*. While epistemic values help us choose theories and classifications, only action-guiding values help us determine what to do.

The debate does not end here, and in may turn out that the class of factual claims which do not contain any action-guiding values is smaller that we intuitively think.[1] Still, when keeping in mind that the border between facts and (action-guiding) values may be vague and contestable or that a conceptual distinction between facts and values does not imply full independence of factual claims and value judgment,

[1] One often-mentioned complication is the class of concepts labeled 'thick concepts' in moral philosophy. Thick concepts such as *courage* or *cruelty* are traditionally conceived of as *both* having descriptive content and being evaluatively loaded. By being evaluative, they differ from purely descriptive concepts such as *water* and *red*, which have no such evaluative quality. But they differ also from the thin evaluative concepts such as *good* and *right*, since they have a more specific descriptive content. This intermediate position has been seen as problematic for theorists who have relied on a sharp distinction between facts and values. It would take us too far to go into the details in this debate, but the interested reader should look into Väyrynen (2013), Dancy (1995), Williams (1985), and McDowell (1978, 1979, 1981).

it is hard to deny that distinguishing some questions as factual questions and others as value questions is *useful*. It captures categories in which we perceive the world and, as we will see in this chapter, keeping separate, as far as possible, matters of value and matters of fact helps us situate the problem we confront of as well as suggest ways of moving forward.[2]

2.2 *Subjective or Objective Values?*

When we talk about preferences, we typically mean *someone's* preferences. *I* prefer apples to oranges, whereas you may not. Preferences understood this way are then *subjective* in that they essentially relate to a subject.[3] Talk about values and norms, on the other hand, is typically ambiguous between *my* subjective values and norms, and values and norms in a more intersubjective or objective sense. Often our value claims express our personal commitments and may thus refer to values in a subjective sense. Expressions such as 'American values', on the other hand, typically refer to intersubjective values, whereas when we talk about 'the unfortunate lack of gender equality in many countries' we are perhaps rather referring to what we take to be an objective value, a value which is valid or correct although perhaps not shared by all.

Sometimes it is of paramount importance to distinguish between subjective, intersubjective and objective values. For one thing, values in the subjective and intersubjective sense obviously exist, since people de facto are committed to certain values, whereas the existence of objective values is a controversial and heavily debated question in moral and political philosophy.[4] For the purpose of this chapter, however, we need not take a stand on matters such as these, and we will not distinguish between subjective, intersubjective or objective values. Value uncertainty is here interpreted in the first instance as a property of a mental state of a person being uncertain about what to do. When I am uncertain about which value to

[2] Note, however, that while the distinction between facts and values utilized here assumes that there is *some* interesting and systematic distinction to be made, rather than a totally gerrymandered one, it does not assume any deeper ontological or metaphysical commitment, such as a denial of truth or objectivity in morality. In moral philosophy, there is an open debate about whether or not there are moral facts, and if so, whether such facts are natural facts in disguise, or constitute some other, non-natural sort of fact. (See footnote 4 for relevant literature.) The distinction between fact and value is well-established, however, and with the now mentioned caveat, we will adhere to this tradition in this chapter. Philosophers subscribing to moral facts may translate what we in the main text label merely 'fact' into 'descriptive fact' or 'non-normative fact'.

[3] For comprehensive accounts of the notion of preferences, see Hausman (2011) and Hansson and Grüne-Yanoff (2006).

[4] In various versions, it is arguably *the* question of the domain within moral philosophy which deals with the status of morality: metaethics. Among the huge literature in the area, recommended modern classics include Blackburn (1998), Smith (1994), Brink (1989), and Mackie (1977). For a comprehensive modern overview, see Miller (2013).

apply, or how to weight different values, what matters are the values to which I am *committed* – in other words, values in the subjective sense. These values may also be intersubjective, or even, were there to be such a thing, objective, just as my subjective beliefs may be both intersubjectively shared and objectively true.[5] But unless *I* am committed to these values (or to abide by them in virtue of other values I hold, such as behaving in accordance with whatever the communal values happen to be) they do not enter into my considerations. Similarly for the case of a group decision, what matters are the values to which we are committed, regardless of any further ontological status beyond this fact.[6]

A potential objection to looking at all values from the subjective point of view when discussing value uncertainty would be that it matters for the *justification* of the values we are committed to whether values exists in any objective sense, since it is then important to *discover them* rather than merely *deciding* on a set of values. But for our concerns this objection would only be valid if there were any method of discovering values which were different from any reasonable method of 'deciding' on them. And it turns out that there is not: whether or not values exist objectively in any interesting sense or not, the only method there is for justifying one's values is through *argumentation*, through giving and asking for reasons for being committed to them.[7] I believe in gender equality, say, since I fail to see that the biological differences between men and women provide any good reason for why women should be discriminated against. If I, on the other hand, were to believe in male superiority, I would believe in this value for *some* reason, for example a belief that women are evolutionary fitted to childcare, and that this fit is hardwired and make them less suitable to other tasks. Others – or indeed our introspecting self – may of course object to any consideration brought up in favor of a value commitment, but we never transcend the circle of giving or asking for reasons for our commitments. Consequently, although we may of course *say* that I should not murder innocent people because it is morally bad to do so, it is only a motivating reason to me if there is a reasonable answer to the question *why* it is morally bad, in the same way as the answer 'because it is true' does not really give me a further reason to believe in a claim in which I doubt.[8]

Related to the question of objective and subjective values is the question of moral and other values. In many circumstances, talk of values implies talk of *moral*

---

[5] My belief that there is water in the glass in front of me, for example, may be shared by others as well (intersubjective) and may be true (objective). Similarly, if justice is an objective value it may be acknowledged by me (subjective) as well as others (intersubjective).

[6] I say 'ontological status' here since other statuses, such as whether we disagree on our values, may of course be important for arguments about how to weigh our values.

[7] The central notion of reflective equilibrium will be treated in Sect. 5 below. See further Betz (2016) and Brun and Betz (2016) in the current volume.

[8] We are thus here referring to internal reasons, i.e. considerations which a person *takes* to be a reason. We may also talk about external reasons, considerations that speak in favor for a certain alternative, whether or not the person in fact realizes that this is so. For further discussion of the distinction, cf. e.g. Finlay (2006), Smith (1987), Williams (1981 [1979]).

values – which typically include also human and political values. That an action is just corresponds to a value (justice) in this sense, whereas that an action benefits my interest, some would say, does not. And indeed, sometimes a distinction between moral and other more prudential or self-regarding values may be of interest. Here, on the other hand, values are understood in a broad sense which is neutral to whether or not they are other-directed or self-directed. If Eve is uncertain about whether to give money to the poor woman, the values which are contributing to this uncertainty may be moral (a right not to be poor, for example) as well as totally self-regarding (how giving to the woman makes her feel, say).[9] What matters for value uncertainty is whether she is uncertain about her values and how to weight them, not what type of values they are.

2.3 Agency

As mentioned above, I will treat value uncertainty as relating, in the first instance, to the values held by an agent. While 'agent' is neutral between *individual* or *group* agent, most examples will consider the individual case. The reason for this is not to claim that value uncertainty is only a phenomenon of individuals, denying that group decisions, small or large, may be fraught with value uncertainty as well. To the extent that we may reasonably talk about group agency, that *we* believe, want or decide things, we may certainly talk about *our* value uncertainty as well.[10] When we do, however, all the methods and techniques mentioned throughout in this chapter are equally applicable to the many person case. Naturally, in addition to the *internal,* intrapersonal deliberation of the individual case we have the *external,* interpersonal deliberation of the many person case. Moreover, metaphorical talk such as 'part of me is committed to never lie' may have a fully literal analogue in the many person case, since there may be an actual person being so committed.

Hence, the decision procedure is more complex in the many-person case: in the single-person case there is only one me who is doing the deciding, whereas there are many potential ways of reaching a decision in the many person case. And this is exactly the point of focusing on the individual case in the present chapter: it is sufficient for introducing the basic problem of value uncertainty and the main ways of dealing with it, while at the same time avoiding many further problems, in particular those of justified decision procedures in group decisions. The latter is an important topic, indeed, much theorized and debated, in political theory and other areas, but has little to do with value uncertainty as such; moreover, it would require

[9] The distinction between moral and other types of values is further controversial, in that there are moral theories, such as ethical subjectivism, which count self-regarding values as the correct moral values.

[10] For discussion of group agency, cf. e.g. Pettit (2009), Tuomela (2007), Bratman (1999), and Searle (1990).

far more space that what is presently available (Peter 2009; Rawls 1993, 1999 [1971]; Dworkin 1986; Habermas 1979, 1996; Dahl 1956).

Consequently, we will focus on value uncertainty on the abstraction level of the agent – which typically is an individual but need not be – and disregard the special problems of many-person decision procedures apart from the techniques and considerations brought up below.

3 Kinds of Value Uncertainty

Value uncertainty comes in many forms. I will not try a complete taxonomy here, but a few distinctions may be helpful in order to get a better grip of the phenomenon. Before going on to address solutions, let us therefore distinguish between different varieties of value uncertainty.

3.1 Hypothetical Versus Actual Uncertainty

Let us imagine an agent who is certain about how to how to rank all possible factual states of the world in all possible circumstances.[11] Some such ranking may be expressed in general terms. Let us say, for example, that the agent would always prefer a cup of coffee to a cup of tea, but a cup of tea to a cup of hot chocolate. Other orderings require more detailed state descriptions. Although she has preferred carrots over peas in every actual decision situation she has faced, she knows that were she to have carrots as the only vegetable for a week, she would actually prefer peas over carrots for the next meal. If her mind is totally made up among all such possible preference relations, sufficiently specified, she is in a state of full outcome preference certainty.[12]

It seems reasonable to assume that such full outcome preference certainty is a fiction. Many of us have considered a hypothetical choice in which we were unable to identify some outcome that we considered to be at least as good as any other alternative.[13] But such hypothetical uncertainty is of course compatible with people being certain about what to do in many (indeed even all) *actual* decision situations.

[11] As mentioned in the last section, the phenomenon of value uncertainty can be expressed not only directly in terms of uncertainty about values, but also in terms of uncertainty about preferences, norms, principles or even theories.

[12] C.f. Gibbard (2003) for a similar conceptualization.

[13] This is so even if we are restricting the domain to the – still very large – domain of *physically* possible states, as opposed to the even larger domains of the outcomes which are conceptually, logically or even metaphysically possible (cf. Erman and Möller 2013). If we are unable to decide whether one of two states of affairs is better, worse or equal in value, we commonly call these two states of affairs *incommensurable* (Raz 1986).

That I am uncertain about how to value a hypothetical case may have no bearing on my decisions if this case never actualizes. I may be uncertain of what to do if I face some hard dilemma such as saving thousand people to the expense of several of those near to me, yet (hopefully) live my whole life without having to face that choice.

In the present chapter, the main focus will be on solving *actual* or more *local* cases of value uncertainty. Specifically, I will focus on value uncertainty in relation to a particular situation. If I, *in a given decision situation*, find myself uncertain about what to do, value or prefer, and this uncertainty goes beyond a lack of factual information, in the sense that additional factual information does not solve my uncertainty, I am facing a case of value uncertainty in this actual or local sense on which we will focus. Consequently, removing our uncertainty in such a given decision situation is compatible with the value uncertainty remaining in a similar (but of course not exactly similar) situation. Still, an important goal has been reached.

3.2 Full or Partial Information About Outcomes

The paradigmatic case of value uncertainty is when we do not know what to do given *full information*, i.e. given that all the relevant facts are settled. If we have full information of what will happen on all available alternatives, but are still uncertain, it is a clear-cut case of value uncertainty. But also some uncertainty under situations with less than perfect factual information can reasonably be considered as value uncertainty. In many situations, we do not fully know what the facts are, or will be, given our decisions.[14] This is evident in decision situations both small and large: when deciding which clothes to wear in light of the weather as well as when deciding on different climate strategies, our decisions are fraught with epistemic uncertainty. In such cases, it does not even suffice to know which outcomes we prefer, we need to know how to value the uncertainties we face as well. I like both coffee and tea, and while I prefer coffee to tea, do I prefer a 50 % chance of getting a cup of coffee (risking to get no drink) to a 100 % chance of getting a cup of tea? And what if I do not know the probability of my getting a cup of coffee at all?

The decision theoretical literature typically distinguishes between at least three different levels of epistemic (un)certainty: cases in which I have full deterministic knowledge of which outcome my decision leads to (decision under certainty), where I may assign probabilities to the outcomes (decision under risk), and where

[14] It might be argued that we face epistemic uncertainty in *all* situations. Still, it is often reasonable to approximate certainty in decision-situations: for example, it is typically not necessary to include the possibility that my shirts suddenly have vanished from my closet (perhaps stolen or eaten by a swarm of moths) when thinking about what to wear for work.

I cannot (even) assign probabilities (decision under ignorance).[15] For all these cases, theorists have argued for various decision procedures, given certain assumptions on our evaluations of the available outcomes. Moreover, cases may be mixed as well. One option may give a certain outcome for sure, whereas we may in another option not be able to assign probabilities to the various outcomes. In all of these cases, I may be uncertain which strategy to use. Should I choose a certain, less valuable outcome to an uncertain but potentially more valuable one, or should I take the chance of gaining more at the price of loosing more?

3.3 *Uncertainty in Relation to Strength of Preferences*

Not only the preference orderings between outcomes but also the 'distances' between them often become relevant for whether or not we have value uncertainty. If all I know is that I prefer coffee to tea, I might be uncertain about how to evaluate a situation in which there is, say, an 80 % chance of receiving coffee (but 20 % risk of receiving nothing) to a definite outcome of receiving tea. But if my preference for coffee is only minimally stronger than my preference for tea, I probably value a definite outcome of getting tea more. If on the other hand my preference for coffee is very strong, even a 10 % chance of coffee may be preferable to a definite outcome of receiving tea.

If I know my preference ordering between all available alternatives, my preferences may be measured on what is called an *ordinal scale*. But an ordinal scale says nothing about the strength of the preferences beyond the relative positions of the outcomes. That $A > B > C$ (where ' $>$ ' should be interpreted 'is preferred to') can be true both if the alternatives are almost equivalent to me and if I take A to be much more preferable to B, etc. For an ordinal scale, that is, the only thing that matters in a numerical representation of the outcomes is their order: $(A, B, C) = (53, 52, 51)$ has the same meaning as $(A, B, C) = (1000, 50, -10)$.

In order to capture the relative strengths of my preferences, we need to be able to measure them on an *interval scale*. An interval scale captures the notion we intuitively read into the above ordered lists, namely that A in the latter is much more preferable than B, whereas in the former they are rather close. In decision theory interval scales are of paramount interest, since only when we have them may we construct *utility values* representing our outcomes so that, given that we may also assign probabilities for all outcomes, the notion of expected utility becomes meaningful. The expected utility of an alternative is the probability-weighted sum of that alternative, and a central – one may even say dominant – method in decision theory is that one should choose an alternative that maximizes the expected utility.

[15] See Alexander (1970), Luce and Raiffa (1957) (who use the term 'uncertainty' rather than 'ignorance' for the third level). See Hansson and Hirsch Hadorn (2016) in this volume for comments on different notions of uncertainty.

Consequently, even when we are certain about our preferences we may still be uncertain as to their relative strength. We then face yet another kind of value uncertainty.[16]

3.4 Four Types of Uncertainty About Values

In this section up till now, we have for illustrative purposes mainly expressed value uncertainty in relation to preferences: the preference *relation* or the *property* or *state* we prefer. Let us now turn to value uncertainty expressed directly in terms of values. There are at least four – related but analytically distinct – ways of being uncertain about values. First, we may *be uncertain about which values we endorse*. Some values we are uncertain about whether we endorse at all. Some argue for the value of saving endangered species, for example, while others take there to be no such value, arguing that it is a natural flow of evolution that some species who are not sufficiently fit become extinct, and that this is as it should be. Secondly, even more common is perhaps *uncertainty about the content of values* we endorse. While most people arguably feel certain about fundamental values such as justice and equality at *some level*, they may be unsure about their more exact content. For example, many of us are genuinely uncertain about the limits of equality of welfare. Too much inequality of welfare is not good, but is total equality the goal, or is some inequality as a consequence of different efforts and talents in relation to our contribution to society preferable to total equality?

Third, even when we have a reasonably good grasp of which values we endorse, we may be *uncertain about which values apply* to the problem at hand. Values are more often than not hidden entities of a decision-problem. While we may identify a stream of feelings and desires in a situation, as well as a number of beliefs about the relevant facts, identifying which values apply to the situation may not be transparent. Take Eve, who wonders about whether to give money to the woman outside the supermarket. Eve is conflicted. She feels sorry for the woman, but she is also troubled by the fact that there has been such an influx of beggars from other countries due to the free movement within the European Union. She wishes there were no beggars in the city at all. But she is fundamentally uncertain about her values among all these feelings and wishes.

The problem for Eve here is not that she *has* no values with which to evaluate different potential outcomes. Whether we know how they apply to the situation, we all have values; and in this situation, Eve's feelings and wishes are definite signs of their presence. But she is still unsure about what her values really are amidst her feelings and wishes. The situation is very common. Values are more often than not hidden entities of a decision-problem.

[16] The distinctions introduced here are commonplace in the decision-theoretical literature. For accessible introductions, see Peterson (2009) or Resnik (1987).

Fourth, in analogy with the discussion above about the ranking of preferences, we may be uncertain about how to *weigh different values*. Often, the main source of our value uncertainty may not be which values there are, or even which values we take to apply to a situation, but how to much weight these different values should have. In other words, it is often unclear which values are more important in a particular situation. Take the Parable of the Prodigal Son, where the younger son, after having wasted his share of the father's estate through extravagant living, returns, now poor and miserable, to his father, asking to be hired as his servant. In line with the abovementioned uncertainty, we may ask which values pertain to this situation. Justice, desert, kindness and forgiveness are values which perhaps come to mind. But how are we to decide which value is more important when they point in different directions? The father famously celebrates the return of the lost son, which the older son, who has stayed and helped the father throughout, takes to be a big injustice. The father may not disagree, but clearly thinks kindness and forgiveness to be more important here. Arguably, part of the power of the parable lies in the tension between desert and justice on the one hand, and kindness and forgiveness on the other.

3.5 Uncertainty About Moral Theories

In moral or political philosophy, explicit treatment of value uncertainty is rare. One exception is the debate on what has been called *moral uncertainty*. Moral uncertainty is typically defined in this debate as good reasons for more than one moral theory. More precisely this is typically spelled out in terms of *positive credence in more than one moral theory*. For example, many of us have *both* consequentialist and deontological (duty-based) intuitions. Consequentialism holds that an action is right when it is the action with the best consequences, typically measured in terms of well-being or preference satisfaction. Many of us believe that what made an action right was this feature of having better consequences than other acts. We should not steal other people's property, for example, since a society in which we did so would be worse than a society in which we refrained from doing so. But many of us also take there to be rules we should follow even when they do not lead to the best consequences, such as not to put an innocent person in jail or sacrifice one person to save the lives of others, even if doing so would maximize happiness or preference satisfaction. Moral uncertainty will arise if we put some credence both to consequentialism and to some duty-based theory.

The problem moral uncertainty theorists aim to solve is what to do under moral uncertainty when we have *diverging* recommendations. (Naturally, when all of the moral theories in which we have positive credence recommend the same action, it seems safe to say that we have no problem knowing what to do. The problem enters when one theory says, for example, 'lie' and another 'do not lie', or when one theory treats one alternative as acceptable whereas the other treats it as unacceptable.) We will return to moral uncertainty in the last section of the chapter.

4 Making It Explicit: Methods and Techniques

Until now I have addressed mainly what value uncertainty *is*. Now I will turn to the difficult question of what to do about it. In one sense, of course, we already know what we should do: make up our minds. The straightforward way of solving value uncertainty, on this line of thought, is to make up our minds sufficiently, decide what we really prefer or value or which norms and principles we really should act upon. Arguably, sometimes we may be able to *directly* follow this advice. When you are about to pick your two flavors of ice cream and suddenly do not know which you prefer, that uncertainty typically passes before the parlor staff becomes too impatient. And when Adam's family is uncertain about whether they should go skiing in the alps or sunbathing in Thailand on the holiday, uncertain about whether they value lazy warmth or active cold, perhaps all they have to do is to sit down and think about what they prefer, and their minds are made up without further ado.

However, this advice is only useful if we have some clue about how to make up our minds. Otherwise it is as helpful as the knowledge that we should buy stocks when the price is low and sell when it is high. That is, not helpful at all. I am in a state of value uncertainty *because* I have been unable to make up my mind, and simply ordering myself to do so does not help if I do not know how. Fortunately, in many cases there are some more substantial pieces of advice to put forward.

Generally speaking, there are two main ways of making up one's mind: through clarification and through argumentation. We will return to argumentation, the main theme of the current anthology (Hansson and Hirsch Hadorn 2016), in the next section. In this section we will investigate a number of techniques and methods which may help us solve cases of value uncertainty mainly through clarifying the problem.

The common core in the methods and techniques presented below is that they help us to specify the parameters of the problem. Our values and norms are often vague and unclear to us, and not fully explicit. Only when our underlying values have been made sufficiently explicit, only when their content is sufficiently transparent to us, are we able to appreciate whether they allow for a solution upon reflection.

Returning to Eve, she believes in many values, although she does not often formulate them explicitly. In particular, she believes in fairness, and while she has a distinct feeling that the beggar-situation is unfair somehow, she is uncertain about what fairness entails in this particular instance. Part of her uncertainty, let us say, is due to this vagueness or lack of clarity. In order to get a better grasp, she may then attempt to *further specify* the conception of fairness in which she believes. Specification of one's values generally means to clarify the content of one's commitments in more detail. What does a commitment to fairness really entail?

We will now turn to several analytic techniques which may help us clarify our value commitments and the decision situation as a whole.

4.1 Contextualization

One useful type of specification is *contextualization*: making explicit the relevant context in which the value will be applied. A person may, for example, be truly uncertain whether she prefers coffee or tea, when the question is asked in the abstract. But she may be completely certain that she prefers tea to coffee in the morning, coffee to tea during the workday, and again tea to coffee if it is late in the evening. A decision situation typically comes with contextual factors which suitable incorporated may solve the value uncertainty for the case at hand.[17]

Explicit contextualization is helpful in two ways. First, and most obviously, it may reveal value uncertainty in the particular decision situation for which the contextualized specification has been made. If the initial uncertainty was about preferences for coffee or tea, and the decision turns out to be about breakfast options, the above specification reveals the uncertainty completely for the case at hand. Secondly, the contextualized specification has made a more general ordering available, to be used in different specific situations. While the person may still be uncertain about the choice between coffee or tea in the abstract (and thus perhaps be undecided what to bring to an isolated island), her values – in this case preferences – have been more extensively clarified and she has reached a state in which a more general ordering is available, where at least the following holds:

tea in the morning > coffee in the morning
coffee during the workday > tea during the workday
tea in the evening > coffee in the evening

In this way, contextualization may solve a case of value uncertainty by clarifying the relevant circumstances for the problem at hand. In the above example it meant *localizing* the problem, in that the more abstract problem was turned into a more concrete one by adding contextual information. But it may go the other way around as well. We may be unsure about how to value a specific conflict between welfare and freedom even if we deem welfare to have precedence, generally speaking. Many parents take the welfare of their children to be more important than their freedom to choose for themselves, for example. Still, there may be many situations where they value the children's freedom more, even when they think that their welfare will suffer. Maybe spending all their saved up allowance on that shiny new toy will not be the choice which best furthers their wellbeing; still, we may value their freedom to make the choice. At least sometimes.

[17] Of course, some contexts come close to implying the highly abstract question about tea or coffee, such as if the decision situation is that the person is going to spend a week in an isolated place and may only bring one type of beverage.

4.2 *Hierarchy of Values*

Another analytic tool for solving value uncertainty is to making explicit one's *hierarchy* of values and norms. Much of what we value, we value for *instrumental* reasons, as means to a further end. Other things we perceive as having final or intrinsic values, sometimes called basic values: we value them for their own sake, not only as means for further ends.[18] Many value a good economy, for example, but for most – if not for Scrooge McDuck – it is hardly a final end. Why do we value a good personal economy? Because of the things we may do, such as going on holiday or being able to replace the refrigerator when it breaks down. Arguably, holidays and refrigerators are not final values either. We go on holidays, say, to rest or to explore new exciting places; and we value the refrigerator since it keeps our food fresh. And so on.

Although the notion of intrinsic value is interesting in its own right, in actual decision situations we seldom need to know which values we take as *most* fundamental.[19] The useful point for our purposes is that thinking in terms of what the *more* basic values are can help us realize, when we are uncertain about our values or norms, which of them should matter more in the situation at hand. Thinking in terms of instrumental and more basic values thus helps us clarify what's at stake.

Sometimes clarifying the order of one's values and norms solves the uncertainty completely. Returning to Eve, she feels sorry for the woman, indicating in her mind that she should help her. But Eve is conflicted since she wishes there were no beggars in the city, and she is convinced that helping the woman would provide further incentives for begging in the streets. Thinking further about what values ground these conflicting feelings, let us imagine, she realizes that her care for the wellbeing of the woman outside the supermarket reflects what she takes to be an even more basic value: the right of every person to fundamental goods such as food, shelter and medicine. Moreover, thinking hard about it, she finds that her desire that there were no beggars in the city is not a basic but an instrumental desire, and that the more basic concern really is that no one should have to resort to begging at all. The relevant underlying value is in fact the same basic right to fundamental goods that grounded her concern for the woman's wellbeing in the first place. Consequently, refraining from giving to the woman would only relieve the symptom by fulfilling the instrumental desire alone, not cure the illness itself.

---

[18] Some authors, such as Christine Koorsgard (1996: 111–112, 1983) make much of the distinction between final and intrinsic value – taking the former to mean the value something has for its own sake and the latter the value something has in itself, which is then argued to be different properties – whereas other authors (e.g. Zimmerman 2001, 2014: 25) treat them interchangeably. For the purpose of this chapter, I will choose the latter practice.

[19] The interested reader is directed to e.g. Zimmerman (2001), Rabinowicz (2000, 2001), Korsgaard (1983, 1996), and Broome (1991).

4.3 Strength of Values

Sometimes an attempt to clarify one's values is not as immediately successful as in the example above. Let us now instead assume that upon reflection Eve's desire that there were no beggars in the city turns out to be an expression of the value for her not having to *bear witness* to poor people begging for money. Meeting beggars in subway stations and outside supermarkets makes her feel bad and disturbs her enjoyment of being outside. Here the basic value is self-directed: her own wellbeing. In this case, reflection has revealed a conflict of values which did not immediately resolve when made explicit. Whereas her wish to give money reflected a concern for the rights of another person, her wish for there to be no beggars reflected a concern primarily for herself. And in this case the former, other-directed concern arguably clashed with the self-directed one.

Clarification has thus revealed a *tension* between two values rather than, as in the previous example, two values pointing in the same direction. Still, having thus clarified the underlying values of the case may have contributed to solving the uncertainty. One way in which the uncertainty is solved is when we realize that the values have different *lexical priorities*, by which is meant that one value is more important than another in the sense that it should always be prioritized. Eve might realize, for example, that the cause of her uncertainty is a real tension between her care for the wellbeing of others and her care for her own everyday wellbeing (which we may assume is well above any basic goods-threshold), but that there is no doubt in her mind, when the background values are made clear, that the wellbeing of persons lacking basic goods always trumps her inconvenience of having to witness this need.

Lexical priority is a strong condition, however, and therefore unlikely to hold for many values. Most people are arguably not really committed to give the wellbeing of people lacking basic goods lexical priority to their own above-threshold level of wellbeing under all circumstances. As utilitarians such as Peter Singer famously have argued, were we so committed we should give away the greater part of our salaries to others in need (Singer 2009, 2015). Fortunately, in order to solve value conflicts all we need is a *contextualized priority*. What matters is that we may judge that the values speaking in favor of a particular action or policy are stronger than the values speaking against it *in the case at hand*.[20] Consequently, even if Eve is not committed to always valuing people's need for basic good above her own (above-threshold) wellbeing, she may acknowledge that *in this case*, her particular inconvenience is outweighed by the value of helping people reach a level of basic goods.[21]

---

[20] This is best perceived as *weighing* values rather than as finding a lexical priority among them.

[21] Of course, it *may* come out the other way around as well. Perhaps Eve realizes that when they come into conflict, *her inconvenience* in fact matters more to her than the basic needs of others. Many theorists agree with David Hume's famous statement, "Tis not contrary to reason to prefer the destruction of the whole world to the scratching of my finger" (Hume 2000 [1738]: part 3, sect. 3). Here it is important to differentiate between solving the value uncertainty of a person or group, and solving it in a *satisfactory manner*. Strictly speaking, the value *uncertainty* is solved as soon as the decision-maker has decided which value is more important to her in the case at hand. At least analytically, it is another question whether or not this solution is morally preferable to alternative ways of settling the uncertainty.

4.4 Embedding (framing)

As discussed at length in (Grüne-Yanoff 2016) in the current anthology, a critical factor for a decision problem is the description or re-description of its elements. Here, we will focus on framing in a looser sense which I will call *embedding,* namely how the delimitation of a problem and the set of actions we perceive as available to us may have consequences for value uncertainty.

As was clear above in our discussion of contextualization, our original understanding of a decision situation is often somewhat vague, allowing for different interpretations. In energy debates, for example, it is often not fully clear what the question is, and how it is delimited. Is the question at hand what means of energy production we should endorse given current levels of energy uses, for example, or is the question rather a broader one in which alternative lifestyles utilizing less energy may be considered? Different values may be relevant, or their strength may vary, depending on how we embed a question, and thus a value deadlock in one embedding might turn out to be solved in another.

In addition to being vague, our initial, often implicit embedding of a problem may also be unnecessarily restrictive. It tends to limit our perspective, making us forget that there are always (in principle at least) many alternative ways of perceiving a decision situation. Returning to Eve, the question whether to give money to the woman begging outside the supermarket has been framed as a question of giving money or refraining from doing so. But Eve could entertain many other alternatives which strive to alleviate poverty. Rather than giving money to the woman directly, she could support initiatives benefitting the begging woman. For example, she could help the woman to other, more societally productive activities, such as selling journals or doing handicraft, by supporting organizations engaging in such activities. Other actions would more indirectly benefit the woman, such as support to organizations who fight for the benefit of the minority group to which she belongs. Alternatively, Eve could include action alternatives in support of *other* poor people, such as the extremely poor (a group to which the begging woman, we may assume, does not belong). By such broadening of what she takes to be the available set of options, she might shift her understanding of the salient features of the decision problem so that her value uncertainty is solved. Let us say that by broadening the potential set of actions to take, Eve's indecision between her personal 'feel good'-value of not having to see beggars in the street and the value of helping people in need matters no more, since there is, she takes it, a more fundamental value at stake: Eve realizes that the begging woman reminds her about the existing poverty in our world, and that Eve's fundamental belief in the human right to a decent standard of life is best served by an action which benefits people in the worst circumstances. So she decides to give the equivalent of what she has previously given to the increasing number of beggars in her city to organizations aiming to help the extremely poor.

Eve's re-framing above is an example of a common phenomenon which could be called *the covering solution*: she finds that there is an option which covers both of the values between which she is undecided.[22] Her action to give to the extremely poor is in line with *both* her self-directed value of not having to see beggars in the street and the value of helping people to live a decent life. Her situation has thus become similar to the moral uncertainty case discussed above, in which two in-principle competing moral theories between which a person is uncertain recommend the same action.

4.5 Transforming the Problem

In the next section, we will turn to the question of what to do, or how to think, in cases where mere clarification of our values or the decision problem is not enough. But first we will say something about transforming or changing the problem as a means of solving value uncertainty. So far, the underlying premise in this section has been that a more thorough specification of the context of the decision situation, the available alternatives, the order of our values, etc., corresponds to a deeper insight into what we want, value and believe, and may in this way contribute to solving the value uncertainty for the case at hand. But these techniques do not necessarily just clarify our original intention. They can also modify our conception of the problem, and they can even change our value commitments.

The border between a specification which is a mere clarification of the original question and one which amounts to changing the question is arguably not sharp. The distinction can be elucidated with the example of bringing tea or coffee to a trip. Let us assume that deliberating on the question of bringing coffee or tea has revealed my more contextualized preferences mentioned above:

tea in the morning > coffee in the morning
coffee during the workday > tea during the workday
tea in the evening > coffee in the evening

If my trip lasts only until lunch, my (here admittedly rather artificial) initial value uncertainty is solved: my uncertainty about which beverage I prefer to bring has turned to certainty that it should be tea. Specifying my preferences has clarified the relevant aspects and solved the case. But say the trip lasts a week. The decision problem 'what to bring if the trip lasted one day' would then amount to *changing* the question, not clarifying it. My counterfactual value certainty does not help to solve the present uncertainty. Admittedly, *something* is clarified, but my attitude to the original problem is still as uncertain.

While obviously not a solution in this example, changing the problem may be the best available alternative, and in such cases we may speak about this as one way to solve value uncertainty. Consequently, even if my initial idea was to be away a

[22] In the introduction to her book on incommensurability, incomparability and practical reason (Chang 1997), Ruth Chang calls this the covering value.

week, I may decide to change the scope of the trip, if that option is open to me, even though I consciously change my problem rather than specify it more thoroughly.

Most of us would perhaps not let the impossibility of bringing both coffee and tea on a trip decide its scope, but attempting to change a decision situation is arguably one of the most common strategies to deal with value uncertainty. Often re-framing of a decision situation is more properly described as changing the decision context rather than clarifying it: if my original question was whether to go on a sunny beach vacation or an adventurous mountain trip, but I cannot decide which, the option of aiming for a vacation in which I can do *both* may be the best solution, even if it is a clear change in my original choice situation.

Postponing a decision is another common way to handle value uncertainty by in effect changing the problem. In many large-scale cases, such when dealing with long term storage of nuclear waste, we find it hard to know how to value the many empirical uncertainties involved. We then often postpone the original decision, hoping for a better epistemic vantage point in the future. Postponing is then a way of re-embedding the decision situation from a decision involving a number of long-term solutions, to a situation which also includes the alternative of short-term storage in combination with a later decision about long term solution. Choosing that additional alternative in effect amounts to valuing the better-known risks involved in short-term storage of nuclear waste, in combination with a potentially more informed long-term decision later, as preferable to the more unknown risks of making a long-term storage decision here and now (Hansson 1996). An alternative to postponing the decision in full is to divide it into 'smaller parts', for example by making sequential decisions. See (Hirsch Hadorn 2016) in the current anthology for further discussion on this topic.

5 Beyond Clarification

In the previous section, a number of analytic techniques for solving value uncertainty have been introduced, relying on the possibility of specifying our values or the relevant circumstances of the decision problem. The underlying hope has been that what started out as uncertainty about which values were salient in the case at hand, or how they should be weighted, would change into (a reasonable level of) certainty when properly specified. Of course, that is a possibility rather than a promise. It may turn out that my most fully specified characterization of a decision situation is just as fraught with value uncertainty as my initial understanding. I may wonder whether justice is more important than kindness, lay out all the relevant facts, specify what I mean by justice and kindness in this exact instance, and still be exactly as uncertain about whether this-instance-of-justice should take precedence over this-instance-of-kindness. A deeper level of uncertainty perhaps, but uncertainty all the same. Or perhaps even more uncertainty: in the abstract I tended to go for kindness rather than justice, although I was uncertain; but pondering on the problem has only made me less certain about what to do.

So what are we to do if making the problem as clear as humanly possible does not present itself with a solution to our uncertainty? Your value commitments are as clear as you can make them, but they point in different directions, and the world will not help you: it has arranged the facts so that your value uncertainty matters. There are at least two types of answers in the literature to how we can move beyond clarification of the problem.

5.1 Answer One: Decision Making Under Moral Uncertainty

Some theorists in what has been labeled the moral uncertainty debate insist that there is a rational way forward even when facing persistent value uncertainty. Remember, value uncertainty in the moral uncertainty debate is spelled out in terms of positive credence in more than one moral theory, i.e. the state of the agent who finds several moral theories plausible but cannot decide which to fully believe in. For example, an agent may think that her moral values and intuitions mostly point to utilitarianism, on which the morally right action is the one that would maximize wellbeing. But she is uncertain, since she also finds that there is something to say for a rights-based ethics, on which some action-types such as lying or failing to keep promises are bad in themselves.

Theorists in the moral uncertainty debate have suggested several different decision strategies, but here we will only consider the two most influential kinds: that the recommended action is given by weighing the moral values of the potential alternatives between all theories into which we put some credence, and that we should select the theory in which we believe the most and stick with it.

The former kind of suggestion may intuitively seem like the most plausible candidate, and is the one which many theorists in the moral uncertainty debate argue for (Broome 2010; Sepielli 2009; Ross 2006; Lockhart 2000). The suggestion is grounded in the observation that different moral theories seem to give the moral goodness or badness of an outcome not only different valences, such that something is either right or wrong, but a more fine-grained moral value: an action might be slightly good or bad, just as it might be very good or bad. Suppose that a person is uncertain between utilitarianism in which killing is sometimes obliged (that is, when it is the alternative which maximizes the resulting happiness), and a duty-based theory which considers killing one of the most serious wrongdoings. If she then finds herself in a situation where the utility of killing a person in front of her is only slightly higher than abstaining from it, it seems reasonable to value the fact that since the other theory she partially believes in strongly forbids it weighs much heavier than the only slight utility surplus the alterative has in the former theory. Generally speaking, if an action is considered to be really bad according to one theory an agent partly believes in and only slightly good in her rival theories, she should typically avoid it.

Perhaps the most popular version of the idea that weighing the moral values of the alternatives between one's candidate theories is the rational choice is to recommend the alternative with the *highest expected moral value* (e.g. Lockhart 2000). Consider the following example on this alternative:

	T1 ($p = 0.5$)	T2 ($p = 0.5$)
A	Slightly bad (−1)	Very good (100)
B	Slightly good (1)	Very bad (−100)

Here, option A gets the total moral value 49.5 ($-1*0.5 + 100*0.5$) whereas option B gets −49.5. Consequently, it should rationally be chosen, some theorists argue. (Moreover, option A remains the preferred alternative even when our credence in T1 is much higher than in T2).[23]

Intuitively plausible as the suggestion may seem, there is a rather forceful objection against it: the problem of comparing the moral value *between* different moral theories. Critics argue that all theories which have been suggested for how such intertheoretic comparisons of moral value would work are implausible, which they take to be sufficiently convincing reasons against the idea. Contrary to how it may seem at first glance, they argue, moral values in different theories may not be compared (Gustafsson and Torpman 2014; Sepielli 2009, 2013; Gracely 1996; Hudson 1989).

The second suggestion is that when we have positive credence in more than one theory, we should act on the theory in which we have most, even if not full, credence. The suggestion takes its cue from the skeptical conclusion that intertheoretic comparisons are not possible. Consequently, proponents of this suggestion argue, the main intuition-pump for weighing the moral value of all our potential moral theories into a resulting recommendation has no force. With different theories come different standards of evaluation, and so if one theory labels a particular action as 'horribly wrong' this does not mean that it is worse than something which is labeled 'somewhat wrong' by another theory. All we can say is that both consider the action to be morally wrong.

The upshot according to this suggestion is that even in face of uncertainty, if there is one theory in which we believe more than others, we should act in accordance with that theory.

While this strategy as well faces objections, it would take us too far to consider them here.[24] Instead, we will end this subsection with discussing the potential problem with the moral uncertainty accounts as such: their exclusive focus on moral *theories*. In the debate, moral uncertainty is characterized as credence in more than one moral theory, and the suggested solutions are given by some or another function of this credence and the moral values the different theories assign to the available alternatives. There are several problems with both the characterization and the solution, however. First, the characterization seems too narrow to

---

[23] Indeed, even if $P(T1) = 0.99$, A would still be the better option.

[24] The interested reader should turn to Gustafsson and Torpman (2014) for a recent run-down of the common criticism and some suggested rebuttals (including modifications to the suggestion).

capture the relevant phenomenon properly. Agents who face value uncertainty need not even partially believe in any particular moral theory. It seems reasonable to claim that many people do not believe in any particular moral *theory* at all. Although they are committed to some values and norms, take some features of an action – that it is kind, perhaps, or just, or produces wellbeing – to speak in favor of or against it; but they do not subscribe to any particular *account* of how these features come together which may be called a moral theory. Some may even be moral particularists who *deny* that there are moral theories in any interesting sense (Hooker and Little 2000). There is thus a worry that the debate about moral uncertainty captures only a small part of the phenomenon of value uncertainty. If I am uncertain about whether kindness or justice should be exercised in a particular situation, and this uncertainty is not due to factual concerns, then this is a case of value uncertainty whether or not I have a credence in several moral theories.

The moral uncertainty theorist may of course argue that 'moral theory' should be understood broadly, including cases where we are committed to a set of values and norms rather than to a theory in a stricter sense.[25] But even if we grant this, we run into the second, and more severe, problem: the sought solutions disregard the best available data. Even when it is correct to say that we have positive credence in more than one moral theory, this does not mean that our moral commitments are *reduced* to this credence, that all that matters in determining what to do is the credence we have in theories X, Y, Z etc., and what moral values these theories assign to particular actions. When we form a belief in a moral theory, we do so because, among other things, we take it to fit well with many of our moral judgments in particular cases, the values we take as important, etc. Perhaps I have a strong belief, as in the first example above, in the absolute wrongness of killing. I am uncertain about other aspects of the duty theory which has this as an absolute rule, but I fully believe in this particular prescription. Now if my credence is divided between this duty theory and utilitarianism, and the choice before me is that of killing an innocent man or not, there would be nothing strange about letting this particular conviction play a deciding role in choosing what to do, even if I put *more* credence in utilitarianism overall.

In sum, it seems as if it is *exactly* when we are not fully committed to one single moral theory that it becomes central that our particular values and considered judgments play a role in deciding what to do – that is, the very aspects the debate about moral uncertainty reduces away.

[25] Or she may bite the bullet, of course, arguing that she is interested in a more limited, but still interesting problem. Even so, she faces the second problem in the main text.

5.2 *Answer Two: Back to First Order Normative Argumentation*

The second answer to what to do if clarification of the problem or our values did not provide us with a solution takes as starting point the insight with which we ended the last subsection: that the primary 'data' we have to work with when in value uncertainty is the set of moral values, norms and particular commitments which we hold. And when the previous answer tried to find a rational way forward given the remaining value uncertainty, the second answer insists that the way forward is to make your values hang together. If you value both justice and kindness, and you want to perform the kind action as well as the – in this case incompatible – just action, this is a signal that your values do not cohere sufficiently. When this is the case, you must find a way to handle this incoherence.

What this amounts to is that the general way forward when value uncertainty remains is to engage in the very theme of the present anthology (Hansson and Hirsch Hadorn 2016, Brun and Betz 2016): argumentation. It is only through argumentation, be it introspection or deliberation (and typically a mix of the two), based on the factual as well as normative information we may gain access to, that we may find a solution to our value uncertainty when clarity itself is not sufficient. In this anthology many such argumentative tools are presented. In this chapter I will focus on what I take to be the dominating methodological development of the basic idea of how to reach coherence in moral and political philosophy: *the method of reflective equilibrium.*

Reflective equilibrium is a coherentist method made popular by the political philosopher John Rawls in his seminal book *A Theory of Justice.*[26] While the core idea is arguably as old as philosophy itself, Rawls's illuminating treatment in the context of his theory of justice (and the developments by other philosophers in its aftermath) has become the paradigmatic instance of the method.[27] (For further analysis, see also (Brun and Betz 2016) in the current volume, where the tool of argument maps, strongly influenced by the conception of reflective equilibrium, is used).

When faced with a normative problem – a problem about what we should do, how to act – we come armed with a set of beliefs about how the world is as well as about how it should be. These beliefs can – but need not – be particularly structured or theoretically grounded. Typically however, our arsenal of value commitments contain both more general ones, such as perhaps the equal value of every person or that we should try to behave kindly to others, and more particular ones, perhaps intuitions pertaining to the very problem at hand, 'What happens *right here* is

[26] Rawls (1999 [1971]). For earlier formulations, see Rawls (1951).

[27] For a recent analysis, see Brun (2014). In a strict sense', reflective equilibrium refers to a state of a belief system rather than a methodology. But it has become commonplace to refer to it as the method through which we try to arrive at this state.

wrong!' The basic idea of Reflective Equilibrium is to scrutinise one's set of beliefs, and modify them until our normative intuitions about particular cases (which Rawls called our 'considered judgments') and our general principles and values find themselves in equilibrium.

The idea that we should modify our value commitments until they reach equilibrium is an analogue to how we should modify factual beliefs. As with value commitments, our factual commitments do not always cohere at the outset. Let us imagine that the communist hunting senator McCarthy both believed that the specter of communism haunted the United States and Europe, and also, believed that every statement in the Communist Manifesto is false.[28] So far his beliefs seem to cohere perfectly. But what if he learnt that that the very first sentence of the Communist Manifesto reads "The specter of communism haunts Europe." Now, if he learns this, we expect senator McCarthy to modify his set of beliefs until they reach equilibrium.

In a similar vein, the method of reflective equilibrium demands that we are prepared to abandon specific normative intuitions when we find that they do not fit with intuitions or principles on which we rely more. Likewise for our principles and values: if we find that on closer examination they go against normative intuitions, principles and values that we are simply not prepared to abandon, they too must be modified. The goal is to reach a state of equilibrium, where all relevant normative commitments fit together.

The factual analogy further suggests how we should go about judging which, among competing values, we should put most faith in. McCarthy should find a coherent set of beliefs based on what he has best reason to believe in. He may, for example, revise his belief that the US and Europe are full of communists: perhaps he has only US statistics to go on, and without good justification believed that what goes for the US must go for Europe as well. The stronger his belief in the total falsity of every sentence in the manifesto, the more he must be prepared to find a coherent set of beliefs which includes this belief, no matter the costs. Another option is reinterpretation: as with the value propositions we have discussed above, our factual beliefs are often vague and possible to specify, perhaps in a way which make the set coherent without having to abandon any belief. Senator McCarthy may perhaps remember that the Communist Manifesto was written in 1848, a hundred years before he started his anti-communist crusade. So the factual claim in the book clearly addresses the situation in Europe back then, and not in the 1950s. McCarthy may then believe that Marx and Engels were wrong about communism hundred years earlier, 'they were really very few back then,' but continue believing that absolutely everything in that book is false and that the communists swamp the western world. Similarly, when our values are not in reflective equilibrium, we should scrutinise our reasons for holding on to our value commitments, general or particular. Something must go.

[28] This example is from Brandom (1994: 516).

What does it entail then, to get our bundle of value commitments to cohere (sufficiently) in practice? Reflective equilibrium may properly describe the general process of adjusting our intuitions, value commitments and principles in order to find a coherent whole. But how do we find the proper argumentative structure, how do we weigh, in actuality, between different options which point in different directions or perhaps seem incommensurable, even when we specify and make our value beliefs as clear as possible? My suggestion is that the best general answer to this question is to point to our very practice of normative theory and applied ethics. Normative theory and applied ethics aim to provide us with moral reasons, justification for what we should do, how we should act, in more general terms and in particular circumstances and domains. This justification is typically viewed as aimed at providing arguments for followers and at meeting the arguments of antagonists, i.e. handling disagreement (see Brun and Betz 2016 for the argument analysis of some examples). But it might equally well be viewed as trying to help us *form* our previously undecided positions, or to sort out our inner disagreements – or, for group agency, a combination of intrapersonal and interpersonal disagreement. As Rawls formulates it:

> justification is argument addressed to those who disagree with us, or to ourselves when we are of two minds. It presumes a clash of views between persons, or within one person, and seeks to convince others, or ourselves, of the reasonableness of the principles upon which our claims and judgments are founded. (Rawls 1999 [1971]: 508)

From what we have discussed in this chapter I would like to add the role of convincing not only of the reasonableness of the principles but also of the particular actions from which we may choose in the contexts in which we find ourselves.

It is arguably in normative theory and applied ethics that the most sophisticated arguments are brought forward, but the practice of searching for justification for our value commitments is exercised in many places in the public and private spheres outside of academia as well: governmental bodies, media, trade and industry as well as among friends, family, or in solitude. It is thus to normative deliberation, discourse and introspection wherever it takes place I suggest we should look when value uncertainty persists. Sometimes there is a lively debate within the domain in which our value uncertainty comes to the fore (topics such as abortion, environmental issues), sometimes our input will be limited to more abstract or general ideas (particular normative theories, epistemic methods). The binding thought is that when facing value uncertainty, the only way forward is to help us decide on how to go on using whatever available resources we may find, internal or external. What the relevant reasons for action are, and how they hang together, is essentially contestable, and there is no foreseeable endpoint in which we will be certain about what to do, even in those situations where we know all relevant facts of the matter. Fortunately, through internal and external deliberation, through argumentation, we often find ourselves able to make up our minds.

6 Conclusion

In this chapter, an introduction to the phenomenon of value uncertainty has been undertaken, discussing the many forms it may take as well as several methods of treating it. In Sect. 2, I discussed the central yet controversial distinction between facts and values, and I touched upon the complex question about the status of values, whether they are subjective or in some sense transcend the individual or interpersonal evaluation. Regardless of such ontological status, however, I concluded that what matter for our decision-making are the *actual* commitments we have, and so our subjective values are central for this chapter.

In Sect. 3, I distinguished several important aspects of value uncertainty: whether we referred to hypothetical or actual situations, whether we have full or only partial information, and the difference in strength of our preferences. Four types of uncertainty of values were distinguished: uncertainty about *which* values we endorse, uncertainty about the specific *content* of the values we do endorse, uncertainty about which among our values apply to the problem at hand, and the relative weight among different values we do endorse. Lastly, I mentioned one comparably technical form of value uncertainty, uncertainty about moral theories.

The two following sections discussed various contributions to solving value uncertainty. In Sect. 4, methods of specifying the problem in order to clarify what the salient factors may be was discussed. Contextualization, making explicit the relevant context in which the value will be applied, is an important way of making what is at stake concrete, and thus making it easier to remove uncertainty. Also, clarifying how much weight the value carry is a significant task in situations where there are conflicting values at place. Furthermore, we may sometimes fruitfully change the way in which the problem is framed or embedded in the overall context. We may also sometimes transform or change the problem, such that we postpone our original decision or make the overall problem into sequential decision-points.

In Sect. 5, we discussed what to do if clarifying the problem is not enough. No matter how concrete and specified we make the decision situation, our value uncertainty may remain. We here discussed two approaches to how we then may go on. The first comes from the debate in philosophy about moral uncertainty, where it is argued that there are rational decision methods for what to do even when we remain uncertain about which moral theory we take to be the right one. While some good formal points have emerged from the philosophical debate, I raised skepticism about the viability of these formal solutions, *in particular* where we are uncertain about our values. Rather, I take the second approach to be the viable way forward. This second approach amounts to the overall theme of the present anthology: argumentation (Hansson and Hirsch Hadorn 2016).

This current volume discusses several argumentative methods, and in the present chapter I focused on the method of reflective equilibrium, a very influential method in current normative philosophy. The central conclusion is that we may always continue the deliberative endeavor by engaging in normative argumentation. There is no guarantee of success, of course. Sometimes we will remain uncertain, no matter what. Then either we will become paralyzed or we will force ourselves to make a choice, regardless. Still, many cases of value uncertainty can be traced to a lack of clarity of our own commitments (or the situation at hand), or can be helped with further input, deliberation or introspection. In principle – if not when in a hurry – there is thus always something we can do when we are uncertain about our values: think about them some more. And the best way forward in order to gain ground is to give and ask for further reasons. In other words: argumentation.

Recommended Readings

While the topic of value uncertainty is seldom directly treated in the literature, the rich literature in moral philosophy and decision theory provide many relevant insights into how to handle uncertainty, both by providing ways in which to view the decision situation, by providing methods for how to solve it, and substantive arguments for some endorsing some values rather than others. Rachels (2002) is an introduction to the main questions in moral philosophy, and Hansson (2013) deals specifically with what to do given uncertainty. Hausman (2011) and Peterson (2009) introduce the complex questions of decision-theory in an accessible way, whereas Broome (1991) and Chang (1997) provide challenging but rewarding insights into comparative assessments. Lockhart (2000) is recommended for the reader interested in moral uncertainty proper, and Putnam (2002) provides both insights and background to the fact-value complexities.

References

Alexander, E. R. (1970). The limits of uncertainty: A note. *Theory and Decision, 6*, 363–370.

Betz, G. (2016). Accounting for possibilities in decision making. In S. O. Hansson & G. Hirsch Hadorn (Eds.), *The argumentative turn in policy analysis. Reasoning about uncertainty* (pp. 135–169). Cham: Springer. doi:10.1007/978-3-319-30549-3_6.

Blackburn, S. (1998). *Ruling passions*. Oxford: Clarendon Press.

Brandom, R. (1994). *Making it explicit*. Cambridge, MA: Harvard University Press.

Bratman, M. E. (1999). *Faces of intention*. Cambridge: Cambridge University Press.

Brink, D. O. (1989). *Moral realism and the foundations of ethics*. Cambridge: Cambridge University Press.

Broome, J. (1991). *Weighing goods*. Oxford: Blackwell.

Broome, J. (2010). The most important thing about climate change. In J. Boston, A. Bradstock, & D. Eng (Eds.), *Public policy: Why ethics matters* (pp. 101–116). Canberra: Australian National University E-Press.

Brun, G. (2014). Reconstructing arguments. Formalization and reflective equilibrium. *Logical Analysis and History of Philosophy, 17*, 94–129.

Brun, G., & Betz, G. (2016). Analysing practical argumentation. In S. O. Hansson & G. Hirsch Hadorn (Eds.), *The argumentative turn in policy analysis. Reasoning about uncertainty* (pp. 39–77). Cham: Springer. doi:10.1007/978-3-319-30549-3_3.

Chang, R. (Ed.). (1997). *Incommensurability, incomparability, and practical reason*. Cambridge, MA: Harvard University Press.
Dahl, R. A. (1956). *A preface to democratic theory*. Chicago: Chicago University Press.
Dancy, J. (1995). In defence of thick concepts. In P. A. French, T. E. Uehling, & H. K. Wettstein (Eds.), *Midwest studies in philosophy* (pp. 263–279). Notre Dame: University of Notre Dame Press.
Dworkin, R. (1986). *Law's empire*. Cambridge: Harvard University Press.
Erman, E., & Möller, N. (2013). Three failed charges against ideal theory. *Social Theory and Practice, 39*, 19–44.
Finlay, S. (2006). The reasons that matter. *Australasian Journal of Philosophy, 84*, 1–20.
Gibbard, A. (2003). *Thinking how to live*. Cambridge, MA: Harvard University Press.
Gracely, E. J. (1996). On the noncomparability of judgments made by different ethical theories. *Metaphilosophy, 27*, 327–332.
Grüne-Yanoff, T. (2016). Framing. In S. O. Hansson & G. Hirsch Hadorn (Eds.), *The argumentative turn in policy analysis. Reasoning about uncertainty* (pp. 189–215). Cham: Springer. doi:10.1007/978-3-319-30549-3_8.
Gustafsson, J. E., & Torpman, O. (2014). In defence of my favourite theory. *Pacific Philosophical Quarterly, 95*, 159–174.
Habermas, J. (1979). *Communication and the evolution of society* (T. McCarthy, Trans.). Boston: Beacon Press.
Habermas, J. (1996). *Between facts and norms* (Trans by William Rehg). Cambridge: MIT Press.
Hirsch Hadorn, G. (2016). Temporal strategies for decision making. In S. O. Hansson & G. Hirsch Hadorn (Eds.), *The argumentative turn in policy analysis. Reasoning about uncertainty* (pp. 217–242). Cham: Springer. doi:10.1007/978-3-319-30549-3_9.
Hansson, S. O. (1996). Decision making under great uncertainty. *Philosophy of the Social Sciences, 26*, 369–386.
Hansson, S. O. (2013). *The ethics of risk. Ethical analysis in an uncertain world*. New York: Palgrave Macmillan.
Hansson, S. O. (2016). Evaluating the uncertainties. In S. O. Hansson & G. Hirsch Hadorn (Eds.), *The argumentative turn in policy analysis. Reasoning about uncertainty* (pp. 79–104). Cham: Springer. doi:10.1007/978-3-319-30549-3_4.
Hansson, S. O., & Hirsch Hadorn, G. (2016). Introducing the argumentative turn in policy analysis. In S. O. Hansson & G. Hirsch Hadorn (Eds.), *The argumentative turn in policy analysis. Reasoning about uncertainty* (pp. 11–35). Cham: Springer. doi:10.1007/978-3-319-30549-3_2.
Hansson, S. O., & Grüne-Yanoff, T. (2006). Preferences. Stanford encyclopedia of philosophy. http://plato.stanford.edu/entries/preferences/. Accessed 23 Aug 2015.
Hausman, D. M. (2011). *Preference, value, choice, and welfare*. Cambridge: Cambridge University Press.
Hooker, B., & Little, M. O. (2000). *Moral particularism*. Oxford: Clarendon Press.
Hudson, J. L. (1989). Subjectivization in ethics. *American Philosophical Quarterly, 26*, 221–229.
Hume, D. (2000 [1738]). *A treatise of human nature*. Oxford: Oxford University Press.
Korsgaard, C. M. (1983). Two distinctions in goodness. *Philosophical Review, 92*, 169–195.
Korsgaard, C. M. (1996). *Creating the kingdom of ends*. Cambridge: Cambridge University Press.
Kuhn, T. S. (1962). *The structure of scientific revolutions*. Chicago: University of Chicago Press.
Lakatos, I., & Musgrave, A. (Eds.). (1970). *Criticism and the growth of knowledge*. London: Cambridge University Press.
Lockhart, T. (2000). *Moral uncertainty and its consequences*. Oxford: Oxford University Press.
Luce, R. D., & Raiffa, H. (1957). *Games and decisions: Introduction and critical survey*. New York: Wiley.
Mackie, J. L. (1977). *Ethics: Inventing right and wrong*. London: Penguin Books.
McDowell, J. (1978). Are moral requirements hypothetical imperatives? *Proceedings of the Aristotelian Society, Supplementary Volumes, 52*, 13–29.
McDowell, J. (1979). Virtue and reason. *The Monist, 62*, 331–350.

McDowell, J. (1981). Non-cognitivism and rule-following. In S. H. Holtzman & C. M. Leich (Eds.), *Wittgenstein: To follow a rule* (pp. 141–162). London: Routledge & Kegan Paul.
McMullin, E. (1982). Values in science. *PSA: Proceedings of the Biennial Meeting of the Philosophy of Science Association, 2*, 3–28.
Miller, A. (2013). *Contemporary metaethics: An introduction* (2nd ed.). Cambridge: Polity.
Peter, F. (2009). *Democratic legitimacy*. New York: Routledge.
Peterson, M. (2009). *An introduction to decision theory*. Cambridge: Cambridge University Press.
Pettit, P. (2009). The reality of group agents. In C. Mantzavinos (Ed.), *Philosophy of the social sciences: Philosophical theory and scientific practice* (pp. 67–91). Cambridge: Cambridge University Press.
Putnam, H. (1990). Objectivity and the ethics/science distinction. In J. Conant (Ed.), *Realism with a human face* (pp. 163–178). Cambridge, MA: Harvard University Press.
Putnam, H. (2002). *The collapse of the fact/value dichotomy and other essays*. Cambridge, MA: Harvard University Press.
Quine, W. V. O. (1953). Two dogmas of empiricism. In *From a logical point of view* (pp. 20–46). Cambridge, MA: Harvard University Press.
Rabinowicz, W. (Ed.). (2000). *Value and choice* (1st ed.). Lund: Lund Philosophy Reports.
Rabinowicz, W. (Ed.). (2001). *Value and choice* (2nd ed.). Lund: Lund Philosophy Reports.
Rachels, J. (2002). *The elements of moral philosophy* (4th ed.). New York: McGraw-Hill.
Rawls, J. (1951). Outline of a decision procedure for ethics. *Philosophical Review, 60*, 177–197.
Rawls, J. (1993). *Political liberalism*. New York: Columbia University Press.
Rawls, J. (1999 [1971]). *A theory of justice* (Rev. ed.). Cambridge, MA: Belknap Press of Harvard University Press.
Raz, J. (1986). *The morality of freedom*. Oxford: Clarendon Press.
Resnik, M. D. (1987). *Choices*. Minneapolis: University of Minnesota Press.
Ross, J. (2006). Rejecting ethical deflationism. *Ethics, 116*, 742–768.
Searle, J. R. (1990). Collective intentions and actions. In P. R. Cohen, J. Morgan, & M. E. Pollack (Eds.), *Intentions in communication* (pp. 401–415). Cambridge: MIT Press.
Sepielli, A. (2009). What to do when you don't know what to do. In R. Shafer-Landau (Ed.), *Oxford studies in metaethics* (4th ed., pp. 5–28). Oxford: Oxford University Press.
Sepielli, A. (2013). Moral uncertainty and the principle of equity among moral theories. *Philosophy and Phenomenological Research, 86*, 580–589.
Singer, P. (2009). *The life you can save: Acting now to end world poverty*. New York: Random House.
Singer, P. (2015). *The most good you can do: How effective altruism is changing ideas about living ethically*. New Haven: Yale University Press.
Smith, M. (1987). The humean theory of motivation. *Mind, 96*, 36–61.
Smith, M. (1994). *The moral problem*. Oxford: Blackwell.
Tuomela, R. (2007). *The philosophy of sociality: The shared point of view*. New York: Oxford University Press.
Väyrynen, P. (2013). *The lewd, the rude and the nasty: A study of thick concepts in ethics*. Oxford: Oxford University Press.
Williams, B. A. O. (1985). *Ethics and the limits of philosophy*. Cambridge, MA: Harvard University Press.
Williams, B. A. O. (1981 [1979]). Internal and external reasons. Reprinted in *Moral luck* (pp. 101–113). Cambridge: Cambridge University Press.
Zimmerman, M. J. (2001). *The nature of intrinsic value*. Oxford: Rowman and Littlefield.
Zimmerman, M. J. (2014). Intrinsic vs. extrinsic value. The Stanford encyclopedia of philosophy. http://plato.stanford.edu/entries/value-intrinsic-extrinsic/. Accessed 23 Aug 2015.

Chapter 6
Accounting for Possibilities in Decision Making

Gregor Betz

Abstract Intended as a *practical guide for decision analysts*, this chapter provides an introduction to reasoning under great uncertainty. It seeks to incorporate standard methods of risk analysis in a broader argumentative framework by re-interpreting them as specific (consequentialist) arguments that may inform a policy debate—side by side along further (possibly non-consequentialist) arguments which standard economic analysis does not account for. The first part of the chapter reviews arguments that can be advanced in a policy debate despite deep uncertainty about policy outcomes, i.e. arguments which assume that uncertainties surrounding policy outcomes cannot be (probabilistically) quantified. The second part of the chapter discusses the epistemic challenge of reasoning under great uncertainty, which consists in identifying all possible outcomes of the alternative policy options. It is argued that our possibilistic foreknowledge should be cast in nuanced terms and that future surprises—triggered by major flaws in one's possibilistic outlook—should be anticipated in policy deliberation.

Keywords Possibility • Epistemic possibility • Real possibility • Modal epistemology • Ambiguity • Ignorance • Deep uncertainty • Knightian uncertainty • Probabilism • Expected utility • Worst case • Maximin • Precautinary principle • Robust decision analysis • Risk imposition • Surprise • Unknown unknowns

1 Introduction

A Hollywood studio contemplates to produce an experimental movie with a big budget. Its success: unpredictable. Long-serving staff says that past experience is no guide to assessing the likelihood that this movie flops. Should the management take the risk? (Some wonder: Could a flop even ruin the reputation of the studio and damage profits in the long run? Or is that too far-fetched a possibility?)

G. Betz (✉)
Institute of Philosophy, Karlsruhe Institute of Technology (KIT), Karlsruhe, Germany
e-mail: gregor.betz@kit.edu

S.O. Hansson, G. Hirsch Hadorn (eds.), *The Argumentative Turn in Policy Analysis*, Logic, Argumentation & Reasoning 10, DOI 10.1007/978-3-319-30549-3_6

Another example: A local authority considers to permit the construction of an industrial site near a natural habitat. There's broad agreement that the habitat must be preserved, but it's totally unclear how the ecosystem would react to a nearby industrial complex. Experts say that anything is possible (from no negative effects at all to the destruction of the ecosystem in the medium term).

The objective of this chapter is to show how one can rationally argue for and against alternative options in situations like these. Intended as a *practical guide for decision analysts*, the chapter provides arguably an opinionated introduction to reasoning under "deep uncertainty."[1,2] It is not supposed to review the vast decision-theoretic or risk-ethical literature on this topic. Moreover, readers should be aware that what the chapter says departs from mainstream risk analysis, and that many scholars would disagree with its proposals.[3] However, the *argumentative turn* does not simply dispose of standard decision-theoretic methods (or their application in risk analysis). Rather, it seeks to incorporate these methods in a broader argumentative framework by re-interpreting them as specific (consequentialist) arguments that may inform a policy debate—side by side along further (possibly non-consequentialist) arguments which standard risk analysis does not account for.[4]

Brief outline. Reasons in favor of or against doing something can be analyzed as arguments in support of a normative statement—which, for example, characterizes the corresponding option as obligatory or impermissible (Sect. 2). Section 3 reviews such so-called practical arguments that can be advanced in a policy debate despite deep uncertainty about policy outcomes. These arguments, being partly inspired by the decision theoretic literature, presume characteristic decision principles, which in turn express different, genuinely normative risk attitudes. Reconstructing such arguments hence makes explicit the competing risk preferences—and basic choices—that underlie many policy debates. In the second part of the chapter, beginning with Sect. 4, we discuss the epistemic challenge of reasoning under deep uncertainty: identifying all possible outcomes of the alternative policy options. It is argued that our possibilistic foreknowledge should be described in nuanced terms (Sect. 4) and that drastic changes in one's possibilistic outlook should be reckoned with (Sect. 5). Both the static and the dynamic features of possibilistic predictions compel us to refine and to augment the arsenal of practical arguments discussed in Sect. 3 (Sects. 6 and 7).

---

[1] Like for example Heal and Millner (2013), I use "deep uncertainty" to refer to decision situations where the outcomes of alternative options cannot be predicted probabilistically. Hansson and Hirsch Hadorn (2016) refer to situations where, among other things, predictive uncertainties cannot be quantified as "great uncertainty." Compare Hansson and Hirsch Hadorn (2016) also for alternative terminologies and further terminological clarifications.

[2] This chapter complements Brun and Betz (2016) in this volume on argument analysis; for readers with no background in argumentation theory, it is certainly profitable to study both in conjunction.

[3] I try however to pinpoint substantial dissent in footnotes.

[4] For an up-to-date decision-theoretic review of decision making under deep uncertainty see Etner et al. (2012).

In the remainder of this introductory section, I will briefly comment on the limits of uncertainty quantification, the need for non-probabilistic decision methods and the concept of possibility.

A preconceived idea frequently encountered in policy contexts states: no rational choice without (at least) probabilities. Let's call this view "probabilism."[5] According to probabilism, mere possibilities are uninformative and useless (for, in the end, anything is possible); in particular, it is allegedly impossible to justify policy measure based on possibilistic predictions.[6] One aim of this chapter is to refute these notions, and to spell out how decision makers can rationally argue about options without probabilistic predictions.

But why are non-probabilistic methods of rational choice important at all? Proponents of mainstream risk analysis might argue that decision makers always quantify uncertainty and that they, qua being rational, express uncertainty in terms of probabilities. We do not only need probabilities, they say, we always have them, too.[7] Or so it seems. My outlook on rational decision and policy making departs from that view. Fundamentally, I assume that rational policy making should only take for granted what we know, what we have reason to assume. If there is for example no reason to believe that the movie will be a success, rational decision making should not rely on that prediction. Likewise, only *justified* probabilistic predictions should inform our policy decisions. Rather than building on probabilistic guesswork, we should acknowledge the full extent of our ignorance and the uncertainty we face. We should not simply make up the numbers. And we should refrain from wishful thinking.[8]

At the same time, it would be equally irrational to discard or ignore relevant knowledge in decision processes. If we do know more (than mere possibilities), then we should make use of that knowledge. For example, if some local fisherman has strong evidence that an industrial complex would harm a key species in the ecosystem, then the policy making process should adequately account for this evidence. Generally, we should not only consider explicit knowledge but try to profit from tacit expert knowledge, too.[9] In particular, whenever we have reliable

---

[5] Terminologically I follow Clarke (2006), who criticizes probabilism on the basis of extensive case studies. A succinct statement of probabilism is due to O'Hagan and Oakley (2004:239): "In principle, probability is uniquely appropriate for the representation and quantification of all forms of uncertainty; it is in this sense that we claim that 'probability is perfect'." The formal decision theory that inspires probabilism was developed by Savage (1954) and Jeffrey (1965).

[6] In the context of climate policy making, (Schneider 2001) is a prominent defence of this view; compare also Jenkins et al. (2009:23) for a more recent example. A (self-)critical review by someone who has been pioneering uncertainty quantification in climate science is (Morgan 2011).

[7] Morgan et al. (1990) spell out this view in detail (see for example p. 49 for a very clear statement).

[8] This view is echoed in various contributions to this book, e.g. Hansson (2016, esp. fallacies), Shrader-Frechette (2016 p. 12) and Doorn (2016, beginning). Compare Gilboa et al. (2009) as well as Heal and Millner (2013) for a decision-theoretic defence.

[9] See again Shrader-Frechette (2016).

probabilistic information, it would be irresponsible not to make use of it in decision processes. In sum, this chapter construes reasoning about policy options as a tricky balancing act: it must rely on no more and on no less than what one actually knows.

Because this point is both fundamental and controversial, I wish to illustrate it further.[10] Assume that the outcome of some policy depends on whether a red or a blue ball is (randomly) drawn from an urn. If we know how many red and blue balls there are, we should consider the corresponding probabilistic knowledge in the decision process. However, if we don't know, neither the policy advisor nor the decision maker should *pretend* to know.[11] One might be tempted to argue that, in the absence of any specific information, we should consider both outcomes as equally likely. But then we'd describe the situation as if we knew that there are as many blue as red balls in the urn, which is simply not the case. No probabilistic description seems to capture adequately our ignorance in case we have no clue about the ratio of red and blue balls.

Now, assume we don't get reliable probabilistic forecasts; for practical purposes we have to content ourselves with knowledge about possible intended consequences and side-effects. Yet, what counts as a decision-relevant possibility? That is which possibilities, which "scenarios" should we consider when contemplating alternative options? E.g., is the potential bankruptcy of the Hollywood studio decision-relevant or is it just too far-fetched? That question will occupy us in the second part of this chapter. Here, I just want to make some preliminary remarks.

A first type of possibility to consider are so-called conceptual possibilities. These are (descriptions of) states-of-affairs which are internally coherent. Conceptual possibilities can be consistently imagined (e.g., me walking on the moon). It seems clear that being a conceptual possibility is necessary but not sufficient for being decision-relevant.

Real possibilities (at some point in time t) consist in all states-of-affairs whose realizations are objectively compatible with the states-of-the-world at time t. In a deterministic world, all real possibilities will sooner or later materialize.[12] Epistemic possibilities, in contrast, characterize states-of-affairs according to their relative compatibility with current understanding. Epistemic possibilities hold relative

---

[10] The illustrative analogy is inspired by Ellsberg (1961), whose "Ellsberg Paradox" is an important argument against probabilism.

[11] It has been suggested that decision-makers can non-arbitrarily assume allegedly "un-informative" or "objective" probability distributions (e.g. a uniform distribution) in the absence of any relevant data. However, most Bayesian statisticians seem to concede that there are no non-subjective prior probabilities (e.g. Bernardo 1979:123). Van Fraassen (1989:293–317) thoroughly discusses the problems of assuming "objective priors." Williamson (2010) is a recent defence of doing so.

[12] For a state-of-the-art explication of the concept of real possibility, using branching-space-time theory, see Müller (2012).

to a given body of knowledge[13]: a hypothesis is epistemically possible (relative to background knowledge K) if and only if it is consistent with K.[14]

The following example may serve to illustrate the distinction. An expert team is supposed to defuse a WW2 bomb (i.e., a bomb from World War II). Its explosion is of course a *conceptual* possibility. The team has only limited knowledge of the bomb, it is in particular not clear whether the trigger mechanism is still intact. Against this limited knowledge, it is an *epistemic* possibility that the bomb detonates upon being moved. Now the trigger mechanism is in fact still intact, but the original explosives have undergone chemical interactions and were transformed into harmless substances over the decades. This means that the detonation of the bomb is not a *real* possibility.

I assume that the decision-relevant notion of possibility is a purely epistemic concept. Quite generally, predictions used for practical purposes should reflect our current knowledge and understanding of the system in question. In the *argumentative turn* especially, we're not interested in what is objectively, from a view from nowhere, the correct decision; we want to understand what's the best thing-to-do *given* what we know—and what we don't. For this task, we need not worry about whether some possibility is real or "just" epistemic.[15] In the above example, one should consider the potential explosion as a decision-relevant possibility, as long as this scenario cannot robustly be ruled out. The rather metaphysical question whether it's really possible that the bomb goes off (i.e., is the detonation pre-determined, or is the world objectively indeterministic such that not even an omniscient being would be in a position to predict whether the bomb would detonate?) seems of no direct practical relevance.

Real possibilities are at best of indirect practical significance. Namely insofar as they bear on our expectations concerning the reducibility of (epistemic) uncertainty: ideally, the range of epistemic possibilities approaches the range of real possibilities as our understanding of a system advances; real possibilities represent lower bounds for the uncertainty we will face in the future, no matter how much we will learn about a system.

Relativizing decision-relevant possibility to a body of background beliefs seems to raise the question: What's the background knowledge? Whose background

[13] Or, more precisely, "knowledge claims." In the remainder of this chapter, I will refer to fallible knowledge claims, relative to which hypotheses are assessed, as "(background) knowledge" simpliciter.

[14] There is a vast philosophical literature on whether this explication fully accommodates our linguistic intuitions (the "data"), cf. Egan and Weatherson (2009). Still, it's unclear whether that philosophical controversy is also of decision-theoretic relevance.

[15] On top, that's a question we cannot answer anyway: Every judgement about whether some state-of-affairs S is a real possibility is based on an assessment of S in terms of epistemic possibility. To assert that S is really possible is simply to say that S represents an epistemic possibility (relative to background knowledge K) and that K is in a specific way "complete", i.e. includes everything that can be known about S. Likewise, to assert that S does not represent a real possibility means that S is no epistemic possibility (relative to background knowledge K) and that K is objectively correct.

beliefs? First of all, note that this is a general issue in policy assessment, no matter whether we evaluate options in a possibilistic, probabilistic or deterministic mood. My reading of the *argumentative turn* is that we don't need general rules which determine precisely what counts as background knowledge. If there is disagreement about this question, then make it explicit, analyze the different arguments that can be set forth from the different knowledge bases, identify the crucial items in the background beliefs which are responsible for the practical disagreement! The *argumentative turn* may accommodate dissent on background beliefs and allows for rational and constructive deliberation in spite of such disagreement.

2 Practical Arguments, Preliminary Remarks

In the *argumentative turn*, decision procedures, decision methods, and justifications of policy decisions are construed as arguments which warrant the corresponding policy measure.[16] Such "practical" arguments have a normative—more precisely, prescriptive—conclusion: they warrant that certain policy options are obligatory (ought to be taken), permissible (may be taken) or prohibited (must not be taken).

Valid arguments with prescriptive conclusions require normative and descriptive premisses. The descriptive premisses characterize the alternative options; often they identify consequences that will or may ensue if one such option is taken. The normative premisses value the alternative options in view of their descriptive characterization.

Our first example of a practical argument (under certainty) is a simple, so-called consequentialist argument. It argues that China should reduce air pollution, despite negative side-effects, because this will curb pulmonary diseases, argument A:

(1) The major effects of reducing air pollution in China, compared to status quo, would be (i) a significant reduction of pulmonary diseases and (ii) the acceleration of regional climate change.
(2) Business as usual policy simply sustains status quo.
(3) A significant reduction of pulmonary diseases and the acceleration of regional climate change are preferable to status quo.
(4) If some option leads to a state of affairs that is preferable to the one that would be brought about by an alternative, the former should be taken rather than the latter.
(5) Thus: China should reduce air pollution rather than continue business as usual.

The conclusion (5) is a (comparative) prescriptive statement: It says that some action should be taken rather than another one. Premisses (1) and (2) are descriptive premisses: They forecast the main consequences of two policy options, reducing air pollution and business as usual. These different states-of-the-world, which are

[16] Brun and Betz (2016), this volume, which nicely complements this chapter, provides practical guidance for analyzing and evaluating argumentation.

predicted in (1) and (2), are then normatively evaluated in premiss (3). The normative evaluation of outcomes is based on, or partially expresses an underlying (frequently implicit) "value theory," a so-called axiology. Premiss (4) states a (rather uncontroversial) decision rule: Of two options, choose the one with the better consequences! That is a normative statement, too.

Practical arguments need not be consequentialist. The following simple rights-based argument argues that new polling stations should be constructed, argument B:

(1) Costly constructions of new polling stations are the only way to ensure that the citizens' rights to vote are not infringed.
(2) Such a measure does in turn not lead to violations of rights of similar or higher (normative) significance.
(3) If a measure is required to avoid the violation of some rights and in turn does not bring about the violation of other rights (of similar or higher weight), then the measure ought to be taken.
(4) Thus: New polling stations should be constructed.

In this argument, premisses cannot be neatly separated into normative and descriptive ones. Premisses (1) and (2) characterize (in a descriptive mood) the policy measure in question (and indirectly—n.b. the "only" in (1)—the alternative options). Yet in referring to rights and their potential violation, these premisses have a normative content, too. Premiss (3) in turn is clearly a normative statement—and a substantial one, too: it implies that violations of rights can only be offset by violations of more important rights (not, e.g., by numerous violations of lesser rights or by diminution of wellbeing).

The descriptive premisses in arguments A and B characterize unequivocally, by means of deterministic predictions, the alternative options. Even if there is uncertainty about the effects of measures to reduce air pollution or the construction of polling stations, these uncertainties are not articulated in arguments A and B. The whole point of decision analysis, broadly construed, is to make uncertainties (in descriptive or normative statements) explicit and to investigate how conclusions can be justified *while acknowledging* the uncertainty we face.

In situations under deep uncertainty, we are not in a position to make deterministic predictions as in the arguments A and B. We can't even provide reliable probabilistic forecasts (such as: "business as usual" policy is unlikely to lead to a reduction in pulmonary diseases; construction of polling stations will ensure with a probability of 90 % that voting rights are not infringed). The descriptive premisses merely state possible consequences of alternative actions, they characterize options in a possibilistic mood (like: moving the bomb *possibly* leads to its detonation). The normative premisses will then value the alternative options in view of their possible characteristics, e.g. in view of their possible outcomes. Crucially, reasoning under deep uncertainty relies on other decision principles than arguments under certainty or risk. As will become clear in the course of this chapter, these principles involve substantial normative commitments and reflect different risk attitudes (such as levels of risk aversion) one may adopt.

Sound decision making under certainty requires one to consider all alternative options and all their consequences (to the extent that they are articulated and foreseen). Likewise, sound decision making under deep uncertainty requires one to consider all alternative options and all their *possible* consequences (under the same condition). In other words, practical reasoning under deep uncertainty must reflect one's apprehension of the entire space of decision-relevant possibilities.[17] Arguments that derive policy recommendations in view of *some* possible consequences only, while deliberately ignoring other possibilities, are typically weak, i.e. rely on implausible decision principles and will be given up in the face of conflicting arguments.

Let me flesh that out. The local authority which considers to permit the construction of the industrial site might reason like this: "The industrial complex may destroy our habitat. That would be disastrous. So we must stop the industrial project." Now, this reasoning is faulty. The decision makers have not explicitly considered further possible consequences of constructing the industrial site (maybe this ensures that the company will not construct a factory at another place where an even more valuable ecosystem would be endangered; maybe the site will generate so much tax revenues that another reserve could be environmentally restored), and they have not considered the possible effects of not building the industrial complex (maybe the local authority will lack the financial resources to clean up a contaminated mine, which in turn might cause the medium-term destruction of the habitat, too). To be sure: The point here is not that the local authority cannot reasonably prohibit the construction because of potential ecological adverse effects. The point is only: in order to make this case, *all* (apprehended) possible consequences of the available options have to be considered and assessed.[18]

Let me eventually comment on the relation between formal decision theory and the argumentative analysis of practical reasoning, picking up my brief remarks in the introduction. Decision theory provides a formal model of consequentialist decision making. All decision-theoretic methods can be recast and interpreted as practical arguments. And many important arguments in practical deliberation will be inspired by decision theory. There is however no reason to think that every legitimate argument can in turn be cast in decision-theoretic terminology. One major advantage of argumentative analysis over decision theory is its universality and hence superior flexibility; it can account for consequentialist as well as non-consequentialist reasoning side by side. Decision theory sometimes evokes the impression that there exists an algorithmic method for identifying the optimal

---

[17] On prerequisites of sound decision making under uncertainty see also Steele (2006).

[18] The symmetry arguments Hansson (2016) discusses are another case in point. Suppose a proponent argues that option A′ should be preferred to option A on the grounds that A possibly leads to the disastrous effect E. An opponent counters the argument by showing that A′ may lead to an equally disastrous effect E′. Now, both arguments only draw on some possible effects of A and A′ respectively. They are weak and preliminary in the sense that more elaborate considerations will make them obsolete. Maybe we can construe them as heuristic reasoning which serves the piecemeal construction of more complex and robust practical arguments.

choice. That is certainly how its methods are frequently presented and applied.[19] The argumentative turn is free from such hybris: Rational decision making according to the argumentative turn consists primarily in rational deliberation, in an argumentative exchange, in the process of giving and taking various reasons for and against alternative options.

But haven't decision theorists shown that someone who doesn't maximize expected utility violates basic axioms of rationality? This seems to be a widespread misinterpretation of so-called decision-theoretic representation theorems. Granted: It can be shown that every agent whose preferences over alternative options satisfy certain criteria acts *as if* she were maximizing expected utility according to some hypothetical, personal utility and probability function. But this result entails nothing about how the agent has originally arrived at her preferences, or how she is making her choices. It may very well be that she adheres to a non-consequentialist ethical theory, which determines her choices and preferences. The existence of a hypothetical utility and probability function is then in a way a mere formal artefact, a theoretical epiphenomenon that has no practical bearing on the agent's rational decision making process at all.[20]

3 Arguing with Possibilities For and Against Options for Action

This section reviews practical arguments that can be advanced in a policy debate despite deep uncertainty about policy outcomes. The worst case and robustness arguments developed in Sects. 3.1 and 3.2, respectively, are inspired by the decision theoretic literature; Sect. 3.3 analyzes arguments from risk imposition, which are prominently discussed in risk ethics.

3.1 Arguments from Best and Worst Cases

Example (Local Authority) The local authority organizes a hearing on the planned industrial site. At this hearing, members of an environmental group argue along the following lines: The construction of the industrial complex may destroy the habitat. The worst thing that may happen if the community does not grant the construction permission is, however, that the local economy will miss a growth opportunity and will expand less quickly than otherwise. The latter case is clearly preferable to the

[19] Nordhaus and Boyer (2000) is a (influential) case in point.

[20] For a more detailed discussion of the implications of representation theorems see Briggs (2014: especially Sect. 2.2) and the references therein.

first one. The local authority should err on the safe side and prohibit the construction.

The environmentalists put forward a simple worst case argument, whose core can be analyzed as follows, argument C:

(1) There is no available option whose worst possible consequences are preferable to the worst possible consequences of not permitting the construction.
(2) If there is no available option whose worst possible consequences are [weakly] preferable to A's worst possible consequences, then one is obliged to carry out option A.
(3) Thus: The local authority should not permit the construction of the industrial complex.

Premiss (2) represents the general decision principle which underlies the reasoning. It states that alternative options should be assessed according to their worst possible consequences. In decision theory, this worst case principle is called maximin criterion.[21]

Premiss (1) has case-specific, normative and descriptive content. It typically takes three steps to justify a statement like premiss (1). First, one identifies, for each option, all possible consequences. Second, one locates those consequences in a 'normative landscape,' and identifies, for each option, its worst possible consequences. Third, one compares the worst possible consequences of all options and identifies the option whose worst possible consequences are best.

In line with our general remarks above, the simple worst case reasoning requires one to grasp the entire space of possibilities. Otherwise, one wouldn't be able to correctly identify the options' worst possible consequences.

Example (Local Authority) The hearing continues and members of another environmental group object that without the new industrial project, we're lacking necessary funds to clean up the contaminated mine, which threatens the habitat, too.

This objection challenges premiss (1) in the above argument, in particular the claim that the worst case of not constructing the new industrial complex is preferable to the destruction of the habitat. In fact, the objection goes, not constructing the complex may have the same catastrophic consequences.

Put more generally, all available options seem to possess equally bad worst cases. The antecedent conditions of the worst case principle (2) above don't apply to any available option and the principle hence is of no use in warranting a choice. In view of such situations, the worst case principle is sometimes described as self-refuting[22]; but that seems inadequate, the simple criterion does not give contradictory recommendations, it rather does not justify any prescription at all.

[21] Cf. Luce and Raiffa (1957:278), Resnik (1987:26).

[22] E.g. Elliott (2010).

Example (Local Authority) Charged by their colleagues, the opponents of the new complex refine their original argument. They concede that if the local authority fails to clean up the mine, the habitat may be destroyed, too. But they say: We may fail to clean up the mine no matter whether we build the new industrial complex or not. That's because money is not even the main problem when de-contaminating the mine, we rather face technical and engineering problems. So, yes, a constantly contaminated mine with all its catastrophic ecological consequences, including the total destruction of the habitat, is clearly a worst case to reckon. But that worst case may materialize independently of the choice we discuss today. It's just not relevant for the current decision. What is *relevant*, though, is the second worst case, i.e. the destruction of the habitat through the new industrial complex.

The opponents of the industrial complex now argue with a refined decision principle.[23] We can reconstruct their reasoning as follows, argument D:

(1) The worst possible consequence of not permitting the construction is preferable to the worst possible consequence of permitting the construction—excluding all possible consequences both options have in common (such as failure to de-contaminate the mine).
(2) An option A is to preferred to an option B, if—excluding all common possible consequences of A and B—A's worst possible consequence is preferable to B's worst possible consequence.
(3) Thus: The local authority should not permit the construction of the industrial complex.

This reasoning generalizes the original worst case argument C. I.e., every choice that is warranted by the original argument can also be justified with the refined principle.[24]

Since the argument justifies a comparative prescription, it can be applied iteratively in order to exclude several options one after another.

The decision principles which fuel the worst case argument express an attitude of extreme risk aversion. Any potential benefits (positive possible consequences) are simply ignored. We can easily think of decision situations where such an attitude seems to be inappropriate (a Hollywood studio that would base its management decisions on maximin would simply stop producing any films at all, since every film may flop). Following Rawls (1971), Stephen Gardiner suggests (sufficient) conditions under which such a precautionary attitude seems to be permissible, if not even morally required. These are: (i) some options may have truly catastrophic consequences, (ii) the potential gains that may result

[23] The lexicographically refined maximin criterion is called "leximin."

[24] Moreover, the general premiss (2) can be understood as an implementation of Hansson's symmetry tests (cf. Hansson 2016).

from taking a risky option are negligible compared to the catastrophic effects that may ensue.[25]

These prerequisites can be made explicit as antecedent conditions in the decision principle and, accordingly, as additional premisses in our worst case arguments, e.g., argument E:

(1) Some of the local authority's options may have truly catastrophic consequences.
(2) The potential gains that may result from taking a risky option are negligible compared to the catastrophic effects that may ensue in the local authority's decision to permit or prohibit the construction of the industrial complex.
(3) There is no available option whose worst possible consequence is preferable to the worst possible consequence of not permitting the construction.
(4) If (i) some options may have truly catastrophic consequences, (ii) the potential gains that may result from taking a risky option are negligible compared to the catastrophic effects that may ensue, and (iii) there is no available option whose worst possible consequence is [weakly] preferable to A's worst possible consequence, then one is obliged to carry out option A.
(5) Thus: The local authority should not permit the construction of the industrial complex.

Gardiner (2006) suggests to consider the modified decision principle (4) as an interpretation and operationalization of the notoriously vague precautionary principle.

In many situations it is not outright unreasonable to be highly risk averse—in some it may even be morally required. But what about other situations, and what about agents that are rather willing to take risks? How can they reason about their choices under deep uncertainty? One straightforward generalization of the maximin reasoning is to account for both worst and best possible consequences of each option.

Example (Local Authority) The hearing is broadcast and citizens are invited to comment on the discussion online. One post argues: The worst case of constructing the industrial site is the destruction of the habitat. But what about the best case? Fact is: We'd attract a green tech company that builds highly innovative products. That does not only mean sustained growth but also that our small town will potentially attract further supplying industries, to the effect that a whole industrial cluster will emerge in the years to come. With the help of these industries, we might become, over the next two decades, the first community in this state that fully generates its energy demand in a CO_2-neutral way.

Unlike worst case reasoning, arguments of this sort assess alternative options in view of both their corresponding best and worst case. In order to do so, best and

[25] Gardiner (2006:47); see also Sunstein (2005), who argues for a weaker set of conditions. The general strategy to identify specific conditions under which the various decision principles may be applied is also favored by Resnik (1987:40).

worst cases have to be compounded for each option. Let's refer to the joint normative assessment of a pair of possible consequences (best and worst case) as "beta-balance."[26] The relative weight which is given to the worst case in such a beta-balance is a measure of the underlying degree of risk aversion. A simple way to reconstruct the above reasoning would be, argument F:

(1) There is no available option whose beta-balance (of best and worst possible consequences) is preferable to the beta-balance of permitting the construction.
(2) If there is no available option whose beta-balance (of best and worst possible consequences) is preferable to A's beta-balance, then one is obliged to carry out option A.
(3) Thus: The local authority should permit the construction of the industrial complex.

In order to justify a statement like premiss (1), one has to (i) identify all possible consequences of each available option; (ii) determine best and worst possible cases (for each option); (iii) balance and combine the best and worst case (for each option) in light of one's risk attitude, so that one is finally able to identify the option with the best beta-balance. A proponent of the illustrative argument above would, in particular, have to compare a combination of destroying the habitat (worst case) and greening the local economy (best case) on the one side with a business as usual scenario on the other side (if we disregard uncertainty about the consequences of not building the industrial complex).

Worst case reasoning is just a special case of this sort of argumentation, it merely consists in determining the beta-balance in an extreme way, namely by ignoring the best case and simply identifying the beta-balance with the worst case.

The idea that options are assessed in view of their best and worst possible consequences allows us also to analyze the following line of reasoning.

Example (Hollywood) It turns out that the Hollywood studio has lost a vital legal dispute and is virtually bankrupt anyway. Now the managers reason: There's nothing to loose and it can't really get worse. So we should go for the highly risky film—if it will turn out to be a blockbuster, then our studio will finally survive.

To me, that sounds perfectly reasonable. Under one option, bankruptcy is nearly certain, and bankruptcy is as bad as it can get. Under the other option, there is at least a chance that the company survives. The general decision principle that can be used for reconstructing this argument is: If option A leads, in the worst possible case, to consequences X but may also bring about better consequences and if option B will

[26] In case the (dis)value of the best lcase and worst case is quantifiable, their beta-balance is simply a weighted mean (where the parameter $0 \leq \beta \leq 1$ determines the relative weight of best versus worst case in the argumentation): $\beta \times$ value-of-best-case $+ (1 - \beta) \times$ disvalue-of-worst-case. The corresponding decision principle is called "Hurwicz criterion" in decision theory (Resnik 1987: 32, Luce and Raiffa 1957:282). Hansson (2001:102–113) investigates the formal properties of "extremal" preferences which only take best and worst possible cases into account.

surely bring about consequences X, then option A is preferred to option B.[27] Now, we can also explain why the reasoning appears so plausible: Whatever the exact level of risk aversion, the beta-balance of option A is greater than that of option B and hence A is preferred to B according to best/worst case reasoning, in general.

We've discussed the problem that sometimes all options may give rise to equally bad worst cases. Our solution was to compare 2nd (and if necessary 3rd, 4th, etc.) worst cases in order to evaluate the options. But what if all options essentially give rise to the same possible outcomes? In possibilistic terms, the options are then indistinguishable and any justification of a choice requires further (non-possibilistic) characterizations. Now, this characterization does not necessarily have to consist in precise probabilistic forecasts, as the following example illustrates.

Example (WW2 Bomb) The team has decided to evacuate the borrow. Question is: What can be done to secure the historic Renaissance building nearby? The experts agree: There is no way to guarantee that the building will not be fully destroyed. Whatever the team does, that remains the worst possible case. In the same time, the probability of this happening cannot be assessed, too little is known about the inner life of this bomb and analogous cases are rare. Eventually, the team decides to erect a steel wall between the bomb and the building before trying to defuse it. It reasons: Whatever the specific circumstances (state of the trigger mechanism, degree of chemical transformation of the explosive, degree of corrosion, density of the underground, etc.), the (unknown) likelihood that the historic building will be destroyed is reduced through the erection of the steel wall.

In this reasoning, the team relies on partial probabilistic knowledge. I suggest to analyze the argument as follows: The possible consequences of the alternative options are themselves described probabilistically. They can be seen as alternative probabilistic scenarios. The value theory which assesses the possible consequences does not only consider the physical effects but also their probability of occurrence; the normative assumptions of the reasoning assess the *probabilistically* described scenarios. More precisely, we assume that the negative value of a possible scenario (which may ensue) is roughly proportional to the (scenario-specific) likelihood that the historic building is fully destroyed. As a result, the alternative options may lead to different possible consequences which can be normatively assessed.[28]

Following the overall direction of this section, we can reconstruct the argument as worst case reasoning, argument G:

(1) The greatest possible probability that the historic building is fully destroyed is smaller in case a steel wall is erected (compared to not erecting a steel wall).

[27] This is a version of the dominance principle (Resnik 1987:9).

[28] In the context of climate policy making, an analogous line of reasoning, which focuses on the probability of attaining climate targets, is discussed under the title "cost risk analysis"; see the decision-theoretic analyzes by Schmidt et al. (2011) and Neubersch et al. (2014). Peterson (2006) shows that decision-making which seeks to minimize the probability of some harm runs into problems as soon as various harmful outcomes with different disvalue are distinguished.

(2) The value of a possible consequence of erecting or not erecting the steel wall is roughly proportional the corresponding likelihood that the historic building is not fully destroyed.
(3) Thus: The worst possible consequence of erecting the steel wall is preferable to the worst possible consequence of not erecting the steel-wall.
(4) An option A is to preferred to an option B, if—excluding all common possible consequences of A and B—A's worst possible consequence is preferable to B's worst possible consequence.
(5) Thus: The team should erect the steel wall.

3.2 Arguments from Robustness

The best/worst case arguments discussed above presume that one can determine which of all possible outcomes is best, and which is worst. In this respect, such arguments side with traditional risk analysis, which allegedly identifies the "optimal" choice. Sometimes, however, we are not in a position to say which possible outcome is clearly best. (Maybe some values are incomparable, cf. Hansson (1997) and Möller (2016)). As an alternative to optimization, we may seek options that bring about at least tolerable and acceptable (if not necessarily optimal) results. That's the core idea of so-called satisficing approaches, such as implemented in the tolerable-windows approach (e.g. Toth 2003) or the guardrails approach (e.g. Graßl et al. 2003). As normative premisses, such reasons only require a very simple normative theory, namely a binary demarcation of all possible states into acceptable versus non-acceptable ones. Sometimes, this demarcation can be provided in terms of minimum or maximum (multi-dimensional) thresholds (e.g. technical safety thresholds, social poverty thresholds, or climate policy goals such as the 2-degree-target).

Satisficing approaches do not only address axiological uncertainty, they also provide a suitable starting point to handle predictive uncertainty. Thus, an option is permissible under deep uncertainty just in case all its potential outcomes are acceptable according to the underlying 'normative landscape' (i.e. satisfy certain normative criteria). Permissible options are robust vis-à-vis all different possible states-of-affairs. Hence the notion of "robust decision analysis." (Cf. Lempert et al. 2003)

Like best/worst case reasoning, robust decision analysis requires one to have a full understanding of the alternative options' possible consequences. Lempert et al. (2002) have, however, proposed heuristics which allow one to estimate which options are robust in light of an incomplete grasp of the space of possibilities. These heuristics involve the iterative construction of ever new possible scenarios in order to test whether preliminarily identified options are really robust.[29]

[29] Robust decision analysis à la Lempert et al. is hence a systematic form of "hypothetical retrospection" (see Hansson 2016, Sect. 6).

We will return to the epistemic challenge of deep uncertainty—namely the problem of fully grasping the space of possibilities—in the second part of this chapter. But deep uncertainty also poses a normative challenge for robust decision analysis: the greater the number of possible outcomes and scenarios, the greater the likelihood that no option will eventually satisfy a given set of minimum standards. Put more bluntly: No available option may guarantee that the corresponding outcome is acceptable. Robust decision analysis seems of no avail in situations like these.

Still, I suggest that the diagnosis to the effect that no option is robust given some minimum standards may nonetheless give rise to a meaningful decision analysis. The following example illustrates the structure of the argumentation.

Example (Local Authority) Besides permitting and prohibiting the construction of the industrial site, the local authority considers further measures that could supplement a decision to grant permission. These include additional restrictions on design and use of the industrial site; natural barriers (hills, woods); gradual extension of the habitat through artificial flooding of agricultural land; etc. So the authority has to choose amongst alternative policy portfolios. It is guided by two main criteria: an environmental (protect our unique ecosystems) and an economic one (increase growth and employment). The mayor has provisionally set the following targets: 3 % growth p.a. over the next 10 years without any environmental degradation whatsoever. Experts say that, when taking all contingencies into account, there is no policy portfolio which will guarantee that these targets are met. There exist however robust options for weaker targets. So, the experts say, there are costly measures that will protect the ecological habitat (come what may) while constructing the new site, to the effect that long-term growth equals at least 2 %. The growth target of 3 % can be met while preserving the endangered habitat at the cost of putting another ecosystem at risk.

So the mayor really faces a choice between different sets of normative minimum standards that are "satisfiable," i.e. there exist robust policy options in view of these standards. Frequently, such a choice may involve normative trade-offs, e.g. lowering the ecological *or* the economic guardrail (tolerate more loss in biodiversity or slower GDP growth).

The above example suggest that robust decision analysis should try to identify

- The strictest, multi-dimensional sets of minimum standards such that there is at least one robust option relative to that set of guardrails.

Each set of guardrails will produce a different argument in favor of a policy option. In the WW2 bomb example, the experts may face a trade-off between costs of the operation and protecting the neighbors. Different ways of striking the balance will result in different arguments.[30] For example, argument H:

[30] These different arguments and the coherent position (cf. Brun and Betz 2016: Sect. 4.2) one adopts with regard to them can be understood as an operationalization of Hansson's degrees of unacceptability (cf. Hansson 2013:69–70).

(1) A possible outcome is acceptable if and only if no person is killed and the operation has a total cost of less than 1 million €. [Normative guardrails]
(2) There is no possible consequence of defusing the bomb according to which a person is killed or the operation has total cost greater than 1 million €. [Possibilistic prediction]
(3) An option is permissible just in case all its potential outcomes are acceptable. [Principle of robust decision analysis]
(4) Thus: It is permissible to defuse the bomb.

An alternative set of minimum standards yields another argument, argument I:

(1) A possible outcome is acceptable if and only if no person is seriously harmed and the operation has a total cost of less than 2 million €. [Normative guardrails]
(2) There is no possible consequence of detonating the bomb according to which a person is seriously harmed or the operation has total cost greater than 2 million €. [Possibilistic prediction]
(3) An option is permissible just in case all its potential outcomes are acceptable. [Principle of robust decision analysis]
(4) Thus: It is permissible to detonate the bomb.

3.3 Arguments from Risk Imposition

Let's stay with the WW2 bomb example. Assume the least expensive option (say detonating the bomb) risks to seriously harm people living and working in the neighborhood. When we deliberate about that option, it seems a relevant aspect whether the persons potentially affected have been informed and have given their consent. If not, this may provide a reason against choosing this option.[31] A simple argument from risk imposition can thus be reconstructed as follows, argument J:

(1) To detonate the bomb possibly causes serious harm (injuries) of persons living and working in the neighborhood.
(2) The persons living and working in the neighborhood have not given their consent to being exposed to the possibility of serious harm as a result of the bomb's disabling.
(3) An option that involves risk imposition (i.e. which potentially negatively affects persons who have not given their consent to being exposed to such a risk) must not be taken.
(4) Thus: The expert team must not detonate the bomb.

Arguments like these face different sorts of problems and are probably in need of further refinement. Sometimes it is just physically impossible for those being affected by a measure to provide consent (e.g. future generations). The simple

---

[31] For a detailed discussion of risk imposition and the problems standard moral theories face in coping with risks see Hansson (2003).

principle of risk imposition is hence too strict. It must be limited to cases where those potentially affected are in a position to provide consent, or it must state alternative necessary conditions for permissibility. Another problem is that the simple principle of risk imposition merely regards one specific aspect of the entire decision situation, it does, in particular, take into account neither *all* the alternative options nor *all* the possible outcomes of the different options. What if *every* available option involves risk imposition? What if the alternative options have clearly worse (certain or possible) consequences than merely imposing some risk of being injured without consent? Maybe the principle in premiss (3) is best seen as a *prima facie* principle.[32]

4 The Statics of Possibilistic Knowledge: Four Classes of Possibilistic Hypotheses

We've seen that practical reasoning under deep uncertainty requires grasp of the entire space of possibilities; justifications of policy recommendations presume that one correctly predicts all possible consequences for each available option. And the conclusions one arrives at depend sensitively on the outcomes one considers as possible.[33] In the second part of this chapter, we will discuss the methodological challenge of identifying all possible outcomes of a given option, i.e. all conceptual possibilities whose realization, as a result of implementing the corresponding option, are consistent with the given background knowledge.

It is sometimes straightforward to determine the decision-relevant possibilities.

Example (Pendulum) Consider a well-engineered pendulum in a black box. We know that it was initially displaced by 10°, but we don't know when it was released (a minute ago, a second ago, just now). The task is to predict the pendulum's position (deviation from equilibrium) in one minute. Given our ignorance about the time when the pendulum was released, any displacement between $\pm 10°$ is possible. That's the space of possibilities. In other words, these are precisely the statements about the pendulum's position which are consistent with our background knowledge.

It seems that case is fairly obvious, but it's nonetheless instructive to ask how exactly we arrive at the possibilistic prediction. So, on the one hand, every statement of the form "The pendulum is displaced by x degrees" with x taking a value between -10 and $+10°$ can be shown to be consistent with our background

[32] Brun and Betz (2016), this volume, discuss how such principles and the corresponding arguments can be analyzed. See also Hansson (2013:97–101).

[33] Thus, Hansson (1997) stresses that in decision-making under deep uncertainty the demarcation of the possible from the impossible involves as influential a choice as the selection of a decision principle.

knowledge. (In particular, for any such statement $H_{|x|\leq 10}$ there exists a time t_{rel} such that $H_{|x|\leq 10}$ can be derived from the Newtonian model of the pendulum and the possibility that the pendulum has been released at t_{rel}.) On the other hand, every statement of the form "The pendulum is displaced by x degrees" with x taking an *absolute* value greater than 10° can be shown to be inconsistent with our background knowledge. (Any such statement implies that the total energy in the contained system has increased, in violation of the principle of energy conservation.) In sum, we have completely mapped the space of possibilities by considering every conceptual possibility and either showing that it is consistent with K or showing that it is inconsistent with K. Or, in other words, each conceptual possibility has been "verified" or "falsified."[34]

That's in a way the ideal situation of possibilistic prediction.

Mapping the space of possibilities requires us to verify or falsify each conceptual possibility. Both tasks are tricky. An argument to the effect that a statement is consistent with the background knowledge (possibilistic verification) has to account explicitly for one's entire knowledge; if some item of information is left out, the argument fails to establish relative consistency (unless it is explicitly argued the item is irrelevant).[35] The more diverse, heterogeneous and dappled our understanding of a system, the more challenging this task. (That is the reason why conceptual possibilities are sometimes only "partially" verified in the sense that they are shown to be consistent with a subset of our background knowledge; e.g., technical feasibility studies may ignore economic and societal constraints on technology deployment.) An argument to the effect that a statement is inconsistent with the background knowledge (possibilistic falsification) may in contrast be comparatively simple, it may suffice to find a single known fact that refutes the conceptual possibility. The challenge here rather consists in finding an item in our background knowledge that refutes the conceptual possibility.

We have sketched the epistemic ideal of possibilistic prediction and identified potential challenges. But due to our cognitive limitations, we may fail to overcome these challenges. Our actual epistemic situation may depart from the ideal in different ways.

i. There might be some conceptual possibilities which actually are consistent with the background knowledge, although we have not been able to show this (failure to verify).

[34] In speaking of "verified" and "falsified" conceptual possibilities, I follow a terminological suggestion by Betz (2010). To "verify" a conceptual possibility in this sense does not imply to show that the corresponding hypothesis is true, what *is* shown to be true (in possibilistic verification) is the claim that the hypothesis is consistent with background knowledge. However, to "falsify" a conceptual possibility involves showing that the corresponding hypothesis is false (given background knowledge).

[35] For this very reason, it is a non-trivial assumption that a dynamic model of a complex system (e.g. a climate model) is adequate for verifying possibilities about that system (cf. Betz 2015).

ii. There might be some conceptual possibilities which actually are inconsistent with the background knowledge, although we have not been able to show this (failure to falsify).

In other words: There may be some conceptual possibilities which are neither verified nor falsified. In addition, it is not always clear that we have fully grasped the space of conceptual possibilities in the first place, so

iii. There might be some conceptual possibilities which we haven't even considered so far (failure to articulate).

That brings us to the following systematization of possibilities (see also Betz 2010):

1. Non-articulated possibilities [**Class 1**]
2. Articulated possibilities
    (a) Falsified possibilities (shown to be inconsistent with background knowledge) [**Class 2**]
    (b) Non-falsified possibilities
        i. Verified possibilities (shown to be consistent with background knowledge) [**Class 3**]
        ii. Merely articulated possibilities (neither verified nor falsified) [**Class 4**]

For ideal agents, the dichotomy between conceptual possibilities that are consistent with background knowledge versus those that aren't is perfectly fine and may serve to express their possibilistic knowledge. For non-ideal agents with limited cognitive capacities, like us, this dichotomy is often an unattainable ideal, and hence unsuitable to express our imperfect understanding of a domain. The conceptual distinctions above provide a more fine-grained framework for expressing our possibilistic knowledge at a given moment in time.

Let me illustrate these distinctions with some examples.

Class 1. Examples of non-articulated possibilities—aka "unknown unknowns"—can at best be given in retrospect. One of the most prominent instances is the hypothesis that HCFCs deplete the ozone layer, which was not even articulated in the first half of the twentieth century. Likewise, the possibility that an increased GHG concentration may cause the dry out of the Amazonian rainforest was not entertained in the time of Arrhenius. And that asbestos may cause lung cancer was not considered at all when asbestos mining began (more than 4,000 years ago). Likewise, "Just underneath the bomb lies King John's Treasure, a medieval fortune of immense financial but even infinitely greater historic value" is not even articulated by the bomb experts. While we can't provide specific cases of possibilities we currently haven't even thought about, we may have more or less strong reasons to suspect that such possibilities exist, e.g. when we deal with a complex system which we have only poorly understood so far.[36]

[36] See also the "epistemic defaults" discussed by Hansson (2016: Sect. 5).

Class 2. By summing up the maximum contribution of all potential sources of sea level rise, climate scientists are in a position to robustly refute the conceptual possibility that global mean sea level will rise by 10 m until 2,100 with business as usual emissions.[37] In various safety reports, CERN scientists have argued that the generation of stable microscopic black holes in the large hadron collider is inconsistent with current background knowledge (specifically basic physical theory and cosmic observations).[38] In our fictitious example, the expert team rules out—given its knowledge about the size of the bomb and the most powerful explosive used in WW2—that a detonation affects a cultural heritage site 2 km away.

Class 3. Following various detailed energy scenarios, it is consistent with our knowledge about the future of the energy system (which is mainly of technological nature) that Europe reduces its CO2 emissions by 80 % in 2050 compared to 1990.[39] Climatologists argue, by means of detailed models of ice shelf dynamics and global warming scenarios, or historic analogies, that a sea level rise of 2 m until 2,300 is consistent with current understanding of the climate system.[40] That the US president in 10 years' time will be a democrat is also known to be consistent with our current knowledge, essentially because we know nearly nothing about the specifics of the US political system in the medium-term. The bomb experts have verified the conceptual possibility that no single glass window breaks due to the detonation of the bomb by running computer simulations according to which the steel wall deflects, under favorable conditions, the pressure wave.

Class 4. A run-away greenhouse effect on earth is a conceptual possibility sometimes articulated and seriously considered by climate scientists; yet it seems an open question whether that scenario is consistent with our knowledge about the climate system.[41] Can the world achieve the 2-degree-target with current energy technologies, but without expanding nuclear energy and without substantial reductions in global economic growth? I suspect we have no proof that this conceptual possibility cannot unfold, but in the same time we haven't shown that this scenario is consistent with our heterogeneous background knowledge, either. In our fictitious example, the policy makers wonder whether the ecosystem can essentially survive even if one species of fish is lost; but preliminary investigations by biologists are so far inconclusive. A schoolgirl asks the bomb experts whether the dust cloud of a bomb explosion may shut down the hospital's air conditioning system; the experts concede that they have not checked this yet.

[37] For a discussion of narrower bounds for future sea level rise see Church et al. (2013:1185–6).

[38] See Ellis et al. (2008) and Blaizot et al. (2003).

[39] Compare the EU Energy Roadmap 2050 (European Commission 2011).

[40] Cf. Church et al. (2013:1186–9).

[41] Hansen et al. (2013) distinguish different "run-away greenhouse" scenarios and discuss whether they can be robustly ruled out—which, according to the authors, is the case for the most extreme ones (p. 24).

5 The Dynamics of Possibilistic Knowledge

Our possibilistic foreknowledge is highly fallible. That's already true for the simple notion of serious possibility in the sense of *relative* consistency with the background knowledge. Changes in background knowledge trigger changes is serious possibilities. In particular, possibilistic predictions are fallible to the extent that background knowledge is fallible. Expansion and revision of background beliefs can necessitate a revision of one's possibilistic knowledge. So can the recognition that the inferences drawn from background assumptions were incomplete or incorrect. And conceptual innovations that allow for the articulation of novel hypotheses may have the same effect.

How do these changes affect a nuanced explication of one's possibilistic knowledge in line with the previous section? We distinguish four cases: (a) The addition of novel items of evidence or inferences which do not affect previously held background beliefs (expansion); (b) the withdrawal of previously held background beliefs without acquiring novel ones (pure contraction); (c) the replacement of previously held background assumptions or inferences with novel ones (revision); (d) the modification of old or the creation of new terminology that allows for articulation of novel hypotheses (conceptual change).

Re (a). Assume the background knowledge, or the set of inferences drawn from it, is *expanded* in a conservative way, i.e., without changing previous background knowledge or inferences. As a first point to note, any previously falsified possibility will remain falsified. But the status of formerly verified or merely articulated possibilities may change: All these hypotheses have to be re-assessed and the arguments which establish that a hypothesis is consistent with previous background knowledge don't warrant that it is consistent with broader background knowledge—they don't carry over, that is, to the novel situation. For some previously verified hypotheses, it may not be feasible to show that they are consistent with novel background knowledge; some of these may even be falsified on the basis of novel evidence. That may also happen with some formerly merely articulated hypotheses.

In sum, conservative expansion tends to reduce the number of verified possibilities and to increase the number of falsified ones. And that's how it should be, as increasing the content of one's knowledge means to be able to exclude ever more conceptual possibilities.

Let me illustrate these dynamics with the WW2 bomb example. Suppose the bomb experts get a call from a colleague, who has just discovered a document in a military archive from which it is plain that the particular bomb to-be-defused was produced before 1942. That novel evidence necessitates the re-assessment of non-falsified possibilities. The possibility that the trigger is intact, for instance, had been verified by reference to other WW2 bombs recently found, whose trigger was intact. But these bombs all dated from the last 2 years of the war. So the argument from analogy does not really warrant anymore that the trigger of the bomb to-be-defused may be intact, too. For the time being, the possibility that the trigger is intact has to count as a merely articulated one. The experts had also considered whether the dust cloud of a potential detonation may damage the hospital's air conditioning, without being able to verify or falsify that possibility. But based on the novel information that the bomb was produced in 1942, they can now exclude that possibility: the explosives used in that year degrade relatively quickly, which severely reduces the overall power of a potential explosion. The dust cloud would hence be too small to affect the hospital.

Re (b). In terms of possibilistic dynamics, pure *contraction* is symmetric to conservative expansion of the background knowledge. If some background beliefs are given up, e.g. because the inferences that have been used to establish them are found to be fallacious, without acquiring novel beliefs, then every conceptual possibility that had been shown to be consistent with the background knowledge remains a verified possibility. Merely articulated possibilistic hypotheses are unaffected, too. But the allegedly falsified possibilities have to be re-examined: Some of these may become merely articulated or even verified possibilities relative to the contracted background belief system.

Continuing the previous example, let's assume the bomb experts realize that estimates of the degraded chemical substances' explosive power are highly uncertain. In fact, it seems that a blunt statistical fallacy has been committed in the extrapolation from small-scale field tests to large-scale bombs, such as the one to-be-defused. So the bomb experts retract their belief that the power of a potential detonation can be narrowly confined—despite the bomb being produced in 1942. That in turn broadens the range of possibilities. Specifically, the hypothesis that a detonation will produce a large dust cloud which shuts down the hospital's air conditioning system cannot be falsified anymore; it becomes a merely articulated possibility.

Re (c). When the background knowledge or the inferences drawn are *revised*, all the conceptual possibilities have to be re-assessed. Previously falsified hypotheses may become merely articulated or verified ones. Formerly verified hypotheses may not be verifiable anymore, and may even be falsified. In short, anything goes. There is no stability, no accumulation of any kind of possibilistic prediction.

Let's illustrate this case, again, with the WW2 bomb example. Assume the bomb team realizes that it had committed, early in the mission, a fatal measurement error. They underestimated the length and hence the weight of the bomb by 30 %! All the possibilities, all the scenarios considered have to be re-assessed. For instance, the team formerly argued, based on detailed computer simulation, that it is consistent with their understanding of the situation that no window breaks upon detonation thanks to a steel wall which deflects the pressure wave. But the simulations were based on an erroneous assumption about the bomb's size, and hence don't verify that specific scenario (given the correct assumption). The possibility that no window breaks becomes a merely articulated possibility (unless, e.g., an accordingly modified simulation re-affirms the original finding). Also, the team originally excluded the possibility that the cultural heritage site will be damaged. But the argument which rules out this scenario, too, relied on a false premiss. Given the novel estimate of the bomb's size, that possibility cannot be robustly ruled out anymore. Even more so, analogies to similar cases, based on the correct size of the bomb, suggest that the detonation may very well damage the cultural heritage site. So this previously falsified scenario becomes a verified possibility. And so on.

Re (d). Finally, let us briefly consider the case of conceptual change. New terminology is introduced or the meaning of old terminology is modified. Such conceptual change will typically go along with a revision or a re-statement of the background knowledge. So anything we've discussed under (c) is applicable here, as well. On top of that, the creation of a new terminology affects the set of conceptual possibilities and therefore the set of possibilistic hypotheses considered by the agents—some previously articulated hypotheses may not be conceptually possible anymore (like "that's not consistent with the way we use the words now"), other possibilities might be newly articulated.

We shall illustrate the effect of conceptual change against the background of the advancement of molecular biology and genetic theory in the twentieth century. The progress in these disciplines went along with the development of novel concepts, an entirely new language that allows one to describe a known phenomenon in a new way. For example, only against this novel conceptual framework could scientists articulate a hypothesis like: The exposition to this and this chemical substance affects the DNA of the offspring and alters the genetic pool in the medium term. Or: Radioactive radiation may damage the DNA in a cell.

Non-monotonic changes in the stock of possibilistic predictions, such as discussed hitherto, correspond to potential surprises. Just assume that the bomb experts had not corrected their initial measurement error—they would have been surprised to see the cultural heritage site being nearly destroyed. Likewise, had the schoolgirl not brought up the possibility that the hospital's air conditioning system will break down, the experts might have faced an outcome they hadn't even thought of.

Rational decision making under deep uncertainty requires one to map out, *given* current background knowledge, the possibilistic predictions in line with the previous section. I want to suggest that, on top of this, rational decision making should attempt to gauge the potential for surprise in a given decision situation—specifically the potential for surprise that is linked to the modification of the background knowledge and conceptual change.

What I have in mind is a second order assessment of one's background knowledge, the inferences drawn and one's conceptual frame. The more stable these items, the smaller the potential for surprise. If there's reason to think that one's understanding of a system will change and improve quickly, however, one should also expect the overhaul of one's possibilistic outlook.

Of course, it's impossible to predict *what* we will newly come to know in the future.[42] But it's not impossible to estimate whether our knowledge will change, and how much. So, in 1799 Humboldt had reason to expect that he would soon know much more about the flora of South America; if NASA plans a further space mission to explore a comet, we have reason to expect that our understanding of that comet (and maybe comets in general) will change in the future. However, if, in spite of serious efforts, our understanding of a system has stagnated in the last decades and we even understand why it is difficult to acquire further knowledge about that system (i.e. because of its complexity, because of measurement problems that can't be overcome with available technologies, etc.), we have a reason to expect our background knowledge (and hence our stock of possibilistic predictions) to be rather stable.[43]

6 The Practical Arguments Revisited

I've suggested that our possibilistic foreknowledge should be cast in terms of verified, merely articulated, and falsified possibilities; it should also comprise an estimate of the scope of currently non-articulated possibilities as well as an assessment of the stability of one's background knowledge.

What does this entail for practical reasoning under deep uncertainty?

The decision principles and practical arguments we discussed in Sect. 3 assume that we have knowledge about plain possibilities, without taking further differentiations into account. When different kinds of possibilities are distinguished, these principles are in need of further specification before being applied. As a result, each decision principle discussed above corresponds to several principles, each referring to a different sort of possibility.

Let's explore these complications by means of our examples. We start with worst case reasoning.

Example (Local Authority) The environmentalists cited the destruction of the ecosystem as a worst case in order to argue against the construction of the industrial site. Upon being pressed, they explain their possibilistic outlook: "Why do we think it's possible that the ecosystem will be destroyed? Well, because no one has convincingly argued so far that this won't happen."

This makes it clear that the environmentalists are concerned with non-falsified possibility. The original argument C can now be more precisely reconstructed as, argument K:

---

[42] See Betz (2011), especially the discussion of Popper's argument against predicting scientific progress (pp. 650–651).

[43] See Rescher (1984, 2009) for a discussion of limits of science and their various (conceptual or empirical) reasons.

(1) There is no available option whose worst *non-falsified* possible consequence is preferable to the worst *non-falsified* possible consequence of not permitting the construction.
(2) If there is no available option whose worst *non-falsified* possible consequence is [weakly] preferable to A's worst *non-falsified* possible consequence, then one is obliged to carry out option A.
(3) Thus: The local authority should not permit the construction of the industrial complex.

This clarification also shows that, in order to challenge the argument, it suffices to point out a non-falsified (not necessarily verified) possibility according to which not constructing the industrial complex will have consequences as bad as the destruction of the habitat.

Other worst case arguments may consistently refer to verified possibilities. Next, consider best/worst case reasoning.

Example (Local Authority) One argument in the hearing (argument F) compared the worst case of constructing the site with its best case, that is the attraction of a green industries cluster and CO2-free local energy generation in the medium term.

What kind of possibilities are we facing here? Assuming the argument follows, on the one side, the outlook of the environmentalists, the worst case is a merely articulated possibility. What about the best case? That optimistic prediction is not shown to be consistent with the background knowledge, either (there exists for example no precise energy scenario that spells out that the respective conceptual possibility is consistent with local circumstances such as potentials for solar and wind energy, etc.). The possibilistic prediction is just set forth, it is a merely articulated possibility, too. So the argument really strikes a balance between best and worst non-falsified possible cases.

Other best/worst case arguments may compare the best verified possible case with the worst verified possible case, or even the best verified possible case with the worst non-falsified possible case.

Let's turn to robust decision analysis. An option was said to be robust vis-à-vis certain normative guardrails just in case every possible consequence satisfies these guardrails. We've designed the WW2 bomb example above such that no option is allegedly robust with respect to the minimum aims that no person should be harmed and that the costs of the operation should not exceed 1 million Euro. At least one of these guardrails had to be relaxed so that a robust option exists (cf. arguments H, I). At this point, a team member intervenes.

Example (WW2 Bomb) "We haven't been able to find a robust option that satisfies our original guardrails because we considered any possibility we just came up with. What if we restrict our deliberation to cases that we're pretty sure may happen, because they happened before or because our simulations give rise to corresponding results? It seems to me that the detonation plus small-scale evacuation is robust vis-à-vis our original minimum standards and relative to all such *verified* possibilities."

So the team member explains that arguments H, I should be understood as referring to non-falsified possibilities. In addition, she sets up a further argument which only takes verified possibilities into account, argument L:

(1) A possible outcome is acceptable if and only if no person is seriously harmed and the operation has a total cost of less than 1 million €. [Normative guardrails]
(2) There is no *verified* possible consequence of detonating the bomb plus small-scale evacuation according to which a person is seriously harmed or the operation has total cost greater than 1 million Euro. [Possibilistic prediction]
(3) An option is permissible just in case all its potential outcomes (*verified* possibilities) are acceptable. [Principle of robust decision analysis]
(4) Thus: It is permissible to detonate the bomb after small-scale evacuation.

A police officer has reservations about this argument, and objects:

Example (WW2 Bomb) "But you can't robustly rule out that some people in the neighborhood, which will not be evacuated, will be harmed, right. So we impose a serious risk on these people and we must not do so without their consent. Which in turn is difficult to get given that some of these persons are comatose."

This brings us to risk imposition. Here, the police officer challenges the conclusion of an argument from robustness (with respect to verified possibilities) with an argument from risk imposition (with respect to falsified possibilities).

Of course, arguments from risk imposition may also be articulated in view of verified possibilities.

Such are the differentiations we have to account for. We get, as a consequence of our more fine-grained framework for possibilistic prediction, a further proliferation of the already numerous decision criteria and argument patterns for decision making under deep uncertainty.

Now, which of these criteria, which of these argument schemes should one use in order to justify one's choice?—That is the wrong question! There is no exclusiveness. In a first step, one should consider different arguments, which rely on different decision criteria, side by side. We typically don't have a single plausible argument that tells us what we should do, but we have a complex argumentation that consists in various, partially conflicting arguments. So the question is rather: Which of these arguments (underlying criteria) should we prefer? Or, even better: How should we balance the conflicting arguments?[44]

The answer to this question seems to depend on at least two factors: (a) One's level of risk aversion. Already the original decision criteria expressed different risk attitudes. That's also true for their refined versions. Whether a catastrophic *merely-articulated* possible consequence or only a catastrophic *verified* possible consequence represents a sufficient reason for some agent to refrain from some action is a matter of that agent's risk aversion. Likewise, an agent who seeks robust options

[44] Brun and Betz (2016: especially Sect. 4.2) explain how argument analysis, and especially argument mapping techniques, help to balance conflicting normative reasons in general.

with respect to non-falsified possibilities is more risk averse than an agent who is content with robustness with respect to verified possibilities. (b) The profile of possibilistic predictions on which the decision is based. If, for example, there is a wide range of non-falsified possibilities whereas only very few of these can be verified, then it seems unreasonable to base the deliberation on the verified possibilities only. Doing so would make much more sense, however, if nearly all non-falsified possibilities were actually verified. Balancing the different decision criteria may also depend on the ratio of verified, merely articulated and falsified possibilities (which reflects the breadth and depth of one's understanding of a system).

The distinction between different kinds of possibilities does not just make things more complicated, it may also help us to resolve dilemmas, especially dilemmas that pop up in worst case considerations. The idea is that verified-worst-case-reasons trump—*ceteris paribus*—merely-articulated-worst-case-reasons.

In one of our examples, the local authority faces a dilemma, which can be fleshed out as follows.

Example (Local Authority) If the authority permits construction, then the new industrial site will affect, essentially through traffic noise, species living in the habitat, which may eventually cause its destruction. If the authority does not grant permission, then it won't have the money to thoroughly decontaminate the mine, which may in turn intoxicate groundwater and destroy the ecosystem, too. In an attempt to resolve the dilemma, engineers point out the following asymmetry: "Both cases can't be ruled out. But the intoxication scenario is really spelled out in detail and on the basis of extensive knowledge about the mine, its status, the effects of contamination on groundwater, the toxic effects on species living in the ecosystem, etc. This is all well understood and we know that it may happen. We have however no comparable knowledge about the precise effects of traffic noise."

The asymmetry consists in the fact that the worst case of one option is a merely articulated possibility whereas the worst case of the other option is even a verified possibility. This information could be used to resolve the dilemma in favor of the option with the merely-articulated worst case.

7 Arguments from Surprise

The fine-grained conceptual framework of possibilistic foreknowledge does not only induce a differentiation of existing decision criteria, it also allows us to formulate novel argument schemes for practical reasoning under deep uncertainty, which can not be represented in terms of traditional risk analysis.

These novel argument schemes concern the various options' potential of surprise. Given a possibilistic outlook, a surprise has occurred just in case something has happened which wasn't considered possible (i.e. was not referred to in some non-falsified possibility). Surprises may happen for different reasons. We may in particular distinguish two sorts of surprise, to which we already alluded above: (a) surprises that result from unknown unknowns; (b) surprises that result from the fallibility of and the occasional need to rectify one's background knowledge.[45]

We develop and explore arguments which refer to these kinds of surprise by means of example.

7.1 Arguments from Unknown Unknowns

Arguments from unknown unknowns set forth reasons to suspect that some relevant conceptual possibilities have not even been articulated, and claim that the available options are affected unevenly by this problem.

Example (WW2 Bomb) A member of the expert team proposes to try a brand new method for disarming bombs, which he has only recently heard of and which involves ultra-deep freezing and nano-materials. Computer simulations have so far been promising (cheap and safe!), he lectures, but no field tests have been carried out yet. The other experts worry that they lack the time to thoroughly think through the potential effects. Without having a particular potential catastrophic consequence in mind, they argue that the team should rather go for one of the more costly options, so that they are at least pretty sure to oversee the space of possibilities and minimize the risk of unknown unknowns.

Example (Local Authority) As a follow-up to the public hearing, some citizens raise, in a public letter, the concern that the endangered ecosystem is not isolated but linked, through multiple migratory species, with other ecosystems—both regionally and nation-wide. They argue that we really have *no idea* about what will be the broader consequences of the destruction of the habitat, not only ecologically, but also agriculturally and hence economically.

Example (Geoengineering) The proposal to artificially cool the planet has sparked a public controversy (see also Elliott 2016; Brun and Betz 2016). One argument against doing so stresses that we know, from other technological interventions into complex systems, that things may happen which we haven't even thought of. A similar worry, the argument continues, does not apply to alternative policies for limiting climate change. Emission reductions, for example, seek to reduce the extent of anthropogenic intervention into the climate system. Because of unknown unknowns, we should refrain from deploying geoengineering technologies.

It seems that the above arguments are not outright unreasonable or implausible. The following decision principles could be used to reconstruct these arguments in detail:

---

[45] Basili and Zappia (2009) discuss the role of surprise in modern decision theory and its anticipation in the works of George L. S. Shackle.

- If, considering all relevant aspects except their potential for surprise (i.e., the extent to which an option is associated with unknown unknowns), the options A and B are normatively equally good, and if A has a significantly greater potential for (undesirable) surprise than option B, then option B is normatively better than (should be preferred to) option A.
- If option A has a significantly smaller potential for (undesirable) surprise (i.e., is associated with more unknown unknowns) than its alternatives and if carrying out option A doesn't jeopardize a more significant value (than surprise aversion), then option A should be carried out.

7.2 Arguments from Fallibility and Provisionality

Arguments from fallibility and provisionality call for caution in the light of potential future modifications of our background knowledge and corresponding revisions of our possibilistic outlook.

Example (WW2 Bomb) Physical scientists who have heard of the proposed method for disarming bombs have reservations about its application, too. They stress that the method relies on a novel theory (about nano-materials) in a science that is evolving quickly. The background knowledge against which the experts assess the brand new method is likely to change in the near future. That speaks against its deployment; in any case, the scientists argue, the experts should prepare for the eventuality that something unforeseen happens, i.e., something they had articulated, but had originally not verified, or even ruled out.

Example (Geoengineering) Another objection to geoengineering: Our detailed understanding of the climate system, its complex feedbacks, and its multi-scale interactions evolves quickly. Changes in this understanding will crucially affect our possibilistic assessment of the effectiveness and side-effects of geoengineering—much more than our assessment of adaptation and mitigation. Even if, under current possibilistic predictions, geoengineering deployment seems promising, we should refrain from it in light of its high potential for (catastrophic) *surprise*.

These arguments, too, appear prima facie reasonable, and they could be reconstructed with decision principles similar to the ones used in arguments from unknown unknowns:

- If, considering all relevant aspects except their potential for surprise (i.e., the extent to which relevant background knowledge is provisional and likely to be modified), the options A and B are normatively equally good, and if A has a significantly greater potential for (undesirable) surprise than option B, then option B is normatively better than (should be preferred to) option A.
- If option A has a significantly smaller potential for (undesirable) surprise (i.e., the relevant background knowledge is provisional and more likely to be

modified) than its alternatives and if carrying out option A doesn't jeopardize a more significant value (than surprise aversion), then option A should be carried out.

The available options' potential for surprise may also be referred to in order to resolve dilemmas, as illustrated in the following case, which also provides an example for a positive potential surprise.

Example (Local Authority) The local policy-makers commissioned a scientific study to identify and assess alternative locations for the industrial complex. The scientists have actually found a second location; at each site however, the report argues, a different ecosystem would be put at risk. The report details that the habitat near the original location has been monitored and studied in depth and over decades, it is moreover well documented from a handful of other places that traffic noise *may* cause the destruction of the highly sensitive habitat. The ecosystem near the novel location is very remote and has not been much studied, it is for example not even clear which mammal species exactly are living there. For both options (i.e., locations), the verified worst case is the destruction of the respective ecosystem. For the alternative location, this worst case is verified not because of sophisticated modeling studies, but simply because so little is known about the corresponding habitat. Further studies may revise the limited understanding of the poorly investigated ecosystem, and show that the system is not really put at risk by an industrial complex at all. The local policy-makers understand that its higher potential of surprise seems to speak for the alternative location: The second option has a higher potential for *positive* surprise.

Such an argument from positive surprise may be reconstructed with the following decision principle:

- If the options A and B have equally disastrous non-falsified worst cases and if A has a significantly greater potential for surprise than option B, and if no surprise associated with A implies that A's worst case is even more catastrophic than originally thought, then A should be preferred to B.

8 Summing Up

This chapter discussed and illustrated a variety of arguments that may inform and bear on a decision under great uncertainty, where uncertainties cannot be quantified and decision makers have to content themselves with possibilistic forecasts. It developed, in addition, a differentiated conceptual framework that allows one to express one's possibilistic foreknowledge in a nuanced way, in particular by recognizing the difference between conceptual possibilities that have been shown to be consistent with background knowledge and ones that merely have not been refuted. The conceptual framework also gives rise to a precise (possibilistic) notion of surprise (e.g. unknown unknowns) and triggers an expansion of the arsenal of standard argument patterns for reasoning under great uncertainty.

One major purpose of this chapter has been to refute the widely held prejudice that rational decision making and practical reasoning requires at least probabilities. We have seen that this notion is simply untenable. But in view of the multitude of arguments that can be advanced in deliberation under great uncertainty, the problem seems to be that there will typically be too many (rather than too few) reasonable arguments, none of which however clearly trumps the others, none of which wins the debate by itself. All arguments that potentially justify decisions under great uncertainty seem more or less contestable, as they rely, in particular, on decision principles which express different levels of risk aversion. The real problem of practical reasoning is not to find any arguments at all, but to cope with the abundance of conflicting arguments and to aggregate diverse reasons in a meaningful way.

How can that be achieved? Here, the fact that the *argumentative turn* in risk analysis is backed by argumentation theoretic models of rational controversy comes fully into play. Specifically, the methods of argument analysis and evaluation as introduced in Brun and Betz (2016) provide techniques for aggregating conflicting reasons. In a nutshell, I recommend, as a strategy for handling the variety of practical arguments under great uncertainty,

1. To reconstruct *all* arguments that are (and can) be advanced pro and con the alternative options as well as further considerations that speak for or against those arguments;
2. To identify agreed upon background beliefs (such as scientifically established facts), which fix truth-values of some premisses and conclusions in the debate;
3. To identify coherent positions one may reasonably adopt in view of the arguments and the background beliefs, which in turn pinpoint the normative trade-offs one faces when justifying a choice.[46]

Individual decision makers may then resolve the normative trade-offs by opting for one such coherent position.

Recommended Readings

Betz, G. (2010a). What's the worst case? The methodology of possibilistic prediction. *Analyse und Kritik, 32*, 87–106.
Etner, J., Jeleva, M., & Tallon, J.-M. (2012a). Decision theory under ambiguity. *Journal of Economic Surveys, 26*, 234–270.

[46] So, to give an example, it may be that in a specific debate, say about geoengineering, one cannot coherently accept in the same time (i) the precautionary principle, (ii) sustainability goals and (iii) a general ban of risk technologies. Whoever takes a stance in this debate has to strike a balance between these normative ideas.

Lempert, R. J., Popper, S. W., & Bankes, S. C. (2003a). *Shaping the next one hundred years: New methods for quantitative, long-term policy analysis*. Santa Monica: RAND.
Resnik, M. D. (1987a). *Choices: An introduction to decision theory*. Minneapolis: University of Minnesota Press.

References

Basili, M., & Zappia, C. (2009). Shackle and modern decision theory. *Metroeconomica, 60*, 245–282.
Bernardo, J. M. (1979). Reference posterior distributions for Bayesian inference. *Journal for the Royal Statistical Society. Series B (Methodological), 41*, 113–147.
Betz, G. (2010b). What's the worst case? The methodology of possibilistic prediction. *Analyse und Kritik, 32*, 87–106.
Betz, G. (2011). Prediction. In I. C. Jarvie & J. Zamora-Bonilla (Eds.), *The sage handbook of the philosophy of social sciences* (pp. 647–664). Thousand Oaks: SAGE Publications.
Betz, G. (2015). Are climate models credible worlds? Prospects and limitations of possibilistic climate prediction. *European Journal for Philosophy of Science, 5*, 191–215.
Blaizot, J-P., Iliopoulos, J., Madsen, J., Ross, G. G., Sonderegger, P., Specht, H. J. (2003). *Study of potentially dangerous events during heavy-ion collisions at the LHC: Report of the LHC Safety Study Group*. https://cds.cern.ch/record/613175/files/CERN-2003-001.pdf. Accessed 12 Aug 2015.
Briggs, R. (2014). Normative theories of rational choice: Expected utility. *The Stanford Encyclopedia of Philosophy*. http://plato.stanford.edu/entries/rationality-normative-utility/.
Brun, G., & Betz, G. (2016). Analysing practical argumentation. In S. O. Hansson & G. Hirsch Hadorn (Eds.), *The argumentative turn in policy analysis. Reasoning about uncertainty* (pp. 39–77). Cham: Springer. doi:10.1007/978-3-319-30549-3_3.
Church, J. A., Clark, P. U., Cazenave, A., Gregory, J. M., Jevrejeva, S., Levermann, A., Merrifield, M. A., et al. (2013). Sea level change. In T. F. Stocker, D. Qin, G.-K. Plattner, M. Tignor, S. K. Allen, J. Boschung, A. Nauels, Y. Xia, V. Bex, & P. M. Midgley (Eds.), *Climate change 2013: The physical science basis contribution of Working Group I to the fifth assessment report of the Intergovernmental Panel on Climate Change* (pp. 1137–1216). Cambridge: Cambridge University Press.
Clarke, L. B. (2006). *Worst cases: Terror and catastrophe in the popular imagination*. Chicago: University of Chicago Press.
Doorn, N. (2016). Reasoning about uncertainty in flood risk governance. In S. O. Hansson & G. Hirsch Hadorn (Eds.), *The argumentative turn in policy analysis. Reasoning about uncertainty* (pp. 245–263). Cham: Springer. doi:10.1007/978-3-319-30549-3_10.
Egan, A., & Weatherson, B. (2009). *Epistemic modality*. Oxford: Oxford University Press.
Elliott, K. C. (2010). Geoengineering and the precautionary principle. *International Journal of Applied Philosophy, 24*, 237–253.
Elliott, K. C. (2016). Climate geoengineering. In S. O. Hansson & G. Hirsch Hadorn (Eds.), *The argumentative turn in policy analysis. Reasoning about uncertainty* (pp. 305–324). Cham: Springer. doi:10.1007/978-3-319-30549-3_13.
Ellis, J., Giudice, G., Mangano, M., Tkachev, I., & Wiedemann, U. (2008). *Review of the safety of LHC collisions*. http://www.cern.ch/lsag/LSAG-Report.pdf. Accessed 10 Nov 2012.
Ellsberg, D. (1961). Risk, ambiguity, and the savage axioms. *Quarterly Journal of Economics, 75*, 643–669.
Etner, J., Jeleva, M., & Tallon, J.-M. (2012b). Decision theory under ambiguity. *Journal of Economic Surveys, 26*, 234–270.

European Commission. (2011). *Commission staff working paper. Impact assessment. accompanying the document communication from the commission to the council, the European Parliament, the European Economic and Social Committee and the Committee of the Regions.* Energy Roadmap 2050. COM(2011)885. http://ec.europa.eu/smart-regulation/impact/ia_carried_out/docs/ia_2011/sec_2011_1565_en.pdf. Accessed 12 Aug 2015.

Gardiner, S. M. (2006). A core precautionary principle. *The Journal of Political Philosophy, 14*, 33–60.

Gilboa, I., Postlewaite, A., & Schmeidler, D. (2009). Is it always rational to satisfy Savage's axioms? *Economics and Philosophy, 25*(Special Issue 03): 285–296.

Hartmut, G., Kokott, J., Kulessa, M., Luther, J., Nuscheler, F., Sauerborn, R., Schellnhuber, H-J., Schubert, R., & Schulze, E-D. (2003). *World in transition: Towards sustainable energy systems.* German Advisory Council on Global Change Flagship Report. http://www.wbgu.de/fileadmin/templates/dateien/veroeffentlichungen/hauptgutachten/jg2003/wbgu_jg2003_engl.pdf. Accessed 12 Aug 2015.

Hansen, J., Sato, M., Russell, G., & Kharecha, P. (2013). Climate sensitivity, sea level and atmospheric carbon dioxide. *Philosophical Transactions of the Royal Society A-Mathematical Physical and Engineering Sciences, 371* (20120294).

Hansson, S. O. (1997). The limits of precaution. *Foundations of Science, 1997*, 293–306.

Hansson, S. O. (2001). *The structure of values and norms. Cambridge studies in probability, induction, and decision theory.* Cambridge: Cambridge University Press.

Hansson, S. O. (2003). Ethical criteria of risk acceptance. *Erkenntnis, 59*, 291–309.

Hansson, S. O. (2013). *The ethics of risk: Ethical analysis in an uncertain world.* New York: Palgrave Macmillan.

Hansson, S. O. (2016). Evaluating the uncertainties. In S. O. Hansson & G. Hirsch Hadorn (Eds.), *The argumentative turn in policy analysis. Reasoning about uncertainty* (pp. 79–104). Cham: Springer. doi:10.1007/978-3-319-30549-3_4.

Heal, G., & Millner, A. (2013). *Uncertainty and decision in climate change economics.* NBER working paper No. 18929. http://www.nber.org/papers/w18929.pdf. Accessed 12 Aug 2015.

Jeffrey, R. (1965). *The logic of decision.* Chicago: University of Chicago Press.

Jenkins, G. J., Murphy, J. M., Sexton, D. M. H., Lowe, J. A., Jones, P., & Kilsby, C. G. (2009). *UK climate projections: Briefing report.* Exeter: Met Office Hadley Centre.

Lempert, R. J., Popper, S. W., & Bankes, S. C. (2002). Confronting surprise. *Social Science Computer Review, 20*, 420–440.

Lempert, R. J., Popper, S. W., & Bankes, S. C. (2003b). *Shaping the next one hundred years: New methods for quantitative, long-term policy analysis.* Santa Monica: RAND.

Luce, R. D., & Raiffa, H. (1957). *Games and decisions: Introduction and critical survey.* New York: Wiley.

Möller, N. (2016). Value uncertainty. In S. O. Hansson & G. Hirsch Hadorn (Eds.), *The argumentative turn in policy analysis. Reasoning about uncertainty* (pp. 105–133). Cham: Springer. doi:10.1007/978-3-319-30549-3_5.

Morgan, M. G. (2011). Certainty, uncertainty, and climate change. *Climatic Change, 108*, 707–721.

Morgan, M. G., Henrion, M., & Small, M. (1990). *Uncertainty: A guide to dealing with uncertainty in quantitative risk and policy analysis.* Cambridge: Cambridge University Press.

Müller, T. (2012). Branching in the landscape of possibilities. *Synthese, 188*, 41–65.

Neubersch, D., Held, H., & Otto, A. (2014). Operationalizing climate targets under learning: An application of cost-risk analysis. *Climatic Change, 126*, 305–318.

Nordhaus, W. D., & Boyer, J. (2000). *Warming the world: Economic models of climate change.* Cambridge, MA: MIT Press.

O'Hagan, A., & Oakley, J. E. (2004). Probability is perfect, but we can't elicit it perfectly. *Reliability Engineering & System Safety, 85*, 239–248.

Peterson, M. (2006). The precautionary principle is incoherent. *Risk Analysis, 26*, 595–601.

Rawls, J. (1971). *A theory of justice.* Cambridge: Harvard University Press.

Rescher, N. (1984). *The limits of science. Pittsburgh series in philosophy and history of science.* Berkeley: University of California Press.
Rescher, N. (2009). *Ignorance: On the wider implications of deficient knowledge.* Pittsburgh: University of Pittsburgh Press.
Resnik, M. D. (1987b). *Choices: An introduction to decision theory.* Minneapolis: University of Minnesota Press.
Savage, L. J. (1954). *The foundation of statistics.* New York: Wiley.
Schefczyk, M. (2016). Financial markets: The stabilisation task. In S. O. Hansson & G. Hirsch Hadorn (Eds.), *The argumentative turn in policy analysis. Reasoning about uncertainty* (pp. 265–290). Cham: Springer. doi:10.1007/978-3-319-30549-3_11.
Schmidt, M. G. W., Lorenz, A., Held, H., & Kriegler, E. (2011). Climate targets under uncertainty: Challenges and remedies. *Climatic Change, 104*, 783–791.
Schneider, S. H. (2001). What is 'dangerous' climate change? *Nature, 411*, 17–19.
Shrader-Frechette, K. (2016). Uncertainty analysis, nuclear waste, and million-year predictions. In S. O. Hansson & G. Hirsch Hadorn (Eds.), *The argumentative turn in policy analysis. Reasoning about uncertainty* (pp. 291–303). Cham: Springer. doi:10.1007/978-3-319-30549-3_12.
Steele, K. (2006). The precautionary principle: A new approach to public decision-making? *Law, Probability, and Risk, 5*, 19–31.
Sunstein, C. R. (2005). *Laws of fear: Beyond the precautionary principle.* Cambridge: Cambridge University Press.
Toth, F. L. (2003). Climate policy in light of climate science: The ICLIPS project. *Climatic Change, 56*, 7–36.
van Fraassen, B. C. (1989). *Laws and symmetry.* Oxford: Oxford University Press.
Williamson, J. (2010). *In defence of objective Bayesianism.* Oxford: Oxford University Press.

Chapter 7
Setting and Revising Goals

Karin Edvardsson Björnberg

Abstract If goals are to fulfil their typical function of regulating action in a way that contributes to an agent's long-term interests in getting what he or she wants, they need to have a certain stability. At the same time, it is not difficult to imagine situations in which the agent could have a reason to revise his or her goals; goals that are entirely impossible to achieve or approach to a meaningful degree appear to warrant some modification. This chapter addresses the question of when it is rationally justified to reconsider one's prior goals. In doing so, it enriches the strictly instrumental conception of rationality. Using Bratman's (1992; 1999) theory of intention and Edvardsson and Hansson's (2005) theory of rational goal-setting, the chapter critically analyses the steps in the argumentative chain that ought to be considered before it can be concluded that a decision maker has sufficient reason to reconsider her goals. Two sets of revision-prompting considerations are identified: achievability- and desirability-related considerations. It is argued that changes in the agent's beliefs about the goal's achievability and/or desirability could give her a *prima facie* reason to reconsider the goal. However, whether there is sufficient reason—all things considered—to revise the goal hinges on additional factors. Three such factors are discussed: pragmatic, moral and symbolic factors.

Keywords Goal-setting • Goal revision • Reasons • Justification • Evidence • Belief change • Intentions

1 Introduction

Goals are typically adopted on the assumption that goal setting will further goal achievement. By setting a goal, it is assumed, it will become easier to deliberate, plan and act—over time and collectively—in ways that are conducive to goal realisation. Moreover, goals are typically adopted on the assumption that goal

---

K. Edvardsson Björnberg (✉)
Division of Philosophy, KTH, Stockholm, Sweden
e-mail: karine@kth.se

S.O. Hansson, G. Hirsch Hadorn (eds.), *The Argumentative Turn in Policy Analysis*, Logic, Argumentation & Reasoning 10, DOI 10.1007/978-3-319-30549-3_7

achievement will be considered valuable when it occurs and that the goal will be sustained unless special circumstances apply. This holds true for goals set by individuals, groups of individuals and organisations.

In several of his works, Bratman (1992, 1999) argues that if intentions are to fulfil their typical function of guiding action and deliberation, they

> will need to have a certain stability: if we were constantly to be reconsidering the merits of our prior plans they would be of little use in coordination and in helping us cope with our resource limitations. (Bratman 1992: 3)

There is reason to believe that the same holds true for goals. If goals are to fulfil their typical function of regulating action in a way that contributes to the satisfaction of the agent's interests in getting what she wants, they need to have certain stability. Frequent goal revision not only makes it difficult for the agent to plan her activities over time; it also makes it more difficult for the agent to coordinate her actions with other agents upon whose behaviour the good outcome of her plans and actions are contingent. Thus, there are reasons to endorse Bratman's view that non-reconsideration of prior intentions (and goals) ought to be the default.

Yet it is not difficult to think of situations in which the agent could have reason to revise her goals.[1] Anna's realization that her teenage goal to become a top diplomat is inconsistent with the goals and plans that she has adopted at a later stage in life gives her a reason to reconsider her prior goal. A government that realizes that its goal to increase energy efficiency by 95 % in 10 years will most likely be impossible to achieve given the means available, is well advised to lower its ambition. Rationally justified non-reconsideration is not the same thing as sheer stubbornness. However, where to draw the line between the two remains to be settled, in theory and in concrete decision situations.

In decision theory, goals are commonly treated as mere inputs to the analysis, which is instead framed in terms of finding the best means to given goals. Admittedly, in a strict 'instrumental' framework there is little room for rational deliberation about how to set and revise goals (Simon 1983; Russell 1954). The aim of this chapter is to enrich the traditional instrumental conception of rationality by shedding light on the issue of when an agent has reason to (set and) reconsider her goals. As in life, goals often have to be set and revised under conditions of uncertainty; at the time of goal-setting, the agent seldom has perfect knowledge about whether she will be able to reach her goal or even how valuable goal achievement will be when (and if) it occurs. Therefore, the chapter will build on insights and arguments presented elsewhere in this anthology, particularly the chapters on the argumentative turn (Hansson and Hirsch Hadorn 2016), evaluating the uncertainties (Hansson 2016), temporal strategies

[1] In the following, the terms "goal revision" and "goal reconsideration" are used interchangeably. It could be argued that reconsideration and revision are two different things and that there could be reasons to reconsider a goal that nevertheless do not support goal revision. In this chapter, no such distinction between the two terms will be upheld.

(Hirsch Hadorn 2016) and value uncertainty (Möller 2016).[2] The chapter will not provide an exhaustive account of when goal reconsideration is rationally justified. Instead, it will lay out and critically analyse the steps in the argumentative chain that ought to be considered before goal reconsideration can be considered sufficiently justified (see Brun and Betz 2016 in this anthology on the task of argument analysis). Providing a structured analysis of the arguments that come into play in goal setting and revision will assist decision makers who are faced with the challenges of deciding, for example, which policies to adopt, pursue or overturn.

The chapter is structured along the following lines. Section 2, which builds on previous work by Edvardsson and Hansson (2005) and Edvardsson Björnberg (2008, 2009), explains the role of goals in deliberation and action. It is argued that goals are typically "achievement-inducing"; that is, by setting a goal, it usually becomes easier to achieve it. The mechanisms behind this idea are briefly explained and discussed in light of empirical evidence in psychology and management theory. In Sect. 3, which draws extensively on Bratman's (1992, 1999) theory of intention, it is explained why frequent goal revision is problematic from a planning perspective and why goal stability therefore should be considered the default. Section 4 outlines two sets of considerations that could give the agent a reason to reconsider her goals: ability- and desirability-related considerations. It is argued that changes in the agent's beliefs about goal achievability and/or desirability could give her a *prima facie* reason to reconsider her goal.[3] However, whether there is sufficient reason—all things considered—to revise the goal depends on additional (non-epistemic) factors. Those factors are laid out and discussed in Sect. 6. Section 5, which builds on previous work by Baard and Edvardsson Björnberg (2015), addresses the question of how strong evidential support is needed to justify a belief in a goal's achievability and/or desirability and why ethical values need to be considered as well.

[2] See Hansson and Hirsch Hadorn (2016) for a discussion of different types of uncertainties. A common distinction in decision theory is between decision-making under *risk* and decision-making under *uncertainly*. The former refers to situations wherein the decision-maker knows both the values and the probabilities of the outcomes of a decision, whereas the latter refers to situations wherein the decision-maker can value the outcomes but does not know the probabilities or has only partial information about the probabilities. In addition, the term "decision-making under great uncertainty" is sometimes used to refer to situations wherein the information required to make decisions under uncertainty is lacking. Hansson (1996) identifies several such types of information shortages, including unidentified options or consequences, undecided values and undetermined demarcation of the decision. Goal setting often involves uncertainty about the probabilities of certain outcomes (that is, how likely it is that a certain state of affairs will be achieved given that it is formulated as a goal), but it could also involve more radical types of uncertainties.

[3] The Oxford English Dictionary (2015) defines the adverb "*prima facie*" as "at first sight; on the face of it; as it appears at first". To have a *prima facie* reason to reconsider a goal thus means that in the absence of evidence to the contrary, the agent is justified in reconsidering the goal.

2 The Role of Goals in Deliberation and Action

Goals are important regulators of action in both individual and social contexts. Agents—individuals, groups of individuals and organisations—typically set goals because they want to achieve (or maintain) the states of affairs that the goals describe (henceforth "goal states") and because they believe that by setting goals, it becomes easier to achieve those goal states.[4] Edvardsson and Hansson (2005) use the term "achievement-inducing goal" to refer to a goal that fulfils its typical function of regulating action towards goal achievement.[5]

Goal setting contributes to goal achievement through two mechanisms. First, goals are typically action guiding; they *direct* attention towards actions that will further goal achievement, and they constitute a standard against which performed actions can be assessed and evaluated. Having adopted a goal, an agent will under normal circumstances act to achieve it (McCann 1991). That is, the agent will typically prefer options that she believes could facilitate goal achievement and will avoid options that she believes could have the opposite effect (cf. Bratman 1999, see also Cohen and Levesque 1990 and Levi 1986).[6]

The following example illustrates this point: Greta has fallen behind in her studies due to extensive engagements with the university's Archaeological Society. To make up for these amusements, she adopts the goal to finish the second chapter of her Master's thesis on the Luwian hieroglyphs by next Sunday. Having adopted the goal, Greta proceeds to make plans for the coming week. To save time for her studies, she decides to buy seven ready-to-eat meals from the local grocery shop. She then decides to leave her mobile phone with her landlady for the coming days, knowing this will prevent her from taking any incoming calls. Bearing the goal in mind, she also decides to turn down every proposal that she receives during the week that is likely to be incompatible with her finishing the second chapter of her thesis, including a much-anticipated visit to the British Museum's collection of Hittite artefacts. As a final measure, she decides to operationalise the goal by adopting a set of realistic sub-goals, or targets, for each of the weekdays ahead. For Tuesday, she sets the sub-goal to finish the section on Emmanuel Laroche's decipherment of the hieroglyphs. For Wednesday, she sets the sub-goal to finish the section of "the new readings", a set of corrections to the readings of certain signs given by David Hawkins, Anna Morpurgo Davies and Günter Neumann and so

[4] A goal typically describes a desired state of affairs that is yet to be achieved, although the maintenance of a current state of affairs could also be a goal (Wade 2009). The goal to remain married despite relationship deterioration would be an example of the latter.

[5] As noted by Edvardsson and Hansson (2005), goals could be set for other reasons than to achieve them. An example would be a government that adopts the goal to halt biodiversity loss within its national borders with the sole aim to facilitate business partnerships with environmentally friendly states. Although such uses of goals and goal setting may be frequent in political practice, they will not be discussed in this chapter.

[6] Another way to put it is to say that goals serve as departure points for practical reasoning about what to do.

on. As the example illustrates, Greta's goal to finish the second chapter of her Master's thesis functions as a filter of admissibility in the sense that it narrows down her scope of future deliberations to a limited set of options (actions, plans and further goals/sub-goals), and it provides a reason to consider some of the options but not others.

The action-guidance provided by a goal can also help groups of agents plan and coordinate their actions in a way that contributes to goal achievement (Sebanz et al. 2006). In a situation where the mutually agreed upon goal of special agents A, B and C is to perform a particular covert operation in Beirut within the next 24 h, special agent A's actions become predictable to B and C, at least in the sense that there are some actions that B and C can reasonably expect A not to perform within the next 24 h, such as taking a flight to Honolulu. Because B and C can rely on A's behaviour (at least to some extent), they can themselves perform actions whose outcomes are dependent on A's specific behavior.[7] My stepping into the pedestrian crossing as I see the motor traffic lights turn from green to yellow, while feeling confident that both the approaching driver and I share the goal of not causing any traffic accidents, is another example. As both examples illustrate, a mutually agreed upon goal can provide a basis on which a group of agents can plan and coordinate their actions efficiently and effectively towards goal achievement. This interpersonal coordinative function of goals can be formal (or formalised through legal rules as in the pedestrian case), as in the above-mentioned examples, or informal, as in the case of opera choir singers tuning their respective vocal parts against the other singers to achieve the joint goal of producing a memorable performance.

Second, in addition to being action guiding, goals also typically *motivate* action towards goal achievement. The motivation induced in the agent could contribute to initiating and sustaining action in the face of experienced implementation difficulties. As noted by Edvardsson and Hansson (2005: 349), in many social situations, the action-motivating function of a goal is the main reason for adopting it. In the 2014 general election, the Swedish Green Party's (unsuccessful) goal to become the country's third biggest political party was not set with the primary aim to instruct the party members what to do to reach it, but to excite them and make them intensify their efforts.

There is compelling empirical evidence to suggest that goal-setting techniques work along the lines sketched above, at least when the goals meet certain criteria. In psychological and management research, these criteria are frequently summarised through the SMART acronym, according to which goals should be *S*pecific, *M*easurable, *A*chievable (or *A*ccepted), *R*ealistic and *T*imed (Robinson et al. 2009; Bovend'Eerdt et al. 2009; Latham 2003).[8] Locke and Latham (1990, 2002)

[7] See Nozick (1993: 9–12) for a related discussion on the coordinative function of principles. In game theoretical settings, knowledge of an agent's goal can help other agents to plan in a way that makes it easier to achieve their individual goals.

[8] There is a considerable variation in what the SMART acronym stands for in the literature (Wade 2009; Rubin 2002).

among others cite extensive empirical evidence showing that goals that are precise, measurable and measured in the sense that feedback on progression is provided, and reasonably demanding, generally have the highest chance of contributing to the intended (and desired) goal states. One central finding in this literature is that specific goals lead to a higher task performance by employees than vague, abstract or "do your best" goals (Locke and Latham 1990). Another central finding is formulated through the so-called "goal-difficulty function", which implies that the more challenging a goal is, the greater the effort the agent is likely to put forth to achieve it, at least up to a certain point (ibid.).

Despite considerable empirical support for the goal-setting theory, it is important to bear in mind that there could be situations in which goal setting—the goal itself or the process by which the goal is adopted—has the opposite effect to what is assumed above. Hansson et al. (2016) explore a number of situations wherein goals are self-defeating, that is, situations wherein goal setting makes it more difficult to achieve the desired goal state. One of the most frequently discussed examples in philosophical literature is the "hedonic paradox" (Martin 2008; Slote 1989; Mill 1971), which is used to illustrate that happiness cannot be pursued as a direct goal; the more attention the agent pays to the goal, the further away from it she tends to end up. The goal to become a spontaneous person, or to fall asleep within 10 min from putting one's head on the pillow, are two other examples. In such situations, it is perfectly reasonable for the agent to deliberate about what states of affairs she would like to achieve, but not to formulate those ambitions as goals to be used for planning purposes.

3 Why Goal Stability Ought to Be the Default

The account of goal setting outlined above bears resemblance to Bratman's (1992, 1999) theory of intention.[9] Bratman (1992, 1999) defends a pragmatist account of intention, the ultimate defence of which is grounded in the role played by intentions in furthering people's long-term interests in getting what they want. Intentions are instrumentally valuable because they involve commitment to action. Intentions

[9] Although goals and intentions play a pivotal role in deliberation about what to do, it is important to note that there could be differences in how strongly they influence an agent's actions. Intentions typically involve a stronger commitment to action than goals. When I have a goal or intention to practice on my violin for at least 14 h the coming week, I have a disposition towards actions that will bring me closer to the goal. However, the relationship between my having this disposition and letting it influence my actions is stronger for intentions than for goals and stronger still for goals than for desires. Thus, while it makes sense to say "I desire to practice on my violin for at least 14 h this week, but I shall not (or cannot) do it", it typically does not make sense to say "My goal is to practice on my violin for at least 14 h this week, but I shall not (or cannot) do it". Further to the point, saying "I intend to practice on my violin for at least 14 h this week, but I shall not (or cannot) do it" comes out as being even more inconsistent (modified from Hansson et al. 2016, cf. Bratman 1992 on "strong consistency").

allow present deliberation to shape later conduct; by settling on a course of action now, the agent does not have to re-deliberate unless special circumstances apply. Extending the "influence of Reason on our lives" (Bratman 1992: 2) in this way, Bratman argues, is particularly important for us humans, who are planning creatures, although of an imperfect sort.

If intentions are to guide action towards the achievement of an agent's long-term interests in getting what she wants, they must have a certain stability. Frequent intention revisions could lead to significant efficiency losses. It could make it difficult for the agent to deliberate, plan and act over time in ways that are conducive to the satisfaction of her long-term interests. Moreover, it could make it more difficult for the agent to form collaborative partnerships upon which the satisfaction of her interests is contingent. The reason for this is that under normal circumstances, rational agents tend to avoid including unreliable players in their collaborative schemes:

> Suppose you and I plan to meet today for lunch. It will be important to me to know how reliable you are about such things. If you are rather resistant to reconsidering such prior intentions, and I know this, I will be somewhat more willing to make such plans with you and to go out of my way to keep such appointments with you. My knowledge of your habits of reconsideration will directly affect the extent to which I am willing to be a partner with you in mutually beneficial coordinating schemes. (Bratman 1992: 8)

Consequently, Bratman concludes that non-reconsideration of a prior intention will typically be the default. As the following two examples show, this reasoning applies to goals and goal setting too:

> *Example 1*. Andi's goal is to have a career that will impress her status-minded circle of friends. With this further aim in mind, she adopts the goal to become an anaesthesiologist. She then starts to plan her day-to-day life based on this goal. After having studied medicine for 6 months, she decides to finish her medical studies and instead try to become a lawyer. She then starts to plan her day-to-day life based on this goal. After having studied law for 6 months, she decides to finish her law studies and instead try to become a peace negotiator. She then starts to plan her day-to-day life based on this goal. After having studied political science for 6 months, she decides to finish her political science studies and instead try to become a dentist, and so on. After having spent 5 years at the university without receiving a degree, she runs out of money and is forced to take a low-paid part-time job. Constantly reconsidering her career goals has made it impossible to achieve her further aim of having a high-status career.
>
> *Example 2*. The overall aim of Government G is to keep its national emissions at such a level that they do not contribute to dangerous anthropogenic climate change.[10] With this overall aim in mind, Government G decides to prioritise reductions in carbon dioxide emissions from the national iron and steel industry. According to the best knowledge available, if the Government's overall goal is

[10] This example is modified from Baard and Edvardsson Björnberg (2015).

to be met, emissions should be reduced by at least 70 % compared to the 2010 level by 2050 for this specific branch of industry. Because of the uncertainty concerning climate sensitivity and socioeconomic development, G decides that the best approach towards emissions abatement is to adopt short-term emissions targets on an *ad hoc* basis. In 2015, the government adopts the goal "In 2025, emissions should be reduced by 10 % compared to the 2010 level". In 2025, when this goal is achieved, G feels confident that much stricter emissions targets can and should be set and achieved. It therefore adopts the goal "In 2035, emissions should be reduced by 50 % compared to the 2010 level". In 2035, however, G realises the latter goal is far too difficult to achieve and therefore adopts the goal "In 2045, emissions should be reduced by 30 % compared to the 2010 level", and so on. A possible drawback of this *ad hoc* goal setting and goal revision is that the measures taken by the industry to achieve the first goal (investments in new technologies, etc.) could be sub-optimal in relation to the further ambition expressed through the second and third goal, as well as the further ambition to reduce emissions by at least 70 % by 2050. As the industry plans for the measures that must be taken to achieve the 2025 goal, it could be useful to know that sometime in the future one will be expected to invest in much more effective emissions abatement technologies. To allow the industry to plan for such initially more expensive abatement technologies, G would have to signal its long-term commitment at an early stage.[11] A further drawback is that G's *ad hoc* goal setting and goal revision could render the industry less motivated to work towards the targets (as they know they will likely be revised over and over again) and, by extension, to participate in future public–private partnerships concerning the environment.

To understand the mechanisms at play in the two examples it is useful to contemplate briefly what it means to reconsider a goal, plan or intention. When an agent reconsiders a previously adopted goal, she "re-opens the question" (cf. Bratman 1999: 62 ff.). This involves something more than simply entertaining the thought of what goal revision might possibly involve. Fantasising about what it would be like to give up one's goal to remain faithful to one's partner is not the same as seriously re-opening the question of whether to have an extramarital affair. Only the latter involves withdrawing the goal from the background against which one deliberates about what to do. When a goal is seriously reconsidered, its role as a "filter of admissibility" on options is suspended, which means that options that were previously considered incompatible with that goal might become admissible again.

Sometimes, withdrawal of the goal from the background against which the agent deliberates is an act that is itself the result of conscious deliberation. Reconsidering one's goal to remain a faithful partner could involve not only deliberating about one's reasons for remaining faithful but also "second-order deliberation" (Bratman

[11] As suggested by Hirsch Hadorn (2016), this problem could be avoided if the government partitions the decision problem by adopting a system of goals wherein the 2025 and 2035 targets are set sequentially as sub-goals to the overall goal of reducing emissions by at least 70 % by 2050.

1999: 61) about, for example, the emotional costs of reopening the issue. However, in many cases, reconsideration is much less explicit, such as when the agent considers having an affair with one of her office colleagues but does not pause to reflect on the potential emotional or symbolic costs of reconsideration. In that situation, she implicitly re-opens the question of whether or not to retain her goal of remaining faithful to her partner. In addition, purely non-reflective instances of goal reconsideration could be imagined, such as when out of pure habit, the agent suspends her goal to maintain a healthy lifestyle when on Friday evenings, she invariably engages in binge drinking with her colleagues at work (cf. Bratman 1999: 60). Such habitual goal reconsideration will not be discussed in this chapter.

4 Reasons for Goal Reconsideration

Thus far, it has been argued that goals must have a certain stability to fulfil their overall function of guiding deliberation and action in a way that contributes to the satisfaction of the agent's long-term interest in getting what she wants. Yet, there could be situations in which the agent has reason to reconsider her goals. Goals are set on the assumption that the states of affairs they describe are valuable and that by setting the goal it becomes easier to achieve those states. From this follows at least two sets of considerations that could give the agent reason to reconsider her goals (Baard and Edvardsson Björnberg 2015; cf. Bratman 1999: 67).

Achievability-Related Considerations. Goals are normally adopted on the assumption that they will be possible to reach or at least approach to a meaningful degree. However, as time passes, the world as the agent finds it may differ from the world as the agent expected it to be when setting the goals. The discrepancy between the expected and actual preconditions for goal achievement could give the agent a reason to reconsider her goal.

> Example: In 2008, Seth (who is an avid runner) adopts the goal to win the 2015 London Marathon. Three years after having set the goal, Seth suffers a major stroke, which confines him to a wheelchair for the rest of his life with zero chances of ever recovering. In this situation, it may be argued that the world has changed in such a way that Seth now has a reason to reconsider his goal.

Desirability-Related Considerations. Goals are normally adopted on the assumption that goal achievement will be considered valuable when it occurs. However, as time passes, the agent's desires or values on the basis of which the goal was set may change in a way that gives her reason to reconsider her goals (see also Möller 2016).[12]

[12] This could involve either a total or a partial rejection of the agent's desires or values. A partial rejection of the agent's values could, for example, be the result of her coming to embrace new values, which means her prior values fade into the background.

Example: In 2008, as she turned 20, Anna adopted the goal to become an RAF pilot with the future aim of serving in Afghanistan and Iraq. Five years after having set the goal, she adopts Adam and Albert together with her partner. Becoming a parent changes the structure of the values on which her career goal (and other goals) have been based. She no longer attaches great value to the goal of becoming an RAF pilot. In this situation, it may be argued that Anna's values have changed in such a way that she now has reason to reconsider her goal.

Both achievability- and desirability-related considerations can be framed in ontological or epistemological terms (Baard and Edvardsson Björnberg 2015). Framed in ontological terms, it is the goals themselves (or the values upon which the goals have been based) and the means available for achieving them that change, and this change in the world gives the agent a reason to reconsider her goal. Framed in epistemological terms, it is the agent's knowledge about the goals (or the values upon which the goals have been based) and the means available for achieving them that change, and this belief change gives the agent a reason to reconsider her goals.[13]

To illustrate the difference between the two framings, consider the case of Seth. It could be argued that Seth has reason to reconsider his goal, either because the preconditions for goal achievement have changed significantly (the means available for achieving the goal have changed) or because Seth's knowledge of the preconditions for goal achievement have changed significantly. In the former (ontological) case, the corresponding reason for reconsidering the goal is *agent-independent* in the sense that it refers to changes in the actual world. In the latter (epistemological) case, the corresponding reason for reconsidering the goal is *agent-dependent* in the sense that it includes an essential reference to the agent's beliefs about changes in the actual world (cf. Nagel 1986: 152–153). Regardless of whether one prefers to conceptualise the issue in ontological or epistemological terms, the relevant changes (in the actual world or in the agent's beliefs about the world) can be related to the achievability or desirability of the goal, or both.

In the remaining sections of the chapter, the discussion will largely be framed in epistemological terms. That is, reasons for goal reconsideration will be framed in terms of the agent's beliefs about goal achievability and/or desirability. It will be argued that certain changes in the agent's beliefs about goal achievability and/or desirability give her a *prima facie* reason to reconsider her goal. However, whether or not there is sufficient reason—all things considered—to revise the goal depends on (non-epistemic) factors in addition to those pertaining to the agent's beliefs. These factors will be laid out and discussed in Sect. 6. The next section discusses what changes in the agent's beliefs can support goal reconsideration. This involves saying something about the role of evidence in forming justified beliefs in goal achievability and/or desirability.

[13] Here, it could be objected that cognitive changes, such as a change in belief, are also changes in the world. This would make the distinction between ontological and epistemological interpretations meaningless. This objection will not be addressed in this chapter.

5 Justified Belief in Goal Achievability/Desirability

What an agent has reason to believe regarding a goal's achievability or desirability depends, at least partially, on the evidence that is available to her. Evidence justifies beliefs about the achievability and/or desirability of a goal and can thus normatively support a decision to reconsider a goal.

What does "evidence" mean in the context of goal setting? Evidence is sometimes understood as some physical object, such as the fingerprints on a gun or a document accounting for the course of events preceding the death of Arsinoë IV. In goal setting and goal revision, evidence is more appropriately thought of as some observation statement or known proposition about the world (Kelly 2014). For Verna, who has adopted the goal to get employment as a bus driver before the end of the year, a failed driving test in late December (corresponding to the observation statement "Verna failed her driving test in late December") constitutes evidence that her goal will most likely not be achieved. For the municipality of Östhammar, which has adopted the goal to maintain a thriving marine fauna within its borders, the extinction of the sea eagle (corresponding to the observation statement "The sea eagle has become extinct") would constitute evidence that work is not going in the right direction.

There are two broad ways of framing the issue of when there is sufficient evidence for an agent to be justified in believing in a goal's achievability/desirability.[14] To illustrate the difference between them, consider the following ability-related proposition:

> (P1): In 20 years from now, agricultural biotechnologies will have been developed to increase global yields of soy, maize and rice by 75 % compared to 2010 levels.

Suppose that P1 is believed to be a necessary condition for attaining the further goal of eradicating global malnourishment. There are two ways of specifying what it means to say there is sufficient evidence to believe P1 is true. First, the evidence required could be specified in probabilistic terms. It could be argued that the agent is justified in believing P1 is true if and only if the probability that effective agricultural biotechnologies will be developed is at least 0.5, 0.6 or 0.75. Alternatively, one could use a non-quantitative estimate, such as 'beyond a reasonable doubt' or 'more likely than not', which are used as standards of proof in some jurisdictions, or 'virtually certain', 'very likely', 'likely', 'about as likely as not', 'unlikely', 'very unlikely', 'exceptionally unlikely', which were used by the Intergovernmental Panel on Climate Change in its 2013 report (IPCC 2013: 2).

Second, the issue could be addressed from a procedural viewpoint based on how the belief in question was formed. Suppose that to obtain a probability estimate for P1, the government can choose between consulting a self-proclaimed eco oracle or gather a group of scientists working in the field of agricultural biotechnology and

[14] The examples and discussion below are taken from Baard and Edvardsson Björnberg (2015) with some modifications.

ask each of them to give a probability estimate for P1. Suppose further that, based on her supernatural abilities, the eco oracle maintains there is a 0.95 probability that P1 is true, whereas the experts agree the probability is only about 0.05. Baard and Edvardsson Björnberg (2015) suggest that most people would rightly be reluctant to use the oracle's estimation as evidence to support P1, as it does not represent a reliable belief-forming process.[15] The example shows that both substantive and procedural aspects come into play when determining what constitutes sufficient evidence for a proposition such as P1 and, by extension, when determining whether there is sufficient evidence to support goal reconsideration. Exactly how substantive and procedural aspects are related is a much-debated question that lies outside the scope of this chapter.

When assessing the achievability of public policy goals, scientific evidence, that is, evidence obtained through scientific inquiry, often plays a central role. For example, when assessing progress towards climate change, biodiversity, eutrophication or acidification goals, governments systematically call upon physical, biological and ecological expertise.[16] Experts are expected to be able to deliver informed opinions not only on the appropriateness of certain target levels (e.g. viable population targets) given broader conservation goals, but also on how work is progressing and what policy measures are likely to increase goal achievement.

When evaluating evidence for and against a public policy goal's desirability, an expert opinion does not seem to possess an equally strong foothold (although there are scientific experts working in the field of future studies who try to predict social changes, including changes in people's values). Baard and Edvardsson Björnberg (2015) suggest that evidence that is more reliable concerning the desirability of a public policy goal could be gathered by consulting a broader range of actors, including governmental authorities and local municipalities, non-governmental organisations, private businesses and the general public.

Giving a principled account of what constitutes sufficiently strong evidence for belief formation in the context of goal achievability and/or desirability requires a significantly more developed normative argument than can be offered in this chapter. Before turning to the question of when resistance to reconsideration is rationally justified, two factors that affect the choice of standard of proof will be elaborated on briefly. Both factors are discussed in Baard and Edvardsson Björnberg (2015).

As noted above, there is an endogenous relationship between goal setting and goal achievement; by setting a goal, one typically increases the likelihood the goal

[15] That is, it does not lead to a high percentage of true beliefs (see also Nozick 1993: 64 ff.).

[16] As noted by Hansson (1996), the notion of 'expertise' is vague. There could be uncertainties regarding an expert's knowledge and there could be multiple experts with competing but well-grounded opinions. In the literature on evidence, the question of higher-order evidence has received substantial attention in recent years (Feldman 2014; Kelly 2010).

will be achieved. Indeed, the underlying rationale for goal setting is that the goals will guide and motivate action (including the development of means) towards goal achievement. It could be argued that because goal setting will make it more likely that a goal will be reached, weaker evidence (e.g. 'about as likely as not' rather than 'beyond a reasonable doubt') should be enough for a goal to count as justifiably believed to be achievable.

A similar argument could be made concerning the evidence required for a goal to count as justifiably believed to be desirable. A public policy goal for which there is rather weak support at the time of goal-setting could catch up in terms of desirability as time proceeds and people start to plan their lives using the goal as a 'background assumption'. Stewart (1995) argues against a purely instrumental conception of economic rationality, noting that adopted economic goals often help to shape people's preferences and values (see also Bowles 1998 on endogenous preferences). For instance, goals, such as to increase the percentage of people living in houses and flats owned by themselves (as opposed to public housing) or to create a national pension system that requires people to invest a certain percentage of their income in funds, could alter people's preferences and values concerning the role of the market in providing basic social goods (Harmes 2001).

The second factor that could have some bearing on the choice of a standard of proof concerns the magnitude of the consequences the agent is trying to bring about (or avoid) by setting and working towards a goal. It could be argued that a policy goal that is justifiably believed to be very difficult to achieve (such as the goal to completely halt biodiversity loss) or for which there is weak public support at present (the goal of a zero growth economy might be an example) could nevertheless be considered sufficiently achievable and desirable to motivate goal setting, provided the magnitude of the harm that might occur if no such goal is implemented is sufficiently large. In this way, it could be argued that moral considerations come into play when setting and revising goals, especially when deciding how to act on uncertain information about a goal's achievability/desirability.[17]

[17] The last point touches on one of the central questions in the ethics of belief, namely what norms ought to govern belief formation. A distinction is commonly made between strict evidentialist accounts, according to which an agent should base her beliefs always and solely on relevant evidence, and moderate evidentialist and non-evidentialist accounts, which permit non-epistemic considerations to have some bearing on what should count as a justified belief (Chignell 2013). As an example of the latter, Chignell (2013) mentions William James (1896 [1979]), who emphasises the central roles played by prudential and moral values in the ethics of belief. Allowing the magnitude of the consequences of setting (or not setting) goals to have some bearing on what counts as a justified belief in goal achievability/desirability departs from strict evidentialism.

6 When Is Resistance to Reconsideration Rationally Justified?

In relation to future-directed intentions, Bratman argues there are some changes in beliefs that directly oblige the agent to reconsider her prior intention:

> I cannot rationally intend to A at t2 and also believe that I cannot A at t2. So if I newly come to believe that I cannot A at t2 then I am rationally obliged to reconsider. (Bratman 1992: 4)

The obligation to reconsider follows the requirement of strong consistency, which Bratman (1992) considers a structural constraint on rational intention. In Sect. 3, it was argued that goals too ought to meet at least some form of consistency requirement (see examples in footnote 9). For a goal to fulfil its typical function of regulating action towards goal achievement there must be actions that can be performed at least to approach the goal (Edvardsson and Hansson 2005; cf. Laudan 1984 on 'demonstrably utopian' goals). However, there is reason to believe the 'straightaway reconsideration' cases will be uncommon in actual goal setting. Instead, in most situations, there will be some uncertainty regarding the goal's achievability and/or desirability, and the task consists of determining whether there is reason to reconsider the goal given the present degree of belief in the goal's achievability/desirability and other factors that have a bearing on the issue. This section tentatively discusses what those other factors might be.

Suppose that a government is justified in believing a goal will be very difficult to achieve or even approach. In that situation, the government could be said to have a *prima facie* reason to reconsider the goal. Does the government also have a reason—all things considered—to reconsider the goal? At least three partly interrelated factors have a bearing on whether reconsideration is justified.

Pragmatic Factors. The first factor was indirectly touched upon in Sect. 3. It relates to the 'costs' of reconsideration (or alternatively, the benefits of goal stability). In Bratman's (1999) view, intentions provide 'framework reasons', that is, reasons that shape what it is rational to do, but whose ultimate force rests on the overall contribution of the planning system to the satisfaction of rational desire. Non-reconsideration is normatively justified by reference to its consequences, in particular, the agent's long-term prospects of getting what she wants. As argued in Sect. 3, frequent goal revision tends to produce losses in planning efficiency. It could potentially undermine coordination with the agent's other plans and it could affect the agent's opportunities to be part of collective enterprises from which she may benefit.

Moral Factors. The second factor was discussed briefly in Sect. 5. From a strict evidentialist viewpoint, non-epistemic (moral) considerations should not to be contemplated when addressing what counts as a justified belief in a goal's achievability/desirability. However, based on this alternative view, the magnitude of the consequences that an agent is trying to bring about (or avoid) by setting and working towards a goal could have some bearing on the choice of a standard of proof for a goal's achievability/desirability. Put differently, moral considerations come into play when making a decision for or against a decision (on goals) in conditions of great uncertainty.

Symbolic Factors. In addition to being valuable from a pragmatic viewpoint, non-reconsideration could have a symbolic value for the agent. It could contribute to the agent's sense of integrity and self-appreciation. It could give the agent a feeling of being someone who does not surrender lightly in the face of hardship. Such self-appreciation could be instrumentally valuable in the agent's pursuit of other goals (in which case it would have pragmatic value), but it could arguably also be considered intrinsically valuable. The following case provides an example of a situation in which non-reconsideration of a goal could be rationally justified on symbolic grounds:

Achievement of the overall goal of the United Nations Convention on Climate Change (UNFCCC) to stabilise greenhouse gas concentrations in the atmosphere at a level that would prevent dangerous anthropogenic interference with the climate system is contingent on the cooperation of many states, particularly 'top emitting countries', such as China, the United States, India, Russia and Japan. Suppose that a binding carbon dioxide emission target has been adopted by a majority of the world's nations, including the 'top emitting countries' and that after some time, all of the latter countries decided to give up the target. This means the target will be very difficult, if not impossible, to reach. Are there good reasons why your country, which plays a marginal role in the global emissions game, should reconsider the target? Probably yes, as cutting national emissions on a unilateral basis appears unreasonable. However, in support of non-reconsideration, it could be argued that adhering to the target has a symbolic value in that it makes visible to the government and the other players in the game the firmness and integrity with which the government's actions and plans are carried out.

7 Conclusion

If goals are to fulfil their typical function of regulating action in a way that contributes to an agent's long-term interests in getting what she wants, they need to have a certain stability. Yet, as shown above, it is not difficult to imagine situations in which the agent could have a *prima facie* reason to revise her goals. In this chapter, the arguments that can be put forward to support goal (non-) reconsideration have been critically examined. Using Bratman's (1992, 1999) theory of intention, it has been argued that goal non-reconsideration ought to prevail unless special circumstances apply. Two sets of such circumstances have been analysed—achievability- and desirability-related considerations—and the degree of evidence required for an agent to form justified beliefs in goal achievability and/or desirability has been discussed (although it is acknowledged that several issues pertaining to justified belief formation remain to be settled). Finally, three factors that have bearing on whether goal reconsideration can be said to be justified—all things considered—have been tentatively outlined: pragmatic, moral and symbolic factors. The ultimate challenge for any decision maker involved in setting and in working towards goals consists of weighing the evidential and non-evidential considerations outlined in this chapter (see Brun and Betz 2016 on reconstructing complex argumentation). Further research is needed to identify guidance that is more principled in relation to how to carry out this balancing act.

Acknowledgement The author would like to thank Gertrude Hirsch Hadorn and Sven Ove Hansson and the participants of the workshop in Zürich 26–27 February 2015 for their valuable comments and suggestions on earlier versions of this chapter. Any remaining errors (if any) are my own.

Recommended Readings

Bratman, M. E. (1999). *Intentions, plans, and practical reason.* Stanford: CSLI Publications.

Edvardsson, K., & Hansson, S. O. (2005). When is a goal rational? *Social Choice and Welfare, 24*, 343–361.

References

Baard, P., & Edvardsson Björnberg, K. (2015). Cautious utopias: Environmental goal-setting with long time frames. *Ethics, Policy and Environment, 18*(2), 187–201.

Bovend'Eerdt, T. J. H., Botell, R. E., & Wade, D. T. (2009). Writing SMART rehabilitation goals and achieving goal attainment scaling: A practical guide. *Clinical Rehabilitation, 23*, 352–361.

Bowles, S. (1998). Endogenous preferences: The cultural consequences of markets and other economic institutions. *Journal of Economic Literature, 36*, 75–111.

Bratman, M. E. (1992). Planning and the stability of intention. *Minds and Machines, 2*, 1–16.

Bratman, M. E. (1999). *Intention, plans, and practical reason.* Stanford: CSLI Publications.

Brun, G., & Betz, G. (2016). Analysing practical argumentation. In S. O. Hansson & G. Hirsch Hadorn (Eds.), *The argumentative turn in policy analysis. Reasoning about uncertainty* (pp. 39–77). Cham: Springer. doi:10.1007/978-3-319-30549-3_3.

Chignell, A. (2013). The ethics of belief. In: E. N. Zalta (Ed.), *The Stanford encyclopedia of philosophy* (Spring 2013 Edition). Available at: http://plato.stanford.edu/archives/spr2013/entries/ethics-belief/. Accessed 19 Jan 2015.

Cohen, P. R., & Levesque, H. J. (1990). Intention is choice with commitment. *Artificial Intelligence, 42*, 213–261.

Edvardsson Björnberg, K. (2008). Utopian goals: Four objections and a cautious defense. *Philosophy in the Contemporary World, 15*, 139–154.

Edvardsson Björnberg, K. (2009). What relations can hold among goals, and why does it matter? *Crítica, Revista Hispanoamericana de Filosofía, 41*, 47–66.

Edvardsson, K., & Hansson, S.O. (2005). When is a goal rational? *Social Choice and Welfare, 24*, 343–361.

Feldman, R. (2014). Evidence of evidence is evidence. In J. Matheson & R. Vitz (Eds.), *The ethics of belief* (pp. 284–300). Oxford: Oxford University Press.

Hansson, S. O. (1996). Decision-making under great uncertainty. *Philosophy of the Social Sciences, 26*(3), 369–386.

Hansson, S. O. (2016). Evaluating the uncertainties. In S. O. Hansson & G. Hirsch Hadorn (Eds.), *The argumentative turn in policy analysis. Reasoning about uncertainty* (pp. 79–104). Cham: Springer. doi:10.1007/978-3-319-30549-3_4.

Hansson, S. O., & Hirsch Hadorn, G. (2016). Introducing the argumentative turn in policy analysis. In S. O. Hansson & G. Hirsch Hadorn (Eds.), *The argumentative turn in policy analysis. Reasoning about uncertainty* (pp. 11–35). Cham: Springer. doi:10.1007/978-3-319-30549-3_2.

Hansson, S. O., Edvardsson Björnberg, K., & Cantwell, J. (2016). Self-defeating goals. Submitted manuscript.

Harmes, A. (2001). Mass investment culture. *New Left Review, 9*, 103–124.

Hirsch Hadorn, G. (2016). Temporal strategies for decision making. In S. O. Hansson & G. Hirsch Hadorn (Eds.), *The argumentative turn in policy analysis. Reasoning about uncertainty* (pp. 217–242). Cham: Springer. doi:10.1007/978-3-319-30549-3_9.

Intergovernmental Panel on Climate Change (IPCC). (2013). Summary for policymakers. In T. F. Stocker, D. Qin, G.-K. Plattner, M. M. B. Tignor, S. K. Allen, J. Boschung, A. Nauels, Y. Xia, V. Bex, & P. M. Midgley (Eds.), *Climate change 2013: The physical science basis. Contribution of working group I to the fifth assessment report of the Intergovernmental Panel on Climate Change*. Cambridge: Cambridge University Press.

James, W. (1896/1979). The will to believe. In F. H. Burkhardt, F. Thayer Bowers, I. K. Skrupskelis (Eds.), *The will to believe and other essays in popular philosophy* (pp. 1–31). Cambridge, MA: Harvard University Press.

Kelly, T. (2010). Peer disagreement and higher order evidence. In R. Feldman & T. A. Warfield (Eds.), *Disagreement* (pp. 111–174). Oxford: Oxford University Press.

Kelly, T. (2014). Evidence. In E. N. Zalta (Ed.), *The Stanford encyclopedia of philosophy* (Fall 2014 Edition). Available at: http://plato.stanford.edu/archives/fall2014/entries/evidence/. Accessed 19 Jan 2015.

Latham, G. P. (2003). Goal setting: A five-step approach to behavior change. *Organizational Dynamics, 32*(3), 309–318.

Laudan, L. (1984). *Science and values*. Berkeley: University of California Press.

Levi, I. (1986). *Hard choices: Decision making under unresolved conflict*. Cambridge: Cambridge University Press.

Locke, E. A., & Latham, G. P. (1990). *A theory of goal setting and task performance*. Englewood Cliffs: Prentice-Hall.

Locke, E. A., & Latham, G. P. (2002). Building a practically useful theory of goal setting and task motivation: A 35-year odyssey. *American Psychologist, 57*, 705–717.

Martin, M. W. (2008). Paradoxes of happiness. *Journal of Happiness Studies, 9*, 171–184.

McCann, H. J. (1991). Settled objectives and rational constraints. In A. R. Mele (Ed.), *The philosophy of action* (pp. 204–222). Oxford: Oxford University Press.

Mill, J. S. (1971). *Autobiography*. Edited with an introduction and notes by J. Stillinger. London: Oxford University Press.

Möller, N. (2016). Value uncertainty. In S. O. Hansson & G. Hirsch Hadorn (Eds.), *The argumentative turn in policy analysis. Reasoning about uncertainty* (pp. 105–133). Cham: Springer. doi:10.1007/978-3-319-30549-3_5.

Nagel, T. (1986). *The view from nowhere*. New York: Oxford University Press.

Nozick, R. (1993). *The nature of rationality*. Princeton: Princeton University Press.

Oxford English Dictionary (OED). (2015). *"prima facie, adv."*. Oxford University Press. http://www.oed.com. Accessed 24 Aug 2015.

Robinson, C. J., Taylor, B. M., Pearson, L., O'Donohue, M., & Harman, B. (2009). The SMART assessment of water quality partnership needs in Great Barrier Reef catchments. *Australasian Journal of Environmental Management, 16*, 84–93.
Rubin, R. S. (2002). Will the real SMART goals please stand up? *The Industrial-Organizational Psychologist, 39*, 26–27.
Russell, B. (1954). *Human society in ethics and politics*. London: Allen and Unwin.
Sebanz, N., Bekkering, H., & Knoblich, G. (2006). Joint action: Bodies and minds moving together. *Trends in Cognitive Sciences, 10*, 70–76.
Simon, H. A. (1983). *Reason in human affairs*. Oxford: Basil Blackwell.
Slote, M. (1989). *Beyond optimizing: A study of rational choice*. Cambridge, MA: Harvard University Press.
Stewart, H. (1995). A critique of instrumental reason in economics. *Economics and Philosophy, 11*, 57–83.
Wade, D. T. (2009). Editorial: Goal setting in rehabilitation: An overview of what, why and how. *Clinical Rehabilitation, 23*, 291–295.

Chapter 8
Framing

Till Grüne-Yanoff

Abstract The concept of framing, experimental evidence supporting framing effects, and models and theories of decision-making sensitive to framing play important roles in policy analysis. First, they are used to caution about various elements of uncertainty that are introduced through framing into policy interventions. Second, framing is often referred to in order to justify certain policy interventions, as framing effects are often seen as sources of irrationality in need of correction. Third, framing effects are often used as instruments for policy-making, as they are seen as effective ways to influence behaviour. This review discusses the different concepts of framing, surveys some of the experimental evidence, describes the dominant descriptive theories and the main attempts to assess the rationality or irrationality of behaviour sensitive to framing in order to clarify how exactly framing is relevant for policy making.

Keywords Framing • Preferences • Lotteries • Uncertainty • Behavioural economics • Mechanisms • Descriptive decision theory • Normative decision theory • Bounded rationality • Behavioural policy • Nudge • Boost

1 Introduction

There are usually many different ways in which we can frame a decision. This chapter clarifies what is meant by framing, why it is important for decision-making and how we can argue rationally about the choice of frames. Specifically, I briefly survey the history of the technical term in psychology (Sect. 2) and then illustrate the use of the term at the hand of various experimental studies in psychology and economics (Sect. 3). Sections 4 and 5 survey attempts to produce descriptively adequate accounts of the thus elicited phenomena, in terms of mechanistic models and more abstract theory, respectively. Section 6 focuses on the philosophical discussion to what extent framing phenomena are irrational, and why they should or should not be. Section 7 discusses some normative theories of framing, which

T. Grüne-Yanoff (✉)
Royal Institute of Technology (KTH) and University of Helsinki, Stockholm, Sweden
e-mail: gryne@kth.se

S.O. Hansson, G. Hirsch Hadorn (eds.), *The Argumentative Turn in Policy Analysis*, Logic, Argumentation & Reasoning 10, DOI 10.1007/978-3-319-30549-3_8

seek to provide some room for rational choice being influenced by frames, and at the same time impose constraints on what "rationally framed" decisions could be. Section 8, finally, addresses how the scientific discussion of framing has led to different policy proposals how to mitigate framing effects, and how framing effects should be used to influence people's decision.

Framing relates to uncertainty in multiple ways. First, the effect of framing on decisions is often observed in contexts involving uncertainty. For example, it matters sometimes whether an uncertain outcome is differentiated into some very unlikely events and some more likely outcomes, or whether this outcome is described as one bundle with a mean probability of all its events. Second, frames also create uncertainty, for example with respect to an individual's preferences. If an agent changes preferences over options under seemingly irrelevant changes of the frame, the uncertainty about that individual's preferences (their authenticity, or their relevance for welfare properties) increases. Furthermore, the fact that frames affect decisions also creates uncertainty about the rationality of these decisions: they might be unduly influenced by these frames, and alternative ways how to arrive at these decisions might be required instead. Overall, these considerations provide arguments against an algorithmic perspective on decision-making (see Hansson and Hirsch Hadorn 2016). Such an algorithmic perspective claims that with sufficient information, decision-making consists in the application of a fully specified procedure (an algorithm), which yields an unambiguous outcome. Contrary to that, framing yields uncertainties that limit the straightforward application of algorithms. Furthermore, deliberation requires reconstruction and analysis of different framings of a decision problem, and this is the task of argumentative methods (see Brun and Betz 2016). Hence, considerations of framing support the argumentative turn of policymaking.

2 History and Taxonomy of the Term "Framing"

In the context of decision theory, Tversky and Kahneman (1981) were the first to propose the term "framing". They define a "decision frame" as:

> the decision maker's conception of the acts, outcomes, and contingencies associated with a particular choice... controlled partly by the formulation of the problem, and partly by the norms, habits, and personal characteristics of the decision maker. (Tversky and Kahneman 1981:453)

Crucial for the understanding of decision framing is the claim that one and the same element of a decision problem, when considered from different frames, might appear in different ways, and these appearances might be decision-relevant. For example, a glass can be described either as half-full or as half-empty, and people might consider these two descriptions of the same outcome as the descriptions of two different outcomes. Similarly, a body movement like forming a fist can be described as single act, or as the sequence of movements that constitute that act.

Finally, the relevant future states of the world can be described in more or less detail. When describing tomorrow's possible states of the weather, for example, I might distinguish (i) "sunshine" or "no sunshine" or I might distinguish (ii) "sunshine", "clouds", "rain", "snow" and "other". Framing in the wide sense refers to the fact that in order to analyse a decision, one always needs to delineate a decision problem or embed it in a particular context (see Doorn 2016; Elliott 2016; Grunwald 2016). This is of course related to a more general attitude towards or thinking about the world (e.g. Goffman 1974), as for example expressed in various forms of discourse analysis. Framing in the narrower sense only concerns how the conception (description and structuring) of the specific decision problem has an effect on decision-making. Of course, because this effect is often not known in advance, the wide and the narrow notion of framing are sometimes not clearly separated.

To distinguish framing with respect to what is framed, Tversky and Kahneman (1981) characterize three kinds of framing:

(A) framing of outcomes,
(B) framing of acts, and
(C) framing of contingencies.

Of these three types, framing of outcomes has received most attention in the literature and is the form most closely associated with the term "framing." As in the glass half-full/half-empty example, outcome framing is typically taken to affect the decision maker's evaluation of the outcome. Therefore, this type is also known as "valence framing" (Levin et al. 1998), which often is differentiated into three sub-types:

(A1) risky choice framing
(A2) attribute framing
(A3) goal framing

Risky choice framing is performed by re-describing the consequences of risky prospects, for example by re-describing a 70 % post-surgery survival chance as a 30 % chance of dying from this surgery. Tversky and Kahneman seem to be the first to describe this type. *Attribute framing* is achieved by re-describing one attribute of the objects to be evaluated, for example by re-describing a glass that is half-full as a glass that is half-empty. This type of framing has been investigated before Tversky and Kahneman, for example by Thaler (1980). *Goal framing*, finally, consists not in a re-description of the outcome directly, but rather in a re-description of the goal by which outcomes are evaluated. For example, one can evaluate monetary outcomes of one's acts either with the goal of "maximizing wealth" or with the goal of "avoiding any unnecessary losses". Note that a goal framing only concerns a redescription, but not a revision of the goal (see Edvardsson Björnberg 2016).

The types of framing discussed so far all concern the conception of a decision problem "controlled ... by the formulation of the problem", as Tversky and Kahneman put it in the above quotation. Here framing is constituted by the description or re-description of elements of a decision problem. Partly because this description-factor can be experimentally manipulated with relative ease, most of the literature has focused on these types (as will become clear in the description of the different experimental designs used). However, framing is not restricted to this, as Tversky and Kahneman themselves acknowledge: framing is affected "partly by the norms, habits, and personal characteristics of the decision maker" (ibid.). Kühberger (1998) stresses this aspect of framing when he distinguishes between a "strict" and a "loose" sense of the framing concept. The strict sense corresponds to those types of framing that are affected by redescription. The loose definition, however,

> refers to framing as an internal event that can be induced not only by semantic manipulations but may result also from other contextual features of a situation and from individual factors, provided that problems are equivalent from the perspective of economic theory. Describing equivalent dilemmata as a give-some vs. as a take-some dilemma is an example of this type of framing. (Kühberger 1998:24)

This introduces elements of the wide sense of framing back into the picture: any delineation and structuring of the decision problem might have an effect on decision-making, even if these are hard to categorise with the tools of decision theory. Unsurprisingly, such cases have been far less discussed in the literature. The following taxonomy therefore cannot be considered comprehensive. Nevertheless, the following distinctions might be useful:

(D) procedural framing
(E) ethically loaded frames
(F) temporal frames

Gold and List (2004) argue that the ways how mental attitudes are elicited or measured constitutes procedural framing. For example, Lichtenstein and Slovic (1971) devised different ways how to elicit people's preferences over the same prospects. They found that the elicited preferences strongly depended on the elicitation procedure, up to the point where the differently elicited preferences over the same prospects became inconsistent. Gold and List therefore argue that such elicitation procedures constitute a kind of framing.

In social dilemma and coordination games, Bacharach et al. (2006) identify different ethically loaded frames that a player may adopt, namely the *I-frame* and the *we-frame*. Standard game theory implicitly assumes that a player in cases like the Prisoners' Dilemma always adopts an I-frame (asking "What should I do?"), leading to the dominant reasoning ("whatever others do, I will be better off defecting"). But she could be adopting, argue Bacharach et al. (2006), a we-frame (asking "What should we do?"). Players who adopted a we-frame will choose to cooperate in social dilemmas, as this contributes to the strategy profile that maximizes the group's payoff. Bacharach explicitly calls such cases "framing"; research on these phenomena however predates the framing terminology (e.g. Evans and Crumbaugh 1966). Some authors seek to subsume ethically loaded frames under goal framing (Levin et al. 1998:168).

Tversky and Kahneman (1981) briefly mention another kind of framing, namely the changing of temporal perspectives.

> The metaphor of changing perspective can be applied to other phenomena of choice, in addition to the framing effects with which we have been concerned here. The problem of self-control is naturally construed in these terms....an action taken in the present renders inoperative an anticipated future preference. An unusual feature of the problem of intertemporal conflict is that the agent who views a problem from a particular temporal perspective is also aware of the conflicting views that future perspectives will offer. (Tversky and Kahneman 1981:457)

In cases of intertemporal conflict – for example doing things now or later – a decision maker can assume the respective perspectives of her different temporal selves. Assuming today's perspective will let the decision maker decide according to her current preferences, while assuming her future self's perspective will give her future preferences an influence (see Hirsch Hadorn 2016; Möller 2016). Tversky and Kahneman seem to suggest that these perspectives correspond to different temporal frames, although this language has not been widely adopted in the literature.

Clearly, other applications of framing in this loose sense are possible, but because they are not widespread in the literature, I will not discuss them here. Instead, I will briefly sketch three motivations that led Tversky and Kahneman to introduce the concept, and that contributed to its pervasive adoption in the literature.

First, before the presentation of the framing concept in 1981, Tversky and Kahneman had developed a new research paradigm in psychology, that sought to document systematic deviations of experimental subjects from the prediction of the standard rational choice model (Heukelom 2014). The experimental elicitation of framing phenomena stands in this tradition, as standard rational choice models descriptively and normatively assume that people's decisions are invariant under alternative descriptions of the same decision elements (I will discuss the normative assumption of these standard models in Sect. 6). As part of this broader research effort, other researchers experimentally investigated behaviour that conceptually is very close to framing, although they did not use this terminology (e.g.; Lichtenstein and Slovic 1971; Thaler 1980).

Second, Kahneman and Tversky (1979) famously proposed "prospect theory" in order to model the systematic deviations that they and other researchers had elicited. Although there is no terminological reference to framing in prospect theory, the theory relies on evidence that conceptually is very close to cases of valence framing. Unsurprisingly, Tversky and Kahneman (1981) then propose prospect theory as an explanation of the framing effects they describe.

Third, many researchers who seized on the framing concept, including Tversky and Kahneman, claim it as a model for understanding anomalous economic phenomena in the real world that cannot be explained with standard economic models. Kahneman and Tversky (1984:347), for example, claim that framing is the factor underlying the observation "that the standard deviation of the prices that different stores in a city quote for the same product is roughly proportional to the average price of that product (Pratt et al. 1979)." Bacharach (2001:4) argues that framing lies at the bottom of the "Money illusion", and Kahneman and Tversky (1984:349) argue that observations of inconsistent choices of gambles and insurance policies (as described e.g. by Hershey and Schoemaker 1980) are driven by framing.

To conclude this section, I would like to point out a certain tension in the research on framing. On the one hand, sustained research activity has produced a manifold of experimental designs (surveyed in Sect. 3) and mechanistic models (Sect. 4). These findings correspond well with the multitude of framing concepts that I discussed in this section, and which seem to suggest that framing should not be treated as a very unified concept. On the other hand, however, the continued use of the term 'framing' for all these seemingly diverse concepts suggests that its users see a deeper unity in the concept of framing. On an abstract level, all these concepts are seen as closely interlinked. As Bacharach put it: "A frame is the set of concepts or predicates an agent uses in thinking about the world... One does not just see, but one sees as" (Bacharach 2001:1). This has given rise to a tendency to seek unified theories of framing (as discussed in Sects. 5 and 7) and derive general claims about when framing effects justify policy interventions or which framing effects can be exploited for policy purposes. One of the purposes of this review is to represent this tension and its determinants appropriately, which hopefully might contribute to its solution.

3 Experimental Elicitation of Framing Phenomena

Framing is fundamentally an experimentally identified phenomenon. Only the presentation of re-described acts, states or outcomes under highly controlled conditions have yielded behavioural evidence for the systematic deviation from standard rational choice models. Because of this strong dependence on experiments, understanding the concept (or the concepts) of framing requires looking into the details of the experiments that elicited this behavioural evidence.

Many hundreds of experimental studies on framing have been published since 1981. It is not the purpose of this section to provide a systematic review of these. The interested reader might instead consult extant reviews (Levin et al. 1998) and meta-analyses (Gallagher and Updegraff 2012; Gambara and Pinon 2005; Kühberger 1998). The overall tenor of these is that the framing effect is a robust phenomenon:

> A meta-analysis of 136 research reports yielded 230 single effect sizes, which, overall, corroborated the framing effect. (Kühberger 1998:47)

Yet this conclusion disguises an important heterogeneity. Not only do such meta-analyses draw on substantially different experimental designs, they also disclose a heterogeneity of effect sizes, depending on the respective experimental designs. I will come back to this at the end of this section. First, I will describe some experiment types, in order to make obvious the heterogeneity in design.

Tversky and Kahneman's (1981) "Asian disease problem" is clearly the prototypical and most-cited example of a framing experiment. They presented two separate groups of experimental subjects with one of the following decision problems. Number of participants and response frequencies are described in rectangular brackets (Tversky and Kahneman 1981:453):

> *Problem 1* [N = 152]: Imagine that the U.S. is preparing for the outbreak of an unusual Asian disease, which is expected to kill 600 people. Two alternative programs to combat the disease have been proposed. Assume that the exact scientific estimate of the consequences of the programs are as follows:
>
> - If Program A is adopted, 200 people will be saved [72 percent]
> - If Program B is adopted, there is 1/3 probability that 600 people will be saved, and 2/3 probability that no people will be saved. [28 percent]
>
> Which of the two programs would you favor?
>
> *Problem 2* [N = 155]:
>
> - If Program C is adopted 400 people will die. [22 percent]
> - If Program D is adopted there is 1/3 probability that nobody will die, and 2/3 probability that 600 people will die. [78 percent]
>
> Which of the two programs would you favor?

The experiment poses two discrete choices between a risky and a riskless option of equal expected value. In one problem, the options are described in positive terms (i.e., lives saved); in the other in negative terms (i.e., lives lost). Because the experimental manipulation consists in a re-description of a consequence of a risky choice, this is a framing of type (A1), as described in the previous section.

Tversky and Kahneman observed a "choice reversal," where the majority of subjects who were given the positively framed problem 1 chose the option with the certain outcome, whereas the majority of subjects who were given the negatively framed problem 2 chose the risky option.

Despite its prototypical status, following framing experiments have often deviated substantially from the Asian disease design. This has led some authors to question whether these experiments provide evidence for the same phenomenon:

> many recent studies of valence framing effects have deviated greatly from the operational definitions and theoretical concepts used in the original studies, thus stretching the limits of Kahneman and Tversky's initial theoretical accounts. (Levin et al. 1998:151)
>
> Diverse operational, methodical and task-specific features make the body of data heterogeneous to a degree that makes it impossible to speak of 'the framing effect.' (Kühberger 1998:43)

To make these worries more salient, let me summarize some of the main differences in experimental designs (in this I largely follow Kühberger 1998:32–33). The first difference concerns the nature of the options. In some experimental designs, one option is riskless and the other is risky – for example in the Asian disease design described above. In others, both options are risky, as for examples when subjects are asked to choose between therapies that are risky to different degrees. The second difference concerns the degree of partitioning of risky option. In many designs, each risky option only consists of a dual partition, with an event either occurring or not occurring. In other designs, for example bargaining tasks, options might be partitioned more finely. A third difference concerns the nature of the framing manipulation. Framing can be manipulated either by explicit labelling (e.g. "win" vs. "loose"; "gain" v. "pay") or by implicitly describing the task in value-relevant ways (e.g. by describing a situation either as a commons-dilemma or a public goods problem). A fourth difference concerns the subjects' responses: they might be asked to choose between options, as in the Asian disease design, or only to rank the different options. A fifth difference between designs concerns the comparison of choices: are choices of the same person in the two different situations compared, or are the compared choices those of different people (as in the Asian disease problem)? Finally, designs vary in the domain of their choices, involving either economic, social, medical or gambling decisions. Thus, the design of experiments that all are supposed to provide evidence for or against framing effects substantially differs.

Furthermore, framing phenomena have also been elicited in inferential tasks, which do not involve the choice between acts, but rather the choice of theoretical conclusions. Many studies in this area have concluded that laypeople and professionals alike (see Koehler 1996; Berwick et al. 1981) make poor diagnostic inferences on the basis of statistical information. In particular, their statistical inferences do not follow Bayes' theorem—a finding that prompted Kahneman and Tversky (1972:450) to conclude: "In his evaluation of evidence, man is apparently not a conservative Bayesian: he is not Bayesian at all." The studies from which this and similar conclusions were drawn presented information in the form of probabilities and percentages. From a mathematical viewpoint, it is irrelevant whether statistical information is presented in probabilities, percentages, absolute frequencies, or some other form, because these different representations can be mapped onto one another in a one-to-one fashion. Seen from a psychological viewpoint, however, representation does matter: Some representations make people more competent to reason in a Bayesian way in the absence of any explicit instruction (Hoffrage et al. 2000; Gigerenzer and Hoffrage 1995).

That the experimental designs for the elicitation of framing differ substantially perhaps would not be a problem if these designs all yielded comparable effects – indeed, such a result would even support the robustness of the framing effect. Unfortunately, this does not seem to be the case. Rather, effect sizes obtained from different experimental designs systematically differ:

> The more experiments differ from the original Asian disease problem, the lesser the reference point effect.... Overall, 4 of 10 procedural designs are ineffective: the Clinical reasoning design is ineffective, and, to make things worse, is used relatively frequently. Further ineffective designs are Escalation of commitment, Message compliance, and Evaluation of objects. (Kühberger 1998:45)

> the likelihood of obtaining choice reversals was directly related to the similarity between features of a given study and features of Tversky and Kahneman's (1981) original 'Asian disease problem.' (Levin et al. 1998:157)

This of course does not invalidate the framing concept altogether, but it should caution against its context-free use: the phenomenon of framing in some important way depends on the design of the manipulation and the environment in which it is elicited. Because the determining factors of this elicitation are not yet fully understood, it is difficult to extrapolate from the laboratory conditions to other contexts. To progress in this matter would require knowing more about the underlying mechanisms through which these environmental factors influence framing (Grüne-Yanoff 2015). I will discuss this topic in the next section.

4 Possible Mechanisms of Framing

Evidence for framing phenomena typically comes in the form of effect sizes – a measure of the correlation between framing manipulation and behavioural changes. These relations are captured by some of the theories discussed in Sect. 5. What remains often opaque is the process through which the framing produces the change.

> Cognitive processes are another stepchild of framing research. Taken the effect for granted (what can safely be assumed), we would be well advised to probe for the cognitive processes and structures that are responsible for it. (Kühberger 1998:47)

This is of particular relevance given the heterogeneity of effect sizes and their seeming dependence on experimental design. One possible explanation for this dependence is that different framing manipulations in different circumstances trigger different cognitive mechanisms, which then consequently produce different effects and different effect sizes.

There is very little research on the cognitive mechanisms underlying framing. Mechanisms typically only appear as mere speculations and ad-hoc how-possibly explanations of observed phenomena. Nevertheless, it is informative to discuss some of these speculations in order to gain an understanding of their diversity.

For the framing of outcomes, for example, Tversky and Kahneman propose *contextual referencing* as a cognitive mechanism:

> There are situations, however, in which the outcomes of an act affect the balance in an account that was previously set up by a related act. In these cases, the decision at hand may be evaluated in terms of a more inclusive account, as in the case of the bettor who views the last race in the context of earlier losses. (Tversky and Kahneman 1981:457)

For the framing of contingencies, multiple cognitive processes have been proposed. For example, Tversky and Kahneman (1981) propose a *pseudocertainty effect*, which consist of an illusion of certainty. Options that are certain, they suggest, are preferred to options that are uncertain. If now an uncertain option is divided into two sequential steps, one of which incorporates all uncertainty, then the decision maker might take the appearance of certainty from the second step as relevant for the whole option, and prefer it as if it were certain.

Another possible cognitive mechanism behind the framing of contingencies might be *limited imagination*:

> the fundamental problem of probability assessment [is perhaps] the need to consider unavailable possibilities... People... cannot be expected... to generate all relevant future scenarios. Tversky and Koehler (1994:565)

That is, because people are unable to imagine relevant possible scenarios, they do not partition contingencies finely enough. But when they are given such scenarios from external sources, they incorporate them into the decision problem and decide accordingly, thus leading to framing effects.

A further possible cognitive mechanism behind the framing of contingencies might be *limited memory*. Even if they have already heard about possible contingencies, they might have forgotten about them again. Provision of more detailed descriptions then might help in remembering such contingencies (and their relevance), leading to framing effects.

Yet another possible mechanism of framing effects is that different descriptions alter the *salience* of events. For example, by re-describing a week either as a single event or as a sequence of 7 days, Fox and Rottenstreich (2003) elicited substantially different answers from subjects asked to report the probability that Sunday would be the hottest day of the coming week. In such cases, descriptions produce framing effects without fostering imagination or recall.

5 Descriptive Theories of Framing

Despite the diversity in concepts, elicitations and mechanisms of framing, various general theories of behaviour have been proposed that claim to adequately describe the framing phenomenon. None of these theories have mechanistic or procedural content; rather, they aim to capture the systematic relationship between framing manipulation and behavioural changes only. This section briefly reviews four such attempts, namely Prospect Theory, Cumulative Prospect Theory, Support Theory and Partition-dependent Expected Utility Theory. Notably, these theories seek to *describe* actual behaviour, influence, amongst other factors, by framing, while refraining to judge whether this behaviour is rational or not.

Prospect theory (Kahneman and Tversky 1979) describes behaviour as influenced by the decision maker's evaluation that is generated relative to a certain reference point. The theory proposes a two-step decision process: in the editing phase, a reference point is set. In the evaluation phase, outcomes are evaluated either as gains or losses, relative to the set reference point. Specifically, people evaluate gains (i.e. outcomes above the reference point) differently than losses (i.e. outcomes below the reference point) and care generally more about potential losses than potential gains. Prospect theory predates the explicit conceptualization of framing, but it clearly captures its main idea: namely, that the presentation of the outcomes of a decision problem systematically influences the decision maker's choice. That the glass is half-full rather than half-empty makes sense only against changing reference points – people consider it half-empty if their reference point was (the expectation of) a full glass, while they consider it half-full if their reference point was an empty glass. Similar with outcomes of medical interventions that are described either as a chance of death or of survival – people will focus more on the chance of death caused by a medical intervention if their reference point is the certain expectation of surviving, while they focus more on the chance of survival if their reference point is the certain expectation of dying.

In 1992, Tversky and Kahneman proposed a new theory, *cumulative prospect theory*, replacing the 1979 approach. In the new model, the editing phase of prospect theory was renamed "framing phase" (Tversky and Kahneman 1992). Furthermore, people tend to overweight extreme yet unlikely events, but underweight "average" events.

While the above versions of prospect theory describe evaluations of outcomes as dependent on reference points, hence focusing on framing of outcomes, the following theories focus on the framing of contingencies and acts. Tversky and Koehler's (1994) *support theory* describes how probability judgments are affected by whether propositions are presented as explicit or implicit disjunctions. For example, subjects are asked to judge how probable it is that a randomly selected person "will die from an accident". Subjects tend to give a lower probability to this implicit conjunction, than they give to an explicit conjunction consisting of "a randomly selected person will die from a car crash", ". . . a plane crash", ". . . a fire", ". . . drowning", etc. Support theory accounts for this phenomenon by describing agents as assigning subjective probability to hypotheses. Subjective probability increases as hypotheses are "unpacked" into more explicit disjunctions.

Ahn and Ergin's (2010) *partition-dependent expected utility theory* allows discriminating between different presentations of the same act. Starting from the standard subjective model of decision-making under uncertainty, they distinguish different expressions for an act as distinct choice objects. Specifically, lists of contingencies with associated outcomes are taken as the primitive objects of choice. Choices over lists are represented by a family of preferences, where each preference is indexed by a partition of the state space. The respective partitions are interpreted as descriptions of the different events.

6 Normative Assessment of Framing

The concept of framing is inextricably linked to normative judgment. Tversky and Kahneman argued that framing leads to preference reversals, violating consistency requirements of standard decision theory:

> we describe decision problems in which people systematically violate the requirements of consistency and coherence (Tversky and Kahneman 1981:453)

Upon closer inspection, however, it isn't entirely obvious which consistency requirements of standard decision theory framing supposedly violates. None of the axiomatisations of von Neumann and Morgenstern (1944), Savage (1954), Anscombe and Aumann (1963) or Jeffrey (1963) contain any explicitly formulated axiom that the standard framing cases would violate.[1]

Instead, the formulation of the framing effect led to the explicit formulation of a rationality axiom that previously had been implicitly assumed. This requirement has been variably called the *principle of invariance* or the *principle of extensionality*. Kahneman and Tversky formulate it thus:

> Invariance requires that the preference order between prospects should not depend on the manner in which they are described. In particular, two versions of a choice problem that are recognized to be equivalent when shown together should elicit the same preference even when shown separately. (Kahneman and Tversky 1984:343)

Arrow formulated the principle of extensionality thus:

> A fundamental element of rationality, so elementary that we hardly notice it, is, in logicians' language, its extensionality. The chosen element depends on the opportunity set from which the choice is to be made, independently of how that set is described (Arrow 1982:6)

Arrow makes explicit reference to extensionality as a principle of logic. In logic, the principle of extensionality requires of two formulas that have the same truth-value under any truth assignment to be mutually substitutable *salva veritate* in a sentence that contains one of these formulas. Thus, "the glass is half-full" and "the glass is half-empty" have the same truth-value in all possible worlds, because they refer to the same fact of the matter. An agent whose choice is affected by how this same fact is described violates extensionality. In the following discussion, I will reserve *extensionality* as the principle based on logical equivalence in this sense; it is determined by the semantic characteristics of the explicit formulations only. In contrast, I will be using *invariance* for the principle based on non-logical versions of equivalence; it is determined by implicit suggestions, that trigger pragmatic inferences, e.g. on expectations. So, two different formulations are invariant, if they implicitly suggest the same pragmatic inferences.

Thus defined, the two principles differ substantially: two descriptions might be semantically identical and yet differ pragmatically – I will discuss an example later in this section. However, two descriptions might be pragmatically identical and yet differ semantically – for example when the semantic differences are pragmatically irrelevant. That this distinction is relevant will (hopefully) become clear in this

[1] A qualification is necessary here. Kahneman and Tversky for example argue that specific kinds of act-framing violate the principle of dominance: "the susceptibility to framing and the S-shaped value function produce a violation of dominance in a set of concurrent decisions" (Kahneman and Tversky 1984:344). Clearly, dominance is an explicitly formulated requirement in these standard axiomatisations. However, because only special cases of framing violate dominance, and because the normative judgment apparently goes beyond these cases, it cannot be dominance violation that lies at the basis of judging framing to be irrational.

section. Unfortunately, the distinction isn't always so clear in the literature. Because the extensionality principle is the much clearer concept, I will discuss its relation to rationality first, and then focus on the invariance principle later.

Tversky and Kahneman (1986) considered invariance (here understood as extensionality) as a tacit axiom of rationality:

> This principle of invariance is so basic that it is tacitly assumed in the characterization of options rather than explicitly stated as a testable axiom. (Tversky and Kahneman 1986: S253)

Indeed, it has been formally shown recently that Jeffrey-Bolker decision theory (Jeffrey 1963) contains extensionality as an implicit axiom (Bourgeois-Gironde and Giraud 2009:391). For explicit formulations of this axiom, see e.g. Rubinstein (2000), and Le Menestreland and Van Wassenhove (2001).

Given the either implicit or explicit assumption of extensionality in most accepted normative decision theories, framing phenomena seem to be clear violations of rationality:

> The failure of invariance is both pervasive and robust. It is as common among sophisticated respondents as among naive ones, and it is not eliminated even when the same respondents answer both questions within a few minutes. . . .In their stubborn appeal, framing effects resemble perceptual illusions more than computational errors. . .. The moral of these results is disturbing: Invariance is normatively essential, intuitively compelling, and psychologically unfeasible. (Kahneman and Tversky 1984:343–4)

Those, like Tversky and Kahneman, who consider the extensionality normatively necessary, but who see its violation as pervasive, distinguish between normatively valid theories of decision making – which adhere to the invariance principle – and descriptively adequate theories of decision making – which describe the ways how people systematically violate extensionality. Theories of the first kind include von Neumann and Morgenstern (1944), Savage (1954), Anscombe and Aumann (1963) or Jeffrey (1963), while theories of the second kind were described in Sect. 5.

However, is the principle of extensionality really a defensible rationality requirement? This question really has two parts. The first concerns extensionality as a requirement for *full rationality*. The second concerns whether some violations are *compatible* with a normatively valid model of *bounded rationality*. In the remainder of this section, I will discuss some criticisms of the validity of extensionality as a requirement of full rationality. In the next section, I will review some normative theories of bounded rationality that allow limited violations of invariance.

Tversky and Kahneman early on acknowledged that cognitive effort considerations might mitigate the irrationality of framing effects:

> These observations do not imply that preference reversals [arising from framing] are necessarily irrational. Like other intellectual limitations, discussed by Simon under the heading of 'bounded rationality,' the practice of acting on the most readily available frame can sometimes be justified by reference to the mental effort required to explore alternative frames and avoid potential inconsistencies. (Tversky and Kahneman 1981:458)

However, this argument relies on a contested narrow interpretation of Simon's concept of bounded rationality (Gigerenzer and Brighton 2009). Tversky and Kahneman in the above quote clearly consider the validity of bounded rationality models to depend on an accuracy-cost trade-off: not-too-catastrophic inconsistencies are justifiable if the costs of avoiding them would be unreasonably high. In contrast, Gigerenzer and Brighton argue that the validity of bounded rationality models depends on the reliability of the models in performing well for their designated tasks in the designated environments.

In the context of framing, we find such arguments at various places. For example, Sher and McKenzie (2006) argue that the framing of an outcome encodes relevant additional information, which most people intuitively understand. They show experimentally that subjects systematically distinguish between "half-full" and "half-empty" glasses. A full glass of water (A) and an empty one (B) are put on the table. The experimenter asks the participant to pour half of the water into the other glass, and then to place the "half-empty glass" at the edge of the table. Most people choose glass A, the previously full glass.

Such violations of extensionality are rational responses when the goal is e.g. to avoid regret, because the different descriptions of the same fact might convey different information about the expectations of the chooser. In the glass example, if the glass was originally full, the resultant regret from obtaining one-half the water is different from the case where the glass was originally empty. Note that distinguishing between "half-full" and "half-empty" glasses violates extensionality, because the semantic properties of any sentence remains unaffected when one replaces one formulation with the other. Instead, the relevant information is obtained through pragmatic inferences, not logical ones.

Such pragmatic inferences often depend on surprising detail. For example, it seems that incomplete specifications are often interpreted as implicit recommendations. In the Asian disease case, described in Sect. 3, the riskless options are not fully specified, stressing only the amount of survivors or fatalities, respectively. When researchers completely specified the riskless options, the framing effect in the Asian disease problem disappeared (Mandel 2001; Kühberger 1995). If subjects interpret incomplete specification as implicit recommendations, then again, it is perfectly rational for them to take this additional information into account.

Another argument against the necessity of extensionality as a rationality criterion comes from the observation of people's ability to solve coordination problems by exploiting 'focal points'. Bacharach (2001) provides a game-theoretic analysis of such coordination problems, in which players have to coordinate on one out of many possible equilibria. This, Bacharach argues, depends on players being able to identify one strategy profile as 'focal'. In a problem where to meet in a big town, such a focal point might be the most notable monument of that town; in a problem when to simultaneously perform a certain action, such a focal point might be 12 o'clock at noon; in a problem to independently choose the same number between 0 and 100, such a focal point might be 0, or 50, or 100. It is an empirical fact that people often are able to solve such coordination problems, without being able to communicate with each other. Instead, they exploit the fact that within a particular way of describing a town, the time or a numerical interval, certain elements "stick out": these elements appear more salient than others under that description, and consequently draw the players focus onto themselves. Of course such salience varies with the descriptive frame – it is for this reason that Bacharach identifies the violation of extensionality as a success condition for coordination on focal points:

> Human framing propensities stand behind the well-known ability of people to solve coordination problems by exploiting 'focal points'. Ironically, it is precisely their incompleteness that we can thank for this. . . .The partiality and instability of frames or 'conceptual boundedness' disables human agents in certain tasks — in particular, it makes them manipulable by framers. However, the sharedness of frames enables them to do well in other tasks, and in some cases it is important for this that the shared frame is partial. (Bacharach 2001:7–9)

The first lesson to learn from these arguments is that the rationality of framing effects cannot be decided on a logical principle of extensionality. In decision-theoretic contexts, it is not relevant whether alternative descriptions are semantically equivalent (i.e. whether they have the same truth-value in all possible worlds), but rather whether they are informationally equivalent. In the above two cases, different frames of decision problems, although semantically equivalent, carried different decision-relevant information with them, and therefore it was rational for the agents to choose differently under these different frames. Sher and McKenzie (2006), for example, separate the issue of informational relevance from that of extensionality:

> There is no normative problem with logically equivalent but information non-equivalent descriptions leading to different decisions. (Sher and McKenzie 2006:487)

To the contrary, rational agents should be indifferent between two co-reportive propositions if and only if the frames in which their common reference is expressed convey exactly the same information about choice-relevant pieces of information.

While this rejects the logical notion of extensionality as a rationality criterion for decision making, it leaves open the possibility of invariance, suitably defined with respect to irrelevant information, as such a criterion. This possibility depends, however, on finding a sufficiently robust delineation of informational relevance. This is a formidable problem, which to my knowledge has not been solved as of now. Recall Kahneman and Tversky's characterization, cited above: "two versions of a choice problem that are recognized to be equivalent when shown together should elicit the same preference even when shown separately." (Kahneman and Tversky 1984:343). Recognized by whom? By the experimenter? By the decision maker herself? And under what conditions? Whether invariance will be a suitable rationality criterion will depend a lot on how these questions are answered. As Bacharach reminds us, this is a metatheoretical question that cannot be answered within a theory of rational decision making:

> whether there is a violation of [extensionality] (and so of rationality) depends on how we, the theorist, 'cut up the world'. . . . The criterion [extensionality] can only be applied after resolving a question about what it is rational to care about. (Bacharach 2001:3)

Various attempts at answering these questions have been provided, yet none has so far won general acceptance. Sen (1986: Chap. 2) introduced the idea of an *isoinformation set* containing objects of choice taken to be similar in terms of relevant information and which will be consequently treated in the same way in actual choices and judgements. Similarity in terms of relevant information here is an intersubjectively defined notion, for which it is difficult to give clear criteria. Broome (1991) discusses invariance a matter of classifying outcomes: two outcomes belong to the same class if it is irrational to have different preferences for both. Here the criterion is subjective, as it is conditional on an agent's subjective preferences. However, it isn't very useful for the present purposes (which are different from Broome's), as the invariance criterion, which is supposed to explicate rationality, would itself depend on a notion of rationality.

Sher and McKenzie (2006) recently proposed a criterion of informational relevance of different formulations as licensing different inferences:

> When there is no choice-relevant background condition C about whose probability a listener can draw inferences from the speaker's choice between frames A and B, we say that A and B are "information equivalent". Otherwise, we say that there has been information leakage from the speaker's choice of frame, and that the frames are therefore information non-equivalent. (Sher and McKenzie 2006:469)

Yet while one might use this criterion to ascertain whether in particular situations, a certain formulation was informationally relevant – and Sher and McKenzie indeed employ it in this way for assessing experimental situations – this criterion does not lend itself for a general assessment of informational relevance, as there is no clear specification when an agent is licenced to draw inferences from the speaker's formulation.

To conclude, the currently extant literature shows that the logical notion of extensionality cannot be a necessary rationality criterion for decision-making. A notion of invariance – suitably defined on informational irrelevance – might be, yet no clear delineation of informational irrelevance has as of yet found wide acceptance. That *some* framing effects – defined on extensionality or some available notion of invariance – are rational therefore seems a plausible conclusion; yet which specific framing effects are rational and which are not remains shrouded in the ambiguity of the underlying criterion.

7 Normative Theories That Model Framing

Normative decision theories prescribe how a rational decision should be made. Most of the standard normative decision theories, as described in the previous section, at least implicitly assume a relatively strong invariance requirement. Consequently, they preclude framing effects from the set of rational decisions: if descriptions of acts, states or outcomes are equivalent (typically understood as semantic identity) then the differences between these descriptions should have no influence on a rational decision. To the extent that defenders of such theories accept the existence of framing phenomena at all, they therefore propose a distinction between theories of actual behaviour and theories of rational decisions.

In contrast to this, others argue that limited violations of invariance are *compatible* with a normatively valid model of *bounded rationality*. That is, even if most people violate invariance some of the time, some of these violations might be less problematic than others, allowing for a normatively valid model of core rationality requirements. Such theories oppose the distinction between normatively valid and descriptively adequate theories of framing. Instead, they propose that one and the same theory can describe how people actually choose under framing effects, while maintaining that such choices are in fact rational. In this section, I discuss two kinds of such theories: first, those that expand standard expected utility approaches to include legitimate invariance violations, and second those that choose a reason-based account, showing how reasoning *processes* constitute legitimate violations of invariance.

Standard expected utility theories typically exclude framing effects as irrational. Savage (1954) and Anscombe and Aumann (1963), for example, did not explicitly distinguish different presentations of the same act, state or outcome. This is why they are typically interpreted as assuming extensionality. Savage, however, discusses the *small world problem*: that people do not form *one* decision problem for their whole life *at one moment in time*, partitioning the world into all relevant contingencies then – but rather divide this big world decision into a sequence of small world decisions, each of which concerning only a much rougher partitioning of the world into states (see Hirsch Hadorn 2016). People should follow the principle

> to cross one's bridges when one come to them [which] means to attack relatively simple problems of decision by artificially confining attention to so small a world that the [expected utility] principle can be applied here. (Savage 1954:16)

Because partitioning the future states of the world differently is an important form of framing, Savage here acknowledges the potential influence of framing on decision making. This conclusion is further supported by the fact that Savage explicitly excludes certain kinds of partitions as not suitable for his prescription how to make rational decisions. For example, act-dependent state partitions are excluded from a proper decision-problem set-up (as e.g. Jeffrey 1963:8–10, points out). Yet by acknowledging the possibility of different partitions, Savage also raises the possibility that such different partitions influence rational decisions in different ways. Take two different partitions, S and T, where T is a more fine-grained partition than S. If preferences over acts in T satisfy the Savage axioms, there is a probability function defined over states of T and a utility function over outcomes of T. Now can we calculate utilities and probabilities for S from those in T? Savage discusses two methods of doing so, and admits that these methods do not necessarily yield the same probability assignments on states in S (Savage 1954:89, for further discussion, see Shafer 1986:480–484). Thus, although a partition satisfies the Savage axioms, this does not guarantee that the probabilities calculated in this partition do not change when the partitioned is refined (or reduced). This is Savage's *small world problem*. Clearly, it is a particularly striking case of framing of contingencies.

Savage sought to resolve the small world problem by reference to "the grand world", i.e. an ultimately detailed refinement. This device, as he admits himself, is somewhat "tongue-in-cheek" (Savage 1954:83): it posits an atomistic view of the world, although no justification is forthcoming. Only by using the grand world as a reference point, and insisting that that probability assignment is correct which is calculated from the grand world, can Savage solve the small world problem. Without it, framing effects remain possible within his theory. To the extent that Savage's theory is interpreted as a valid normative theory, it follows that these framing effects are rational.

In contrast to the partition dependence, Jeffrey's (1963) decision theory explicitly seeks a partition-invariance calculation of the expected utility of acts. He conceives of acts, outcomes and states as propositions, and calculates the expected value of acts as the sum of values of outcomes, weighted by the *conditional* probability of outcomes, given acts. As Joyce (1999:212) shows, this approach allows us to express the utility of any disjunction as a function of the utilities of its disjuncts. Thus, the partition of acts, states or outcomes has no influence on rational decision, and framing, understood in this sense, cannot be rational. Amongst decision theorists, this is commonly seen as an advantage:

> In Jeffrey's theory ... there is guaranteed agreement between grand- and small-world representations of preferences. This guarantee is precisely what Savage could not deliver. The partition invariance of Jeffrey's theory should thus be seen as one of its main advantages over Savages' theory. (Joyce 1999:122)

Scholars who do not agree with Joyce on the advantages of Jeffrey's theory have introduced modifications to allow for invariance violations that might be pragmatically, if not semantically justified (e.g. Bourgeois-Gironde and Giraud 2009). However, these extensions typically do not themselves provide a criterion to distinguish between admissible and non-admissible invariance violations (as discussed in the previous section).

An alternative route of re-introducing framing into the normative framework is to deny that the Jeffrey's notion of partition invariance can exclude all relevant cases of framing. This would require that there are partitions of the world, which do not stand in the required relationship – one partition is not the disjunct in another partition. Bacharach (2001) seems to hint at such a possibility. On the one hand, he wrote, most partitions exhibit this relationship – for example, partitions with respect to

> shape, colour and position: we can easily see a mark as a triangle, as a blue triangle, as a blue triangle on the left,. . . on the other hand. . . a person can see the marks as letters and as geometric shapes, but not at the same time ... you can't integrate these two perception. (Bacharach 2001:6)

Fig. 8.1 An example of frame ambiguity

By integration, Bacharach means that two existing partitions – e.g. F={triangle, non-triangle} and G={blue, not blue} – are combined to a new partition, e.g. $H = F \bigoplus G$ = {blue triangle, blue non-triangle, non-blue triangle, non-blue non-triangle}. But he argues that not all sets of partitions can be thus integrated. A simple example, which he mentions in the quotation above, is depicted in Fig. 8.1.

One can either see the three marks as (Greek) letters or alternatively as geometric shapes, but one cannot see them as both at the same time. Other examples that Bacharach proposes include ambiguous images like Rubin's vase or the duck/rabbit image, as well as seeing outcomes either from an "I" or a "we" perspective (Bacharach 2001).

If not all frames can be integrated, then the question is how to choose when the tension between such alternative frames cannot be resolved. This is where Bacharach's *variable frame theory* applies. It suggests that in coordination games, players should select strategies by choosing their best reply in each available frame. More specifically, there is an exogenous probability measure $V(F)$ defined on frames F. $V()$ is common knowledge. A strategy profile (s_i, s_{-i}) *is a variable frame equilibrium* if, *for each frame F*, the option expected from playing s_i is subjectively best from the perspective of F against s_{-i} as perceived in F (Bacharach 2001:8–9). The optimality judgment for s_i then depends on the expected utility of playing s_i against s_{-i} in each frame F, weighted by the probability of F, $V(F)$. This theory, amongst others, explains why "conceptual boundedness" of human agents, to the extent that it results in the sharedness of frames, positively contributes to people's ability to coordinate.

The above theories show how framing effects can be incorporated into expected-utility accounts of rational decision-making. An alternative, reason-based, account seeks to identify how reasoning processes rationally influence choice. Let me briefly address how extensions of this account lead to rationalization of framing, by describing Gold and List's (2004) *path-dependent decision-making*. Their account starts from the assumption that particular presentations of decision problems lead agents to consider relevant background propositions in a particular sequence, so that different presentations lead to different consideration sequences and hence to different decision paths. Such a model produces framing effects if (i) different decision paths produce different choices, and (ii) different decision problem presentations lead to such different-choice producing paths.

To give an illustrative example, let's consider Kahneman and Tversky's Asian disease problem again (see Sect. 3). The first, "lives saved", presentation, may induce a decision path starting with factual and normative propositions about saving lives, including normative propositions like "It is not worth taking the risk that no one will be saved" – leading the agent to choose the certain option. In contrast, the second, "lives lost", presentation, may induce a decision path starting with factual and normative propositions about losing lives, including normative propositions like "It is unacceptable to consign some people to death with certainty" – leading the agent to choose the uncertain option.

In cases like the Asian disease problem, agents have *dispositions* both to accept propositions like "It is not worth taking the risk that no one will be saved" as well as "It is unacceptable to consign some people to death with certainty". Yet depending on the decision path taken, only some of these dispositions get actualized and consequently influence decisions. As Gold and List point out, while the propositions that the agent is disposed to accept might be inconsistent (as they are in the Asian disease case), the propositions that the agent accepts on the specific decision path taken are not. Thus agents violating invariance need only suffer from implicit inconsistencies (i.e. inconsistencies regarding propositions that the agent is disposed to accept) while avoiding explicit inconsistencies between actually accepted propositions. Because such reason-based models propose specific reasoning processes, their validity (including their normative validity) will depend on what the actual mental mechanisms are that people make use of when dealing with framed acts, states or contingencies. As I argued in Sect. 4, however, research on mechanisms has been rather neglected with respect to framing.

8 Policy Relevance: How Should Decisions Be Framed?

The literature on framing discussed in the previous sections has inspired many policy proposals for intervening in human behaviour. Three key influences on policy must be distinguished. First, framing is used to *caution* policy interventions based on the reductive approach to policy analysis. Framing, as we saw, introduces various kinds of uncertainty into decision-making, including uncertainty about people's preferences, about the effect of changing the descriptions of a decision problem, and about the rationality or irrationality of observed choices. Consequently, considerations of framing might provide support for argumentative methods to deal with uncertainty in policy analysis.

Second, framing had been used to *justify* such interventions. The basic idea here is that the various framing phenomena show people to behave irrationally in a systematic way, and therefore need help from the policymaker. Third, framing has been used as the *instrument* by which various policies propose to intervene on people's behaviour. The basic idea here is that framing is an important factor that influences behaviour, and that policy interventions can make use of it in order to achieve their ends.

Those who stress the justificatory role of framing generally agree that (i) framing phenomena are widespread and (ii) framing effects are results of irrational decision-making.

> ... research by psychologists and economists over the past three decades has raised questions about the rationality of many judgments and decisions that individuals make. People ... exhibit preference reversals ... and make different choices depending on the framing of the problem.... (Sunstein and Thaler 2003:1168)

> So long as people are not choosing perfectly, it is at least possible that some policy could make them better off by improving their decisions.(Sunstein and Thaler 2003:1163)

That is, framing is a systematic behavioural phenomenon that is accurately described by some descriptive theory (discussed in Sect. 5). However, there is a normatively valid theory of behaviour, which excludes framing effects (as described in Sect. 7). Due to the difference between actual systematic behaviour and rationally required behaviour, policy interventions that make actual behaviour more rational might be justified (for similar arguments, see Conly 2013; Ariely 2008; Trout 2005; Camerer et al. 2003).

More specifically, framing plays an important role in the justification of *nudge policies* (Thaler and Sunstein 2008). Nudges are interventions on the context in which people make decisions with the aim of steering people's behaviour into specific directions. Proponents of nudges often argue that people do not have well-defined preferences, because they change their preferences in the light of rationally irrelevant frame changes. Because people often do not have clear preferences over options, welfare assessments should take into account different criteria than their preferences. Thus the justification of nudge interventions is often supported with framing phenomena: people's preferences are variant under changing descriptions of the same choice situations.

Not everybody agrees with this argument. Critics point out, with arguments related to those reviewed in Sect. 6, that framing phenomena need not be irrational, and that the irrationality judgment is often based on an overtly narrow consistency criterion (Berg 2014; Berg and Gigerenzer 2010). Other concerns, in line with those discussed in Sect. 3, might question the prevalence of framing phenomena and consequently the need for interventions. Finally, some critics wonder whether framing effects really justify interventions on behaviour, and suggest instead that education can prepare people to deal with frames better on their own (Gigerenzer 2015).

This debate about whether framing justifies policy interventions is quite separate from the ways that framing has been proposed as a tool for policy interventions. One can well imagine that even if the justificatory project failed (but some other justification of policy interventions succeeded), that such policies might still employ framing as a means of influencing people's choices, if framing should prove to be an effective means for that purpose.

Three such instrumental uses of framing can be distinguished. First, policy interventions might exploit the effect of framing in order to make people choose an option the policy maker deems optimal.

> A physician, and perhaps a presidential advisor as well, could influence the decision made by the patient or by the President, without distorting or suppressing information, merely by the framing of outcomes and contingencies. Formulation effects can occur fortuitously,

> without anyone being aware of the impact of the frame on the ultimate decision. They can also be exploited deliberately to manipulate the relative attractiveness of options. (Kahneman and Tversky 1984:346)

Such exploitations of framing effects have been proposed, amongst others, by the *Nudge* program (Thaler and Sunstein 2008). Examples of nudging with frames include suggestions to apply lessons from the Asian disease case to the descriptions of medical treatment alternatives, so that patients are more likely to choose that option that the policymaker considers superior (Thaler and Sunstein 2008:36–37). Another example is the recent proposal by Slovic and Västfjäll (2013) how to increase charitable giving through framing. Slovic and Västfjäll diagnose a systematic "insensitivity to mass tragedy" (94) in people's behaviour: when faced with suffering of large groups of victims, for example from genocide or natural disasters, people feel comparatively less compassion and give less aid than when confronted with individual victims. They propose a psychophysical model of *psychic numbing* that describes an inverse relationship between an affective valuation of saving a life and the number of lives at risk. They also argue that this affective valuation is the basis for most intuitive moral judgments about how much effort or how many resources to devote to saving lives. Consequently, they propose corrective interventions on these moral intuitions through framing the plight of many as the many plights of different individuals, each of who deserves compassion and support. Framing, as these two examples show, has become an important argument for nudge policies, as well as one of their chief policy intervention tools.

Note that these interventions might be motivated very differently. One possibility is that people go against their own preferences and do not choose what they judge best (perhaps even due to existing framing effects). In this case, (re-)framing as policy intervention is motivated by the goal to get people to choose what they really want. Another possibility is that people act according to their own preferences, but that the policymaker would prefer if they chose differently. In that case, (re-)framing is motivated to make people choose against their own wishes.

This ambiguity in the use of framing as an instrument of influence is present even in the everyday notion of framing. In colloquial English, the notion of framing has two rather disparate meaning. On the one hand, framing means "the action, method, or process of constructing, making, or fashioning something", or the result of this activity or process (OED). On the other hand, framing can also mean "the action or process of fabricating a charge or accusation against a person; an instance of this" (OED). The crucial difference here is that between a construction *simpliciter* and a construction *with deceptive intention*. It is therefore difficult to say something general about the moral evaluation of framing policies, but it is obvious that at least *some* uses of framing in this way are not compatible with liberal values (Grüne-Yanoff 2012).

Another use of our knowledge of framing effects as a policy tool is to design choice environments in such a way that framing effects are neutralized or eliminated whenever possible. This requires the idea that some frames exert less strong influences on reasoning and decision than others – i.e. that there is a canonical frame. Kahneman and Tversky suggest something along these lines, when they recommend to

> adopt a procedure that will transform equivalent versions of any problem into the same canonical representation. This is the rationale for the standard admonition to students of business, that they should consider each decision problem in terms of total assets rather than in terms of gains or losses. Such a representation would avoid the violations of invariance illustrated in the previous problems, but the advice is easier to give than to follow. (Kahneman and Tversky 1984:344)

One possible basis for such a neutrality argument is the hypothesis that human cognition is well adapted to certain kinds of representations, but not to others. With respect to statistical inference, for example, some have argued that our cognitive algorithms are not adapted to probabilities or percentages, as these concepts and tools have been developed only rather recently. Consequently, policies should aim to design inference or choice tasks with representations that people are most adapted to. In the case of statistical inference, Gigerenzer and Hoffrage (1995) and Hoffrage et al. (2000) showed that statistics expressed as *natural frequencies* improve the statistical reasoning of experts and non-experts alike.[2] For example, advanced medical students asked to solve medical diagnostic tasks performed much better when the statistics were presented as natural frequencies than as probabilities. Similar results have been reported for medical doctors (in a range of specialties), HIV counsellors, lawyers, and law students (Anderson et al. 2012; Akl et al. 2011; Lindsey et al. 2003; Hoffrage et al. 2000).

Bacharach seems to consider a similar idea when he suggests that many frames might be integrable: by providing a finer partition, two seemingly conflicting perspectives on the world can be combined in a more detail-rich frame. However, it remains unclear why this frame should be considered more 'neutral' than either of the original ones. What remains true is that "one does not just see, but one sees as" (Bacharach 2001:1); hence the neutral frame might remain a chimera.

A third use of our knowledge of framing effects as a policy tool – particularly if the first one is ethically questionable and the second one unachievable – is to elicit reflection through reframing. That is, the policy maker might present decision makers who are prone to framing effects with relevant information in *different* formats at the same time. In effect, this seeks to test the robustness of preferences by deliberate attempts to frame a decision problem in more than one way (cf. Fischhoff et al. 1980). Such an approach, instead of nudging or neutralising, seeks to *boost* people's abilities to deal with informationally and representationally challenging situations (Grüne-Yanoff and Hertwig 2015). The boost approach aims to enhance people's ability to understand and see through confusing and misleading

[2] Natural frequencies refer to the outcomes of natural sampling — that is, the acquisition of information by updating event frequencies without artificially fixing the marginal frequencies. Unlike probabilities and relative frequencies, natural frequencies are raw observations that have not been normalized with respect to the base rates of the event in question.

representations by making those representations less manipulative and opaque, rendering them less computationally demanding (Gigerenzer and Hoffrage 1995), and making them semantically and pragmatically less ambiguous (Hertwig and Gigerenzer 1999). From the boost perspective, difficulties understanding statistical information are seen not as an incorrigible mental deficiency of, say, doctors or patients, but as largely attributable to poor or intentionally misleading information. Moreover, the goal is not to push people toward a particular goal (e.g., to seek or not seek a particular treatment), but to help everybody (e.g., doctors and patients) to understand statistical information as the first critical step toward figuring out one's preference.

9 Conclusion

Framing is an important set of phenomena that challenges the standard theories of rational decision making and the notions of rationality they propose. Because framing seemingly drives a wedge between actual behaviour and normative standards imposed on behaviour, it has been used as a justification for policies intervening in behaviour. Nevertheless, many questions remain. From the survey of experimental elicitation, it isn't obvious how unified the notion of framing is, nor is it obvious that it is as prevalent as sometimes claimed. From the survey of mechanistic models and descriptive theories it appears that many questions when and how framing effects behaviour are not fully settled. Furthermore, there is considerable controversy to what extent the sensitivity of decisions to framing is irrational. Finally, consideration of framing might provide support for argumentative methods in policy analysis. All these questions have import on whether policies intervening on framing are justifiable, as well as whether framing is an effective and morally permissible tool of policy making.

Recommended Readings

Arrow, K. J. (1982). Risk perception in psychology and economics. *Economic Enquiry*, *20*, 1–9.

Hertwig, R., & Gigerenzer, G. (1999). The "conjunction fallacy" revisited: How intelligent inferences look like reasoning errors. *Journal of Behavioral Decision Making*, *12*, 275–305.

Sher, S., & McKenzie, C. R. M. (2006). Information leakage from logically equivalent frames. *Cognition*, 101, 467–494.

Tversky, A., & Kahneman, D. (1981). The framing of decisions and the psychology of choice. *Science* (New Series), *211*, 453–458.

References

Ahn, D., & Ergin, H. (2010). Framing contingencies. *Econometrica, 78*, 655–695.

Akl, E. A., Oxman, A. D., Herrin, J., Vist, G. E., Terrenato, J., Sperati, F., Costiniuk, C., Blank, D., & Schünemann, H. (2011). Using alternative statistical formats for presenting risks and risk reductions. *Cochrane Database of Systematic Reviews*. doi:10.1002/14651858.CD006776.pub2.

Anderson, B. L., Gigerenzer, G., Parker, S., & Schulkin, J. (2012). Statistical literacy in obstetricians and gynecologists. *Journal for Healthcare Quality, 36*, 5–17.

Anscombe, F. J., & Aumann, R. J. (1963). A definition of subjective probability. *Annals of Mathematical Statistics, 34*, 199–205.

Ariely, D. (2008). *Predictably irrational: The hidden forces that shape our decisions* (1st ed.). New York: HarperCollins.

Arrow, K. J. (1982). Risk perception in psychology and economics. *Economic Enquiry, 20*, 1–9.

Bacharach, M. O. (2001). Framing and cognition in economics: The bad news and the good. Lecture notes for the *ISER Workshop, Cognitive Processes in Economics*. http://cess-wb.nuff.ox.ac.uk/documents/mb/lecnotes.pdf.

Bacharach, M., Gold, N., & Sugden, R. (2006). *Beyond individual choice: Teams and frames in game theory*. Princeton: Princeton University Press.

Berg, N. (2014). The consistency and ecological rationality approaches to normative bounded rationality. *Journal of Economic Methodology, 21*, 375–395.

Berg, N., & Gigerenzer, G. (2010). As-if behavioral economics: Neoclassical economics in disguise? *History of Economic Ideas, 18*, 133–166.

Berwick, D. M., Fineberg, H. V., & Weinstein, M. C. (1981). When doctors meet numbers. *American Journal of Medicine, 71*, 991–998.

Bourgeois-Gironde, S., & Giraud, R. (2009). Framing effects as violations of extensionality. *Theory and Decision, 67*, 385–404.

Broome, J. (1991). *Weighing goods: Equality, uncertainty and time*. Oxford: Wiley-Blackwell.

Brun, G., & Betz, G. (2016). Analysing practical argumentation. In S. O. Hansson & G. Hirsch Hadorn (Eds.), *The argumentative turn in policy analysis. Reasoning about uncertainty* (pp. 39–77). Cham: Springer. doi:10.1007/978-3-319-30549-3_3.

Camerer, C., Issacharoff, S., Loewenstein, G., O'Donoghue, T., & Rabin, M. (2003). Regulation for conservatives: Behavioral economics and the case for "Asymmetric Paternalism". *University of Pennsylvania Law Review, 151*, 1211–1254.

Conly, S. (2013). *Against autonomy: Justifying coercive paternalism*. Cambridge: Cambridge University Press.

Doorn, N. (2016). Reasoning about uncertainty in flood risk governance. In S. O. Hansson & G. Hirsch Hadorn (Eds.), *The argumentative turn in policy analysis. Reasoning about uncertainty* (pp. 245–263). Cham: Springer. doi:10.1007/978-3-319-30549-3_10.

Edvardsson Björnberg, K. (2016). Setting and revising goals. In S. O. Hansson & G. Hirsch Hadorn (Eds.), *The argumentative turn in policy analysis. Reasoning about uncertainty* (pp. 171–188). Cham: Springer. doi:10.1007/978-3-319-30549-3_7.

Elliott, K. C. (2016). Climate geoengineering. In S. O. Hansson & G. Hirsch Hadorn (Eds.), *The argumentative turn in policy analysis. Reasoning about uncertainty* (pp. 305–324). Cham: Springer. doi:10.1007/978-3-319-30549-3_13.

Evans, G. W., & Crumbaugh, C. M. (1966). Effects of prisoner's dilemma format on cooperative behavior. *Journal of Personality and Social Psychology, 3*, 486.

Fischhoff, B., Slovic, P., & Lichtenstein, S. (1980). Knowing what you want: Measuring labile values. In T. Wallsten (Ed.), *Cognitive processes in choice and decision behavior* (pp. 117–141). Hillsdale: Erlbaum.

Fox, C. R., & Rottenstreich, Y. (2003). Partition priming in judgment under uncertainty. *Psychological Science, 14*, 195–200.

Gallagher, K. M., & Updegraff, J. A. (2012). Health message framing effects on attitudes, intentions, and behavior: A meta-analytic review. *Annals of Behavioral Medicine, 43*, 101–116.

Gambara, H., & Piñon, A. (2005). A meta-analytic review of framing effect: Risky, attribute and goal framing. *Psicothema, 17*, 325–331.

Gigerenzer, G. (2015). On the supposed evidence for libertarian paternalism. *Review of Philosophy and Psychology, 6*, 361–383.

Gigerenzer, G., & Brighton, H. (2009). Homo heuristicus: Why biased minds make better inferences. *Topics in Cognitive Science, 1*, 107–143.

Gigerenzer, G., & Hoffrage, U. (1995). How to improve Bayesian reasoning without instruction: Frequency formats. *Psychological Review, 102*, 684–704.

Goffman, E. (1974). *Frame analysis: An essay on the organization of experience*. Cambridge, Mass: Harvard University Press.

Gold, N., & List, C. (2004). Framing as path-dependence. *Economics and Philosophy, 20*, 253–277.

Grüne-Yanoff, T. (2012). Old wine in new casks: Libertarian paternalism still violates liberal principles. *Social Choice and Welfare, 38*, 635–645.

Grüne-Yanoff, T. (2015). Why behavioural policy needs mechanistic evidence. *Economics and Philosophy*. doi:http://dx.doi.org/10.1017/S0266267115000425.

Grüne-Yanoff, T., & Hertwig, R. (2015). Nudge versus boost: How coherent are policy and theory? *Minds and Machines*. doi:10.1007/s11023-015-9367-9.

Grunwald, A. (2016). Synthetic biology: Seeking for orientation in the absence of valid prospective knowledge and of common values. In S. O. Hansson & G. Hirsch Hadorn (Eds.), *The argumentative turn in policy analysis. Reasoning about uncertainty* (pp. 325–344). Cham: Springer. doi:10.1007/978-3-319-30549-3_14.

Hansson, S. O., & Hirsch Hadorn, G. (2016). Introducing the argumentative turn in policy analysis. In S. O. Hansson & G. Hirsch Hadorn (Eds.), *The argumentative turn in policy analysis. Reasoning about uncertainty* (pp. 11–35). Cham: Springer. doi:10.1007/978-3-319-30549-3_2.

Hershey, J. C., & Schoemaker, P. J. H. (1980). Risk taking and problem context in the domain of losses: An expected-utility analysis. *Journal of Risk and Insurance, 47*, 111–132.

Hertwig, R., & Gigerenzer, G. (1999). The "conjunction fallacy" revisited: How intelligent inferences look like reasoning errors. *Journal of Behavioral Decision Making, 12*, 275–305.

Heukelom, F. (2014). *Behavioral economics: A history*. Cambridge: Cambridge University Press.

Hirsch Hadorn, G. (2016). Temporal strategies for decision making. In S. O. Hansson & G. Hirsch Hadorn (Eds.), *The argumentative turn in policy analysis. Reasoning about uncertainty* (pp. 217–242). Cham: Springer. doi:10.1007/978-3-319-30549-3_9.

Hoffrage, U., Lindsey, S., Hertwig, R., & Gigerenzer, G. (2000). Communicating statistical information. *Science, 290*, 2261–2262.

Jeffrey, R. C. (1963). *The logic of decision*. Chicago: University of Chicago Press.

Joyce, J. M. (1999). *The foundations of causal decision theory*. Cambridge: Cambridge University Press.

Kahneman, D., & Tversky, A. (1972). Subjective probability: A judgment of representativeness. *Cognitive Psychology, 3*, 430–454.

Kahneman, D., & Tversky, A. (1979). Prospect theory: An analysis of decision under risk. *Econometrica, 47*, 263–291.

Kahneman, D., & Tversky, A. (1984). Choices, values, and frames. *American Psychologist, 39*, 341–350.

Koehler, J. J. (1996). The base rate fallacy reconsidered: Descriptive, normative and methodological challenges. *Behavioral and Brain Sciences, 19*, 1–53.

Kühberger, A. (1995). The framing of decisions: A new look at old problems. *Organizational Behavior and Human Decision Processes, 62*, 230–240.

Kühberger, A. (1998). The influence of framing on risky decisions: A meta-analysis. *Organizational Behavior and Human Decision Processes, 75*, 23–55.

Le Menestrel, M., & Wassenhove, L. V. (2001). The domain and interpretation of utility functions: An exploration. *Theory and Decision, 51*, 329–349.

Levin, I. P., Schneider, S. L., & Gaeth, G. J. (1998). All frames are not created equal: A typology and critical analysis of framing effects. *Organizational Behavior and Human Decision Processes, 76*, 149–188.

Lichtenstein, S., & Slovic, P. (1971). Reversals of preference between bids and choices in gambling decisions. *Journal of Experimental Psychology, 89*, 46.

Lindsey, S., Hertwig, R., & Gigerenzer, G. (2003). Communicating statistical DNA evidence. *Jurimetrics: The Journal of Law, Science, and Technology, 43*, 147–163.

Mandel, D. R. (2001). Gain-loss framing and choice: Separating outcome formulations from descriptor formulations. *Organizational Behavior and Human Decision Processes, 85*, 56–76.

Möller, N. (2016). Value uncertainty. In S. O. Hansson & G. Hirsch Hadorn (Eds.), *The argumentative turn in policy analysis. Reasoning about uncertainty* (pp. 105–133). Cham: Springer. doi:10.1007/978-3-319-30549-3_5.

Oxford English Dictionary (OED). Sept 2011. "framing, n.". Oxford University Press. http://dictionary.oed.com/. Accessed 30 Sept 2014.

Pratt, J. W., Wise, D., & Zeckhauser, R. (1979). Price differences in almost competitive markets. *Quarterly Journal of Economics, 93*, 189–211.

Rubinstein, A. (2000). *Modeling bounded rationality*. Cambridge: MIT Press.

Savage, L. J. (1954). *The foundations of statistics*. New York: Wiley.

Sen, A. (1986). Information and invariance in normative choice. In W. P. Heller, R. M. Starr, & D. A. Starret (Eds.), *Social choice and public decision making* (Essays in Honor of Kenneth J. Arrow, Vol. 1, pp. 29–55). Cambridge: Cambridge University Press.

Shafer, G. (1986). Savage revisited. *Statistical Science, 1*, 463–485.

Sher, S., & McKenzie, C. R. M. (2006). Information leakage from logically equivalent frames. *Cognition, 101*, 467–494.

Slovic, P., & Västfjäll, D. (2013). The more who die, the less we care: Psychic numbing and genocide. In A. Olivier (Ed.), *Behavioural public policy* (pp. 94–109). Cambridge: Cambridge University Press.

Slovic, P., Fischhoff, B., & Lichtenstein, S. (1982). Response mode, framing, and information-processing effects in risk assessment. In R. Hogarth (Ed.), *New directions for methodology of social and behavioral science: Question framing and response consistency* (pp. 21–36). San Francisco: Jossey-Bass.

Sunstein, C. R., & Thaler, R. H. (2003). Libertarian paternalism is not an oxymoron. *The University of Chicago Law Review, 70*(4), 1159–1202.

Thaler, R. (1980). Toward a positive theory of consumer choice. *Journal of Economic Behavior & Organization, 1*, 39–60.

Thaler, R. H., & Sunstein, C. R. (2008). *Nudge*. New Haven: Yale University Press.

Trout, J. D. (2005). Paternalism and cognitive bias. *Law and Philosophy, 24*, 393–434.

Tversky, A., & Kahneman, D. (1981). The framing of decisions and the psychology of choice. *Science* (New Series), *211*, 453–458.

Tversky, A., & Kahneman, D. (1986). Rational choice and the framing of decisions. *The Journal of Business, 59*, S251–S278.

Tversky, A., & Kahneman, D. (1992). Advances in prospect theory: Cumulative representation of uncertainty. *Journal of Risk and Uncertainty, 5*, 297–323.

Tversky, A., & Koehler, D. J. (1994). Support theory: A nonextensional representation of subjective probability. *Psychological Review, 101*, 547–567.

Von Neumann, J., & Morgenstern, O. (1944). *Theory of games and economic behavior*. Princeton: Princeton University Press.

Chapter 9
Temporal Strategies for Decision-making

Gertrude Hirsch Hadorn

Abstract Temporal strategies extend decisions over time, for instance by delaying decisions (postponement), reconsidering provisional decisions later on (semi-closure), or partitioning decisions for taking them stepwise (sequential decisions). These strategies allow the decision-makers to use further argumentative methods to learn about, evaluate, and account for the relevant uncertainties. However, temporal strategies also open up opportunities for eschewing the decision problem. I propose four general criteria that serve as a heuristic to structure reasoning for and against the application of temporal strategies to a decision problem: the relevance of considering uncertainties for taking a decision, the feasability of improving information on or evaluating relevant uncertainties; the acceptability of trade-offs related to the temporal strategy, and the maintenance of governing decision-making over time. These criteria need to be specified and weighted in each case of application. Instead of determining a temporal strategy, the criteria provide a framework for systematic deliberation.

Keywords Closure • Postponement • Semi-closure • Sequential decisions • Great uncertainty • Decision procedure • Adaptive governance

1 Introduction

Since we cannot know for sure what will be done or happen in the future, information about policy decision problems regarding the future is always uncertain. This uncertainty, however, can be dealt with: we can intentionally extend decision-making over time in order to learn about, evaluate, and account for the uncertainty of information. In what follows, I call a plan for extending a decision over time a "temporal strategy".[1] Delaying a decision, reconsidering a provisional

[1] The term "strategy" is used in a variety of ways in common language as well as in the sciences. Following the Oxford English Dictionary, entry "strategy", meaning 2d (http://www.oed.com), I use "strategy" here to refer to a plan for successful action, and I extend its application to the

G. Hirsch Hadorn (✉)
Department of Environmental Systems Science, Swiss Federal Institute of Technology, Zurich, Switzerland
e-mail: hirsch@env.ethz.ch

S.O. Hansson, G. Hirsch Hadorn (eds.), *The Argumentative Turn in Policy Analysis*, Logic, Argumentation & Reasoning 10, DOI 10.1007/978-3-319-30549-3_9

decision later on, or partitioning decisions in order to take them stepwise are ways to extend a decision over time. Instead of taking a definitive decision now, temporal strategies in some regard keep the decision open in order to retain some opportunities for learning more before we decide. The extension of decisions over time allows for learning about uncertainties by considering changes in the real world such as events that occur naturally or have been initiated for this purpose, as well as through elaborating on the existing body of uncertain information. Furthermore, temporal strategies facilitate improving the evaluation of uncertainties in decision-making, for instance, if one has to account for additional information on possible outcomes, on further values that are at stake, or on relevant ethical principles that have not been considered so far. Such learning may result in:

- Additional information about options, outcomes, values and modifications of how the uncertainties are characterized and evaluated
- Adaptation or revision of the embedding and structuring of the decision problem as well as the framing of specific components or aspects of the decision problem (options, values, outcomes), the context, the decision-makers, or stakeholders, etc.
- Reconsideration of the arguments for and against the options for choice

By assessing and developing the arguments for and against the available policy options, temporal strategies enable to substantiate the uncertain descriptive and normative knowledge about the decision problem decision-makers are faced with. Core elements of a decision problem include the options for choice, their outcomes, and the values of these outcomes. In order to avoid that postponing and eschewing a decision problem get confused with an explicit decision in favour of the current practice, I suggest that staying with the current practice should count as an option for choice only if it is explicitly listed as such an option.

Although temporal strategies are not unusual in practice, there are only a few systematic analyses of the different strategies regarding the conditions for appropriate application (Hirsch Hadorn et al. 2015; Hansson 1996). This lack needs to be addressed since in the case of great uncertainty[2] about a decision problem, temporal strategies are not a panacea for appropriate decision-making. When taking a temporal strategy into consideration, a careful analysis of the elements of the decision problem as well as of the context of the decision problem is required in order to see whether or not under the given conditions, a certain temporal strategy

---

various conditions of uncertainty. Furthermore, I distinguish strategies from plans on an operative level. Defining "temporal strategy" as a plan to extend a decision over time excludes taking a definitive decision now from the temporal strategies. However, deciding now is a (perhaps not explicitly considered) decision about when to decide. So, I distinguish closure, i.e. taking a decision now, as the default position from alternative strategies.

[2] In this chapter, the term "great uncertainty" is used for "a situation in which other information than the probabilities needed for a well-informed decision is lacking" (Hansson and Hirsch Hadorn 2016). The term "risk" is used to characterise a decision problem, if "we know both the values and the probabilities of these outcomes" (Hansson and Hirsch Hadorn 2016).

should be followed. For instance, such an analysis should clarify whether a certain temporal strategy would allow for providing the required information about the uncertainty related to the elements of the decision problem at hand, and whether this temporal strategy is desirable in face of possible trade-offs. But temporal strategies alone neither provide information on uncertainties (except from what we can learn from just "wait and see") nor do they tell us what can be concluded from such information in order to obtain a reasonable decision.[3] So, for an effective use of opportunities opened up by a temporal strategy, additional considerations are required on feasible means that are appropriate to provide useful information for taking the decision. Finally, in order to prevent eschewing a decision problem by choosing a temporal strategy, one has to establish an appropriate governance structure for decision-making over time, which also accounts for changes in the context of decision-making.

The basic temporal strategies can be distinguished as follows. A typical default strategy is closure that consists in deciding (i) now, (ii) once definitively, (iii) on the whole problem. The extension of the decision into the future is zero, but its consequences can extend far into the future. To create opportunities for learning, evaluating and deliberating, at least one of the three aspects needs to be changed. Instead of deciding now, one could delay the decision taking. Instead of deciding definitively, one could go for a provisional decision to be reconsidered later on. Or, instead of deciding on the whole problem, one could decide stepwise on its parts. The resulting alternative general temporal strategies are called postponement, semi-closure and sequential decisions (see Table 9.1).

Temporal strategies for decision-making that are used as a means to account for uncertainty have to be distinguished from further temporal aspects of a decision problem. For example, long-term and short-term policies differ in terms of the time span in which their intended effects are expected to occur, and consequently also with regard to who will carry the burden and profit from the benefits in each case. For an example from climate policy, see Hammitt et al. (1992). Decision-makers often give more weight to expected near-term effects (time preference) or they

Table 9.1 Differences in decision-making between types of temporal decision strategies

Default strategy	Alternative strategies		
Accept decision problem	Consider revision of decision problem		
Closure	Postponement	Semi-closure	Sequential decisions
Now	Later	Now and later	Now and later
Once	Once	Recurrently	Sequentially
Whole problem	Whole problem	Whole problem	Partitions of problem

[3] I use terms like "reasonable" and "sound" to indicate that the restricted sense of "rational" in traditional decision theory does not apply to decisions under great uncertainty (Hansson and Hirsch Hadorn 2016).

value long term effects less (discounting the future) (Frederick et al. 2003). Because of such biases in the weighing of options of both kinds (Möller 2016), temporal aspects of decisions may give rise to uncertainty of values. Also, uncertainty may arise with regard to the question of how to structure the decision problem or frame the options in order not to mislead the decision-makers (Betz 2016; Grüne-Yanoff 2016). How to account for those uncertainties in decision-making might then be a question of choosing an appropriate temporal strategy of postponement, semi-closure or sequential decisions.

After a short discussion of criteria for and against the default position of closure (Sect. 2), I describe the basic temporal strategies of postponement (Sect. 3), semi-closure (Sect. 4), and sequential decisions (Sect. 5) with reference to some examples of how these are found in practice. Such applications often consist in the specification of one general strategy or a combination of different strategies. As an example, the strategy of just-in-time used in business management combines postponement with sequential decisions (see below). I point to problems that have arisen in the application of such temporal strategies and discuss criteria that have been proposed for or against their application. These criteria may be used as a heuristic for considering which temporal strategies are (in-)appropriate for a given policy decision problem (Sect. 6). To illustrate the use of these criteria, I refer to the example of nutritive options to reduce methane emissions from ruminants (Sect. 7). I conclude by summarizing the specific contribution of temporal strategies to deal with uncertainty in decision-making. Moreover, I emphasise the fact that decisions under great uncertainty force us to make a fundamental shift in conceiving the task of policy analysis (Sect. 8).

2 Closure

"Closure" means to take a definitive decision now. Several kinds of considerations are important for deciding whether closure is appropriate for a reasonable decision from multiple perspectives. First of all, do we need to learn about and evaluate uncertainties? Reasons against closure include uncertainty concerning the embedding of the given policy decision problem on the one hand, and the framing of the policy options, i.e., how the options for choice are formulated, on the other (Grüne-Yanoff 2016). Further reasons speaking against closure are disagreement about the drawbacks of the options as well as about other relevant, but incomplete or unreliable information (Hansson 2016). Closure is recommended if is not possible to learn about relevant uncertainty by extending the decision into the future. This is the case if there is a lack of money or expertise needed to learn about uncertainty, or if properties of the policy options would require a longer time span for learning. Moreover, further aspects such as the severity of the problem and its development in the future as well as the contribution of proposed options to mitigate or solve the problem need to be considered. Finally, it has to be considered whether and how the context, the mandate, as well as the commitment to implement decisions and take action of decision-makers will change over time. If the actual situation is seen as a window of opportunity, this speaks for closure.

3 Postponement

"Postponement" is a way to extend a decision into the future by not deciding now but later on. Postponing a decision about whether to continue or to stop an established activity could either suspend the established activity provisionally or let it go on until a decision is taken. Postponement is also applied in cases of deciding on which of alternative new activities to follow, or when to start with a certain activity. Delaying these decisions serves to get additional information that helps learning more about or better evaluating uncertainties before a decision is taken. There are several ways of postponing decisions. A first choice has to be made between passive and active postponement, which is a choice between just "wait and see" until more information comes in, or starting a search for additional information. A further choice is whether to take specific measures in order to assure that delaying a decision does not end up with eschewing the decision problem or running into obstacles that impede reasonable decisions. This second choice needs to be considered in both, passive and active cases of postponement.

Of course, there are also other reasons that may speak in favour of or against delaying a decision, such as determining the optimal timing of a decision from a cost-benefit perspective. The debate between Nordhaus and Stern on whether to take climate mitigation policies now or later is a well-known case. Because they used different discount rates for valuing future goods as a basis for calculating cost-effectiveness of measures, Stern arrived at the conclusion that an immediate decision would be better, while Nordhaus recommended postponing this decision. See, e.g., Broome (2008) for comments on this debate. The uncertainty of whether or not to postpone a decision from a cost-benefit perspective results from different reasonings about the appropriate discount rate and further assumptions for the calculations. Postponement was not considered as a means to better evaluate and manage these uncertainties. Here, I focus on postponement as a means to account for uncertainty in information about the decision problem.

In business and operation management, "postponement" is used for

> delaying activities in the supply chain until customer orders are received with the intention of customizing products, as opposed to performing those activities in anticipation of future orders. (van Hoek 2001:161)

This is *passive postponement* in the sense of wait and see until uncertainty – in this case about order volumes, specifications of orders and order mixes – is turned into certainty so that the decision of starting some activity can be taken under less uncertainty then. Which of the decisions along the supply chain can reasonably be delayed depends on how the supply chain is managed and what specific technologies are used at each stage. So, the feasibility of postponement for the increase of efficiency depends on the operating characteristics of the process and product design. Furthermore, implementing postponement of decisions on sourcing, fabrication, assembling, packaging or distribution of products may require that one reconsiders and adapts the configuration and management of the supply chain. Such cases may demand a change management regarding the proper reconfiguration and the management of the supply chain for effective implementation of postponement. In order to account for other related decisions in the supply chain and to coordinate with them, the respective actions should be based on considerations from an integral supply chain perspective. As a consequence, it is recommended to consider postponing a decision as a part of a sequence of decisions, see Sect. 5. For details see van Hoek (2001) who gives a review of the literature on postponement in industry, where the information above is taken from. For a survey on decision determinants of the postponement strategy in manufacturing companies in Europe, North America and Asia, see Kiperska-Moron and Swierczeck (2011).

Delaying decisions about public policy is sometimes called "moratorium". Here, too, the purpose is to get more information so that uncertainties can be better characterised and evaluated before a decision is taken. An example would be the moratorium on genetically modified plants in Switzerland since 2005. In this case, voters accepted a 5-year moratorium on the commercial use of genetically modified plants. This moratorium postpones a decision on a possible new option for Switzerland, where the commercial use of genetically modified plants is not allowed. This moratorium is a case of *active postponement*: from 2007 to 2012, the Swiss National Science Foundation funded a National Research Programme (http://www.nfp59.ch/e_index.cfm, 13.11.2014) with the aim of examining the benefits and risks of genetically modified plants under the ecological, social, economic, legal and political conditions of Switzerland. In order to make use of the outcomes of this research programme for the elaboration of a policy on the coexistence of genetically modified and traditional crops in Switzerland, the Federal Council has prolonged the moratorium twice, actually until 2017.

In this example, it seems that the results from active postponement did not enable decision-makers to take a decision by the end of the moratorium. Of course, in issues of public policy, the search for new information by active postponement such as a National Research Programme in the case of genetically modified crops in Switzerland does not turn a decision under great uncertainty into a decision under certainty. Unlike the moratorium on public policy, postponement in the supply chain relates to a type of problems that is largely uncontested and clearly determined, such as customizing products efficiently. This allows for clarifying uncertainty for closure, i.e., to take a definitive decision. Contrary to this, a moratorium on public policy has to deal with a broader range of uncertainties about the problems at stake. In many cases, important uncertainties pertaining to the situation of decision-makers, such as uncertainty of embedding, of consequences, of reliance, or of values (Hansson and Hirsch Hadorn 2016) cannot be clarified in the search for information. Therefore, it seems important to consider whether it is feasible to get the information required for decision-making within the time-span of the moratorium before a decision whether to implement a moratorium is taken. In the case of genetically modified crops in Switzerland, one could have considered whether it is feasible to get the relevant information on ecological consequences from the coexistence of genetically modified and traditional crops within the time span of the moratorium. Or, one could have taken into account to what extent controversy about the embedding and structuring of the decision problem has to be clarified in order to avoid controversy about the reliability of expert knowledge.

If, however, a moratorium on public policy is well targeted to what is relevant for decision-making, it can result in closure after the end of its time despite of enduring further uncertainties. An example would be the moratorium on the runtime of nuclear power plants in Germany (Tigges 2011), which was decided by the German government in the wake of the reactor accidents at Fukushima in 2011 and the subsequent nuclear disaster. In this case, the moratorium was used to suspend an already made decision to prolongate the runtime of the existing nuclear power plants in Germany. The same government, which had taken a definitive decision for prolongation a year before, reopened the decision problem. It was argued that – as a consequence of the information about the reactor accident and subsequent nuclear disaster at Fukushima – the decision on prolongation needed to be reconsidered, based on information from security tests of power plants in Germany. So, a moratorium was put into place. A new decision was taken 3 months later, namely to keep 8 of 17 nuclear power plants in Germany closed, and to install a revised law as the legal basis for a nuclear power phase-out.[4]

Several factors allowed for taking a definitive decision despite of a range of remaining uncertainties. Here, I am not concerned with the question of whether those factors have been explicitly considered by decision-makers, but restrict my comments to their importance for taking a decision at the end of the moratorium. Basically, activities have been focussed on those uncertainties, which needed to be clarified for the decision (Hansson 2016), and which could be clarified to a sufficient degree within the time span. There were no indications for immediate major nuclear risks that would have required a short moratorium. However, a short moratorium is favourable to minimize uncertain planning for business that is related to transitions in energy supply. In addition, a short moratorium in the wake of the nuclear disaster at Fukushima accounted for the political context of decision-making as a window of opportunity. Furthermore, the constellation of the political parties involved favoured the commitment to take a decision. The left-wing party, traditionally critical of nuclear energy, was not expected to oppose the proposal initiated by the right wing government, which was in a position to achieve consent from the right wing party. Appropriate governance of decision-making across the moratorium was accounted for by establishing an ethics commission to work on a political consensus among the parties and organisations involved. With the moratorium, the decision problem on nuclear power plants was

[4] I am grateful to Elmar Grosse Ruse for a helpful discussion of this example.

reframed from a decision on remaining life to a decision about passing security tests. Uncertainties about technical factors that needed to be considered not only included the performance of power plants according to security standards, but also the grid capacity and the security of supply. In addition, legal uncertainty was addressed by elaborating a draft for the regulation of energy supply in future, if nuclear power phase-out would result from the decision process. While the use of established procedures and standards allowed for assessing the safety of nuclear power plants in time, providing the draft for the new law within 3 months was a very challenging task.

Postponement as a temporal strategy allows for learning and improving incomplete or unreliable information. It is suitable if reconsideration of the decision problem and reframing of options should be sought (Grüne-Yanoff 2016). The contribution of passive postponement would be restricted to learning from how the problem develops given the current practice. Active postponement can provide more information than that, but it cannot reduce inherent uncertainties. Furthermore it is not reasonable if the search for new or improved options is costly, takes too much time, or requires expertise that is not available. Also, the severity of the problem and its development in the future, as well as the contribution of proposed options to mitigate or solve the problem, need to be considered in the case of postponement. Finally, it has to be considered whether and how the context and mandate of decision-makers will change as well as their commitment for implementing decisions and taking action. If the actual situation is seen as a window of opportunity, this speaks against long-term postponement. If there is no commitment for taking a decision, passive postponement may run into avoidance of the decision problem by simply continuing with the current practice.

4 Semi-closure

Semi-closure is a way to extend a decision into the future by taking a provisory decision for implementation on a certain option to be reconsidered later on. Terms such as “recurrently”, “recursively”, “iteratively”, or “repeatedly” are used to describe decision-making under a strategy of semi-closure. In this context, all these terms are used to simply indicate that decisions on a certain problem are taken repeatedly. They do not indicate that the same considerations are always applied. To take a provisory decision means that the decision could be corrected based on the outcomes observed, or on the values then attributed to the outcomes, or because of new policy options or changes in the embedding or structuring of the decision problem, etc. So, semi-closure allows for iteratively adapting a measure or even change a measure because of experiences made with its implementation in a given context. Recurrent decisions on the same problem are sometimes called “sequential decisions” (e.g. Parson and Karwat 2011; Gregory et al. 2006:2421). I use “sequential decisions” for successively deciding on different parts of a decision problem, not for reconsidering the same decision several times. However, in a series of decisions, decisions on parts may not be taken in accordance with the original plan, but be adapted to the actual course of events. Therefore, sequential decisions and semi-closure are typically applied in combination, see also Sect. 5. Semi-closure could be used, for instance, as an alternative strategy to postponement, or as a follow-up strategy to postponement, if the information gathered through postponement does not allow for closure. Semi-closure could be used as a permanent strategy, if inherent variability in problems does not allow for a definitive decision on policy options as in adaptive management of natural resources and ecological systems (e.g. Gregory et al. 2006) or in adaptive governance of social-ecological systems (e.g. Folke et al. 2005) and adaptive policymaking more generally (e.g. Van der Pas et al. 2013; Swanson et al. 2010). The broad range of adaptive approaches can be distinguished with regards to whether

- A single option is searched or several options in comparison
- A trial and error procedure is used or a systematic design
- Qualitative methods for data sampling and analysis (e.g. decision seminars) are used or formal ones, (e.g. computer simulations)
- Governance of the policy process is part of the approach or not

In describing some adaptive approaches and problems with application, I will draw on these distinctions.

The inception of *adaptive management* of natural resources and ecological systems, also called "adaptive environmental management", is attributed to Holling (1978) and Walters (1986). The purpose of adaptive management has been to consider the implications of uncertainty about ecological systems for appropriate management options. "Adaptive" refers to (i) the goal of management policies, which is to enhance the capacity of ecological systems to cope with various kinds of impacts called their "adaptive capacity", as well as to (ii) the modification (adaptation) of management policies to meet this goal (see e.g. Pahl-Wostl 2007:52). To use semi-closure in order to account for uncertainties about ecological systems in management, adaptive management is conceived as a cycle of different steps, which includes (re-)designing, deciding, implementing, monitoring and evaluating management policy.

At first glance, the simple core idea of learning by doing for effective environmental management seems appealing when it comes to uncertainty about ecological systems. However, its application to problems of environmental management is not without difficulties. It is broadly conceded in this field that careful assessment is required of the decision problem with regards to whether and how adaptive management could provide information that is useful for decision-making. Otherwise, instead of supporting reasonable decisions, adaptive management would result in unwanted effects, such as that decision-makers eschew the decision problem or that the problem gets worse, for instance, if tipping points for adaptation are ignored (Doorn 2016). Gregory et al. (2006) have analysed some problems that may come along with adaptive management. They have identified

> four topic areas that should be used to establish sensible criteria regarding its appropriateness for the application of AM [adaptive management] techniques. These include (1) the spatial and temporal scale of the problem, (2) the relevant dimensions of uncertainty, (3) the associated suite of costs, benefits, and risks, and (4) the degree to which there is stakeholder and institutional support. (Gregory et al. 2006:2414)

Adaptive management is a strategy that uses semi-closure in order to better characterise uncertainty about how management options affect ecosystems and their development and modify policy accordingly. To this end, one has to consider whether temporal and spatial scales allow for monitoring the effects in question, and whether the design of experiential or experimental intervention and the uncertainties of knowledge about the system allow for attribution of effects to interventions or external changes, among them also surprises. Topic areas (1) and (2) in Gregory et al. (2006) point at these sorts of considerations. In order to improve the information about uncertainties and their consequences, trial and error approaches based on changes ad hoc have been complemented by more systematic approaches, including formal methods such as modelling and simulation (e.g. Schreiber et al. 2004). These formal methods are used as a basis for attributing possible or observed events in passive as well as in active adaptive management. In passive adaptive management, existing information is used to design and implement a management option in order to watch its outcomes for appropriate adaptation. In active (also called experimental) adaptive management, multiple management options are modelled and simulated or implemented in order to compare their effectiveness. However, whether changes of measured indicator values can be attributed to certain causes may remain uncertain in passive as well as active adaptive management. One of the reasons for persistence of uncertainty may be inertia of variables (Parson and Karwat 2011), so that effects are not observable within the given time-span.

Adaptive management has been criticised for its narrow focus on uncertainty about ecological systems and for its narrow goal of enhancing adaptive capacities of ecosystems in the spirit of ecological restoration. To indicate dissent with restoration as the goal, the term "real world experiments" is being used for recursive implementation and modification of measures in ecological management in the spirit of environmental innovation to assure sustainable use for human well-being (Gross and Hoffmann-Riem 2005). Environmental innovation basically questions the conception of the decision problem as an issue of environmental conservation. Others do not criticise this goal, but question that uncertainties about social aspects in a broad sense such as goals, values, perceptions of various stakeholders and practitioners are taken simply as stumbling blocks in environmental management (see e.g. Pahl-Wostl 2007; Gregory et al. 2006; Schreiber et al. 2004). A strategy of semi-closure could, of course, include a systematic approach to learning about uncertainty of social aspects as well. However, the fact that a decision problem is open for reconsideration and change of policy later on also gives rise to uncertainty about further social aspects. For instance, the fact that decisions are taken but not set in stone could weaken or strengthen practitioners' commitment to implementing policy (Edvardsson Björnberg 2016; Doorn 2016). Furthermore, with deciding recurrently, a range of relevant aspects such as the context of the decision problem, the public perceptions of the decision problem or the mandate of decision-makers for future decisions is open to change. This is a source of uncertainty about the governance of the decision process.

Uncertainties about social aspects are explicitly considered in various more comprehensive but quite different conceptions of *adaptive governance*. The term "governance" from political science indicates that actors from public administration, the private sector and the civil society are involved in the design, decision, implementation and evaluation of policy, which could combine a variety of different specific means (see Doorn 2016, also for an example). *Adaptive governance of social-ecological systems* (e.g. Folke et al. 2005) has a broader goal, namely enhancing the adaptive capacity of integrated social-ecological systems. The basic idea is to extend the systems perspective to social aspects of decisions such as the diversity of actors and their networks in order to integrate these as elements of an integrated systems approach. The *institutional approach to adaptive governance* builds on a theory of social institutions as an approach to the governance of the commons such as natural resources:

> We refer to adaptive governance rather than adaptive management because the idea of governance conveys the difficulty of control, the need to proceed in the face of substantial uncertainty, and the importance of dealing with diversity and reconciling conflict among people and groups who differ in values, interests, perspectives, power, and the kinds of information they bring to situations. (Dietz et al. 2003:1911)

The institutional approach uses formal methods to compare possible policies and deals with problems from the local to the global scale. *Policy sciences' conception of adaptive governance* shares with the institutional approach the eminent role of participatory governance for advancing the common interest, but differs from it in other regards. Adaptive governance is proposed

> as a reform strategy, one that builds on experience in a variety of emergent responses to the growing failures of scientific management, the established pattern of governance. (Brunner 2010:301)

The pillars of this reform strategy are (i) to split global problems and downscale them into local ones, (ii) to address policy issues in community based participatory approaches, and (iii) to use interpretative methods to understand local experiences on the ground with policy and adapt policy accordingly. The application of this approach has been extended from ecological and climate change issues to issues of public policy of great uncertainty in a broad range of fields.

This broad range of policy issues and a strong focus on the policy process are shared by *adaptive policymaking* (e.g. Van der Pas et al. 2013). However, the purpose of adaptive policymaking is to gather information about the behaviour of systems in the long-term future, about possible unintended consequences of policy interventions, and about ways of preventing those or modifying the policy. So, the basic idea here is to design adaptable policies together with how to respond to signals from the monitoring of consequences, once policy will have been implemented. Adaptive policymaking could be elaborated by using formal tools such as modelling and simulation together with the help of participatory workshops with decision-makers, practitioners and stakeholders, by using, e.g., decision seminars. Adaptive policymaking is taken to be robust in the sense of being capable to deal with surprises, and it is taken to be dynamic in the sense of being adaptable to changing policy contexts:

> No longer are ex-ante evaluation tools used only to select the optimal or most robust (static) policy option; the tools are now also used to probe for weaknesses in an initial basic policy, and to understand how the system might react to external developments (e.g. in order to search for vulnerabilities and opportunities). This use of futures research allows policy analysts to develop meaningful actions to take to avoid a policy failing due to future external changes. Thus, policymakers can be prepared for the future and will decide in advance when and how to adapt their policy. (Van der Pas et al. 2013:15)

All these adaptive approaches under a strategy of semi-closure indicate a shift away from a "predict-then act" culture in taking policy decisions to a culture of "decisions and revisions", borrowing the term from Levi (1984). A strategy of semi-closure may be appropriate if on the one hand, the severity of the problem or its future development calls for action, while on the other, the problem requires us to learn more about and to evaluate uncertainties of information on the decision problem in order to take a definitive reasonable decision. Or, as a permanent strategy, semi-closure is recommended for problems that cannot be definitely solved, such as problems with relevant inherent variability. However, the various approaches under a strategy of semi-closure that have been described above make it clear that there are several restrictions for reasonable application as a temporal strategy for dealing with uncertainty. Firstly, semi-closure presupposes that options are reversible to a certain extent in order to account for experiences with the provisionally implemented (or simulated) options regarding the outcomes and values. Secondly, reasonable application is restricted to those uncertainties that can be clarified within the given time-span of semi-closure, considering also costs and expertise that would be required. Thirdly, semi-closure also creates the possibility of new uncertainties, for instance if the decision-makers, the goals, or the political agenda will change (Edvardsson Björnberg 2016). Fourthly, proper governance structures are required to assure that semi-closure is not misused for abandoning the decision problem.

Up to now, elaborated systematic approaches in adaptive management, governance or policymaking are rarely implemented (see e.g. Van der Pas et al. 2013; Gregory et al. 2006; Schreiber et al. 2004). This fact indicates that restrictions of time-span, costs and expertise are crucial. Furthermore, it seems that examples of adaptive governance in practice often come along with partitioning a big global problem into a range of smaller regional or local problems, which then are treated on the basis of practical expertise. One may appreciate downscaling of problems as a means to achieve democracy and account for diversity of contexts by a diversity of contextualised policies. However, considerations are also needed about whether, and if so, how dependence of long-term development and global interactions of natural and social processes and their regulations can be accounted for in regional approaches. For policy problems on all scales, a comparative design to learn about different options is important for various reasons. For instance, an evaluation that compares different options or different contexts may help to clarify the causes of the events, produced or simulated with a strategy of semi-closure. Or, if the purpose of semi-closure includes a possible reconsideration of how the policy problem is demarcated, different options and related values and outcomes need to be explored. So, semi-closure could be used to turn unknown unknowns about a decision problem into recognised uncertainty. Uncertainty of events, but also of values related to outcomes, of options to be considered or excluded, and of goals to be pursued, may come up. While these issues may also arise if a policy is implemented after closure, a working governance structure as part of an adaptive approach is an important advantage if upcoming uncertainties call for extending a decision into the future. An institutional framework enables actors from the public and the private sector as well as the civil society to argue about how to react to these uncertainties. Argumentation will be needed for determining relevant uncertainties (Hansson 2016) and respective consequences for the (re-)design of policies and goals (Edvardsson Björnberg 2016), as well as requirements for decisions, implementations and monitoring. So, in order to account for uncertainties of information in a broad sense by using strategies of semi-closure, the design of policies that can be modified and the implementation of a governance framework for the policy process are both crucial requirements.

5 Sequential Decisions

A third way to extend decision-making into the future is to partition a complex decision into a series of decisions on its respective parts so that they can be taken successively. In decision theory, this is called "dynamic choice" (e.g. Andreou 2012; McClennen 1990) or "linked decisions" (Hammond et al. 1999), while "sequential decisions" is more familiar in the field of policy analysis (e.g. Parson and Karwat 2011; Webster et al. 2008). I use "sequential decisions" as an umbrella term for the various approaches to taking decisions in sequences in order to learn about, evaluate and account for uncertainty of information about the decision problem, but I also use "dynamic choice" and "linked decision" when I follow the wording of an author.

In decision theory, the analysis of sequential decisions focuses on how the rationality of a series of decisions over time may be challenged, so that these decisions do not serve their goal well enough (Edvardsson Björnberg 2016). Rationality may be challenged, for instance, because of incommensurable alternatives, because the actual outcomes are preferred to the ones expected in the future while discounting the value of future outcomes, because of intransitive preferences, or because of vague overall goals, for an overview, see Andreou (2012). In such cases, considering each one of the decisions on its own and taking the best option, independently of the ones taken before and the ones to be taken later on, may lead to a final result that is worse than what would have been achievable. Therefore, the

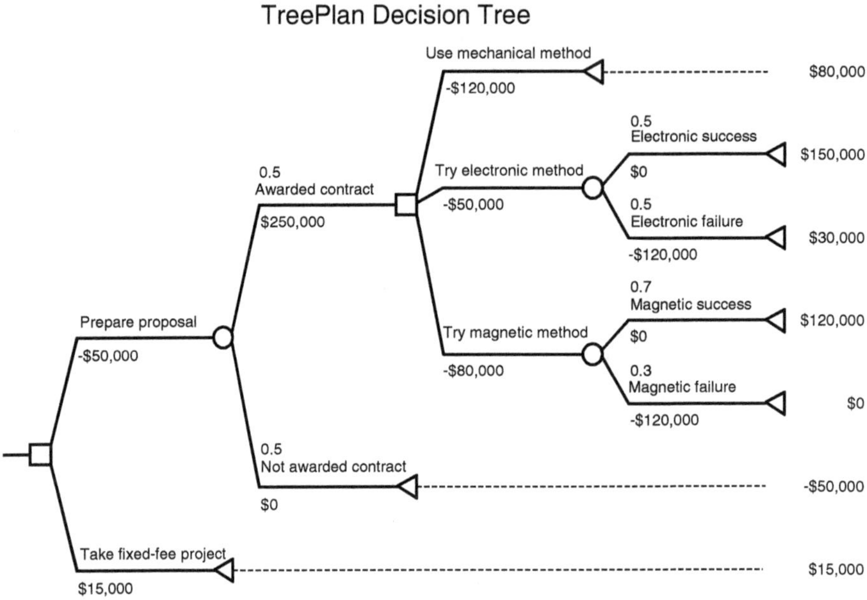

Fig. 9.1 Example of a two steps (□) decision tree with probabilities (○) and outcomes for each decision path (◁) (Source: http://www.treeplan.com/images/treeplan-decision-tree-diagram.gif; accessed 02.01.2015)

partitioning of a complex decision problem should be based on a structure or plan of how the parts relate to each other. A typical approach to partitioning distinguishes a set of subdecisions on alternative options by forming a series of steps to reach the goal. A so-called "decision tree" can be used to illustrate the structure of a complex decision into a sequence of decision points between possible or probable alternatives and the outcomes of each path of the decision tree, see Fig. 9.1.

Treating decisions at each of the steps separated from each other is called "myopic choice". In sophisticated choice, the remaining future plan is reconsidered after having reached the next decision step. In resolute choices, decision-makers are committed to decide in accordance with the plan adopted at the very beginning (see e.g. McClennen 1990). However, if there is no flexibility for changing the original plan as in resolute choice, sequential decisions cannot be used to learn about, evaluate and account for great uncertainty accordingly (Edvardsson Björnberg 2016). The flexibility in deciding on parts that is required for this purpose may include a delay of a certain decision in the series of decisions to be taken or a change of some of its components such as new options or a different evaluation of expected outcomes. So, as a means to account for uncertainty, sequential decisions include postponement or semi-closure, which are considered here as parts of a more complex temporal decision strategy.

The concept of "real options" in investment under uncertainty is an example of combining sequential decisions with semi-closure and postponement as aspects of a sophisticated choice. Capital budgeting for a project typically formulates a plan of investment activities. So, to realise the project, a sequence of decisions on investment actions has to be taken. Real options are options to alter the operating strategy for capital budgeting regarding future actions in order to respond to the actual development of things. Altering the operating strategy may include postponement or adaptation of future decisions on actions foreseen in the plan of investments:

> As new information arrives and uncertainty about market conditions and future cash flows is gradually resolved, management may have valuable flexibility to alter its operating strategy in order to capitalize on favorable future opportunities or mitigate losses. For example, management may be able to defer, expand, contract, abandon, or otherwise alter a project at different stages during its useful operating life. (Trigeorgis 2001:103)

Van Reedt Dortland et al. (2014) discuss the application of real options in combination with scenario planning as a means to flexible management decisions in the design of new healthcare facilities. Among the various uncertainties speaking for flexible decisions are policy and demographic change. They found that

> the real options concept appeared to be too complex to be immediately adopted, although it was recognized as a useful tool in negotiating with contractors over flexibility. (Van Reedt Dortland et al. 2014:27)

They highlight that reasoning about real options to understand possible consequences of future decisions requires respective cognitive capacities, and it may challenge the mindsets of people in organisations. Both factors, if not properly addressed, may work against a successful application of real options.

A second way to partition complex decisions is proposed by Hammond et al. (1999) in their practical guide to smart linked decisions. They use the term "linked decisions" to highlight that what is decided now will substantially affect future decision problems. Therefore, they stress the importance of learning about, evaluate and accounting for uncertainty in planning ahead, be this in personal life, business or public policy. Hammond et al. (1999) distinguish between (i) a decision on the basic decision problem – i.e. the proper embedding and specification of the decision problem -, (ii) an information decision about what one needs to know before taking the basic decision, as well as (iii) considering also future decisions that will be necessarily linked with the basic decision before taking the basic decision. More specifically, they propose the following six steps:

1. Understand the basic decision problem, its embedding and structure, including options and outcomes for whom and when as well as respective values.
2. Identify ways to reduce critical uncertainties related to the decision problem.
3. Identify future decisions linked to the basic decision to be considered in planning ahead.
4. Understand relationships in linked decisions for planning ahead.
5. Decide what to do in the basic decision, which means to work backward in time and consider what speaks for and against each option, based on the embedding and structuring of the decision problem and the information about the decision problem.

6. Treat later decisions as new (basic) decision problems, i.e. understand planning ahead in steps 3 and 4 as a strategy under semi-closure. (see Hammond et al. 1999:168–172)

Basically, the heuristic for linked decisions stresses learning and understanding before deciding in steps 1–4 as well as after step 5 before deciding in step 6. Learning and understanding before step 6 essentially means repeating steps 1–4, which at this stage, is used to prepare the next decision to be taken. Understanding the next decision to be taken as a new decision may also include that goals have to be reconsidered. Hammond et al. (1999) argue that in the case of great uncertainty, flexible plans are needed in order to make action possible that avoids possible or unforeseen negative events. Flexible plans such as all-weather plans, short-cycle plans, option wideners, or "be prepared" plans keep options open (Hammond et al. 1999:173–174). However, treating future decisions as new basic decisions may cause serious problems for socially coordinated activities. Decision-makers that give up their goals too easily appear as unreliable partners, namely when conclusive reasons to do so are missing. In such cases, decisions lack consistency and coherence, which might also be a problem for the (individual or collective) decision-maker him- or herself (Edvardsson Björnberg 2016; Bratman 2012). These are reasons for considering also past decisions in planning ahead.

A third way to partition a decision problem is to separate uncontested from contested parts of a complex decision problem in order to decide now on an uncontested subset while sorting out the unresolved parts later on. However, an agreement to decide sequentially on these parts may be difficult to reach. For instance, while it is uncontested that adaptation measures to protect from climate change impacts are needed, deciding on adaptations measures now while deciding on mitigation measures later on is contested. In this case, deciding sequentially could misdirect future decisions on mitigation measures, since it is unclear to which extent adaptation measures could substitute mitigation measures and vice versa, or how much of available resources should be devoted to each kind of climate policy (Tol 2005). For a more general discussion of empirical findings about partition dependence such as how allocating resources varies with a particular partitioning of a complex decision, see Fox et al. (2005). Therefore, the dependence of future decisions on decisions taken now has to be taken seriously in partitioning between clear and unclear options in order to prevent that decisions on unclear options are not misdirected.

Approaching a goal stepwise by determining interim targets is a fourth way to partition a decision. In the case of utopian goals of long-term character such as sustainable development, determining interim targets, which are measurable in order to monitor the impact of the measures that have been taken, can be used as a means to learn about uncertainties of outcomes and to revise the respective measures (Edvardsson Björnberg 2008; Edvardsson 2004). The goal of sustainable development gives rise to value uncertainty in the sense that it comprises multiple and incommensurable ecological, economic and social subgoals that do not allow for aggregation (Brun and Hirsch Hadorn 2008). Trading for instance performance on ecological indicators for performance on social indicators would be questionable, at least to the extent that thresholds have to be met. So, it is uncertain how alternative policies for sustainable development would compare all subgoals considered. In such cases proceeding sequentially makes it possible to meet thresholds for indicators sequentially. Proceeding sequentially in such cases requires the structuring and monitoring of interim targets for performance on each of the indicators. It is also necessary to consider the whole decision paths and their overall outcomes in order to prevent irrational decisions (Allenspach 2013).

There are further purposes to partition a complex decision in order to learn about, evaluate and account for uncertainty, besides doing so for a temporal strategy. For instance, to partition a global problem into local problems it is necessary to consider the distribution and decentralisation of decision-making and governance. This, for instance, has been proposed as an alternative to the Kyoto Protocol, which was established in 1992 as the global institution for global governance of climate change and policy (Hulme 2009). Partitioning global problems into local ones has been proposed by policy sciences as a general strategy to deal with wicked problems in public policy in order to distribute and decentralise decision-making and governance, see Sect. 4. However, it is unclear how this strategy manages to deal with global interconnections of problems.

To use sequential decisions as a means to learn about, evaluate and account for uncertainty by deciding stepwise, it is required to consider those steps as a series of decisions in combination, i.e. a plan needs to be established for how these steps would contribute to achieve the overall goal of the complex decision problem (see Elliott 2016 for an example). However, learning about, evaluating and accounting for uncertainty requires flexibility to change the original plan based on experience with the steps that have already been taken. Flexibility in deciding on future steps may include a delay of a certain decision in the series of decisions to be taken or a modification of some of its components such as new options or a different evaluation of expected outcomes. So, as a means to account for uncertainty, sequential decisions include postponement or semi-closure on its parts. In such cases, criteria for or against postponement and semi-closure also need to be considered for the respective steps in sequential decisions. These criteria comprise uncertainties related to the information about the decision problem, various aspects related to the options at hand, characteristics of the problem and how it might develop, as well as the context of decision-making and the governance structure. Specific criteria for sequential decisions relate to the partitioning of the complex decision problem in order to avoid biased partition dependence of later steps on earlier ones. Decisions on later steps may be misdirected, for instance, by how the allocation of resources varies with a particular partitioning of a complex decision, by excluding relevant alternative options, or by abandoning the (revised) plan.

6 A Heuristic Method for Deciding on Temporal Strategies

Extending a decision on a policy problem into the future is a means to enable to learn about, to evaluate and to deal with for great uncertainty. Choosing deliberately among alternative temporal strategies for taking policy decisions is not a substantive decision on the alternative options of a given policy decision problem, but a decision about certain procedural aspects of decision-making, namely about those that are related to time. So temporal strategies need to be complemented by further methods for learning about, evaluating and accounting for uncertainty, e.g. methods for assessing arguments (Brun and Betz 2016), considering framings (Grüne-Yanoff 2016), revising goals (Edvardsson Björnberg 2016), evaluating the uncertainties (Hansson 2016), making uncertainties of values explicit (Möller 2016), or accounting for possibilities in practical argumentation (Betz 2016).

A decision on which temporal strategies are (in-)appropriate for a given policy decision problem under great uncertainty needs careful consideration. Such a decision should be based on various criteria that speak for or against closure, postponement, semi-closure or sequential decisions as discussed in Sects. 2, 3, 4, and 5. Here, I summarise the broad range of considerations that may be relevant as criteria and suggest a way to classify them into four broad groups: first, the *relevance* of considering uncertainties for taking a decision; second, the *feasability* of improving information on or evaluating these uncertainties; third, the *acceptability of trade-offs* related to the temporal strategy, and fourth, the *maintenance of governing* decision-making over time (see Table 9.2).

Firstly, if improving information on relevant uncertainties is needed for a better decision, this speaks against closure and for some temporal strategy. These uncertainties may relate to components of the decision problem, i.e. options, values, outcomes that are relevant for the decision to be taken. They may, e.g., arise from lack of knowledge (Hansson 2016), from how these components are framed and perhaps partitioned (Grüne-Yanoff 2016), or from uncertainties about which values to apply to the problem (Möller 2016). In addition, they may also arise from a contested embedding of the decision problem, see the example below.

Secondly, whether it is feasible to learn about or evaluate uncertainties for a better uncertainty management in policy decisions depends to a large extent on aspects related to options and values. Basically, we have to consider whether improving information on uncertainty is feasible (i) within a reasonable timespan, (ii) in view of the state of information and know-how on the problem, (iii) in view

Table 9.2 A heuristics of four guiding questions to cluster criteria for and against the application of a temporal strategy to a decision problem

Criteria	Guiding questions
Relevance	Which uncertainties need further information or evaluation for taking a decision?
Feasibility	Is improving information feasable within the temporal strategy?
Trade-offs	How serious are trade-offs from (not) following the temporal strategy?
Governance	Is appropriate governance of decision-making across time assured?

of conflicting goals, values and norms held in civil society, public bodies and the private sector (Edvardsson Björnberg 2016; Möller 2016), (iv) in view of the costs that would arise from the temporal strategy as compared to closure, and, finally, (v) in view of the possibility of change, e.g. whether options are reversible in case of semi-closure, or, whether misleading dependencies are imposed with partitioning a complex problem in a case of sequential decisions.

Thirdly, regarding the trade-offs that may speak against a temporal strategy, the characteristics of the problem such as how serious it is and whether it will aggravate quickly or slowly in the near future are important for deciding for or against a temporal strategy. Also, whether the contribution of the options at hand to mitigate or solve the problem is expected to be substantial or marginal could make a difference in considering a temporal strategy. Furthermore, possible drawbacks of the problem at hand, further connected problems that would arise from deciding later on, or reconsidering a provisional decision on the options have to be acknowledged.

Fourthly, establishing appropriate measures or institutions to govern the decision process over time seems to be crucial for effective postponement, semi-closure and sequential decisions, see Sects. 3, 4, and 5. However, governance of the decision process should be concerned not only with the commitment of the decision-makers and the organisation of the decision process across time, but also with the broader context in civil society, public bodies and the private sector (Doorn 2016). So, possible future changes of institutions, context and mandate of decision-makers as well as of commitments for implementation of decisions need to be taken into account in order to not miss a window of opportunity for taking a decision.

The four groups of general criteria systematise reasons that may speak for or against temporal strategies. This structuring of criteria is useful as a heuristic that provides guidance for what to consider for deciding on a temporal strategy for decision-making. Considering these criteria may prevent us from inappropriately reducing what is accounted for in the decision. While these criteria primarily work against biases by accounting for the range of relevant considerations, they rarely also work for determining the decision (Betz 2016; Möller 2016). One reason is that criteria are ambiguous and vague. So, they need to be specified for application. In addition, they have to be weighted in relation to the decision problem at hand, since, taken together, they rarely speak unanimously for a certain and against another temporal strategy. Also, because of plural perspectives on a decision problem, there are plural ways to specify and weight criteria with regards to the problem. This does not exclude that some sufficiently specifiable criteria can be turned into an algorithm. However, whether these specifications and weightings are appropriate for the case in question needs to be checked. Furthermore, arguments based on these criteria for and against a temporal strategy are typically non-deductive arguments that support their conclusions conditionally on incomplete information. Therefore, the main value of these criteria is to provide guidance for deliberating on how to proceed with the policy decision problem at hand. To illustrate the use of these criteria as a heuristic for considering postponement, semi-closure and sequential decisions for a given policy decision problem of great uncertainty, I refer to the example of technological options to feed ruminants, which have been proposed as a means to reduce methane (CH_4) emissions in Europe.[5]

7 An Example: Reducing Methane Emissions from Ruminants

Methane is the second most important greenhouse gas (GHG) after CO_2 in terms of radiative forcing (Forster et al. 2007), and at 14.3 % also the second largest source of global anthropogenic GHG emissions. Ruminants account for about 28 % of all anthropogenic CH_4 emissions (Beauchemin et al. 2008). These emissions are caused by digestion processes in ruminants. To mitigate CH_4 emissions from digestion processes in the rumimant, technological options to feed these animals have been developed (UNFCCC 2008; Smith et al. 2007). Within the agricultural system in Europe, these technologies seem to be the only means to mitigate CH_4 emissions from ruminants in Europe without decreasing the production level. These nutritive technologies include two options for diet composition (concentrate rich diets/low roughage diet; increase in dietary fat/lipid), one option for feed plants (legumes), one option for feed quality (improve forage quality: low fiber/high sugar), and two options for extract supplementation (tannins/saponins). Possible outcomes of their application considered by UNFCCC (2008) include the mitigation potential of the respective nutritive option, economic effects such as production level, cost for diets, etc., environmental effects focusing on GHGs which cannot be mitigated, as well as effects on animal health and welfare, such as toxicity. However, there is a lot of uncertainty related to this information, some examples are given in Table 9.3.

Referring to the various exemplary uncertainties mentioned in Table 9.3, closure, i.e. taking a definite decision on the proposed options, is not an appropriate strategy in the case of nutritive options for reducing CH_4 emissions from ruminants. For instance, the nutritive technologies described above promote morally problematic ways of treating animals (Singer and Mason 2006), and they entail a morally questionable trade-off between using crops for the nutrition of animals or of humans, because increasing the level of food consumption is the major driver of increase of water consumption (Steinfeld et al. 2006; Oenema et al. 2005). Since these issues are not considered in the analysis of the nutritive options, the embedding and structuring of the decision problem has to be reconsidered. Because of ethical considerations, further kinds of options such as changes in lifestyle and consumer behaviour should be included.

[5] This example summarises joint interdisciplinary work with Georg Brun (philosophy), Carla Soliva (agricultural sciences), Andrea Stenke (climate science), and Thomas Peter (climate science) on methane emissions, which is published in Hirsch Hadorn et al. (2015).

Table 9.3 Examples of uncertainties in making decisions on how to control GHG emissions from European animal livestock by nutritive technologies (Reprinted with permission from Hirsch Hadorn et al. 2015:115)

Location of uncertainty	Source of uncertainty in CH_4 abatement strategies		
	Incomplete information	Inherent indeterminacy	Unreliable information
Options	Unfinished list of /unclear options: *e.g. if farmers compensated CH_4 emissions by reductions/sinks of other GHGs, so that pressure on CH_4 emission reduction lessens*	Unfinishable list of options: *e.g. unpredictable innovations in animal husbandry or feeding*	Contested framing of decision problem: *e.g. disagreement on the necessity of including life-style changes* (e.g. *less meat*)
Outcomes	Subdivided into statistical uncertainty, scenario uncertainty, ignorance: *e.g. concerning future realizations of abating CH_4 including effects on natural and social systems*	Subdivided into statistical uncertainty, scenario uncertainty, ignorance: *e.g. concerning the prediction of effectiveness of CH_4 abatement in a chaotic system such as the Earth's climate system*	Questionable information base: *e.g. concerning dangers related to mitigation measures addressing CH_4 emissions from ruminants*
Values	Pragmatic incompleteness of rankings: *e.g. concerning present appreciation or ignorance of animal welfare and human health*	Fundamental incompleteness of rankings: *e.g. fundamental lack of appreciation of effects of changed animal feeding practices in different and varying climatic and societal conditions*	Completed rankings despite fuzzy or ambiguous values: *e.g. when experts disagree on reliability of valuation methods of animal husbandry and feeding practices*

Sequential decisions can account for additional options that are still unclear if it is appropriate to partition the options into two subsets, one which can be decided on now, and another to be decided on later. However, understanding the nutritive options as a subset of options which can be decided on now would require firstly that uncertainties of outcomes and related values allow for closure of the subset, which is not the case, see Table 9.3. Secondly, it has to be taken seriously that future decisions on changes in lifestyle and consumer behaviour may be misdirected because they depend on decisions about nutritive technologies taken now. Although both sets of options share the goal to mitigate CH_4 emissions from ruminants, they don't agree both with another goal, namely whether there should be a decrease of the production level or not.

Semi-closure, i.e., a provisory implementation of nutritive technologies, enables learning about or evaluating uncertainties of outcomes and related values. Semi-closure would be feasible, since implementation of nutritive technologies is in principle reversible, and these technologies could be improved, based on experience. There are, however, further properties of these options that need consideration. For a clear case of semi-closure, one should know how nutritive options compare to other kinds of options that mitigate CH_4 emissions: are there better, not necessarily also reversible, ones? Information regarding comparative performance on relevant criteria is missing since there is a lack of search on other kinds of options. As in the case of sequential decisions, it must be taken seriously whether semi-closure on nutritive technologies could lead to eschewing the search for other kinds of options.

Active postponement is a commitment to actively improve information on uncertainties. This includes the embedding of the decision problem to improve as well as complement options accordingly. Since the embedding of the decision problem seems to be a crucial issue with nutritive technologies to mitigate CH_4 emissions from ruminants, this speaks for active postponement. But other points, such as the costly search for new options that have not been undertaken, the severity of the problem, and problem escalation speak against this temporal strategy. However, the role of CH_4 emissions from ruminants in abating climate change as a severe escalating global problem speaks against passive postponement of decision-making. So, whether to go for active postponement or semi-closure as the temporal strategy to decide on nutritive technologies depends on how uncertainties related to nutritive options, their drawbacks as well as the risk of misleading decisions, are judged.

This example shows the value of considering an appropriate temporal strategy for decision-making by using the above-mentioned criteria. Basically, these criteria provide guidance for judging whether a temporal strategy would be conducive for learning about or evaluating uncertainties as required for reasonable decisions from a plurality of perspectives. However, application of the criteria requires clarification whether learning about and evaluating may be restricted to reclassifying decision-relevant uncertainties and acquiring additional information, or, whether reframing elements of the decision problem or even rethinking the embedding of the decision problem is needed. So, considering temporal strategies is a means to identify relevant uncertainties as well as missing or biased information about the decision problem, which is important for a transparent and reliable decision procedure – instead of "muddling through" or abandoning decisions altogether. In cases like the one of mitigating CH_4 emissions from ruminants in which it is contested or unclear what the options and their possible outcomes are, as well as which trade-offs are permissible, temporal strategies are one of the means to guide deliberation in participatory policy processes to achieve a reasonable decision from a plurality of perspectives. However, going for a temporal strategy instead of taking a definitive decision now requires us to establish an appropriate governance structure in order to prevent us from eschewing the decision problem.

8 Conclusion

In the case of great uncertainty about a decision problem, conditions for the application of formal methods from decision theory, decision support or policy analysis to calculate which option would be rational to chose are not fulfilled. If the decision problem cannot be properly defined or important information on options, outcomes and values is missing, decision-makers could be misled when building on results from these kinds of analysis, since relevant aspects must have been ignored in the respective calculations. Using temporal strategies opens up opportunities that enable us to apply further argumentative methods in order to learn about, evaluate, and deal with great uncertainty in taking decisions.

Temporal strategies extend decisions into the future by postponing decisions, recurrently modifying decisions, or taking them sequentially. Temporal strategies enable us to improve the ways in which we deal with uncertainty in the course of decision-making. As a consequence, temporal strategies do not make decisions simpler but more demanding in various regards. These demands pose restrictions on their effective application for a given policy decision problem. To structure reasoning for and against the application of temporal strategies to a decision problem, four general criteria are useful: first, the relevance of considering uncertainties for taking a decision; second, the feasibility of improving information on or evaluating relevant uncertainties; third, the acceptability of trade-offs related to the temporal strategy; fourth, the maintenance of governing decision-making over time. These criteria need to be specified and weighted in relation to the decision problem at hand. Instead of determining a temporal strategy, the criteria provide a framework for systematic deliberation on temporal strategies.

Only rarely will the argumentative methods, which can be applied within the time span of a temporal strategy, turn a decision under great uncertainty into a decision under certainty. So, in most cases, expecting expert advice on definitive solutions is inappropriate for these sorts of problems. Instead, a fundamental shift is needed in how the task of policy analysis is conceived. To account for complex interactions and future development in policy decision problems of great uncertainty, a reasonable strategy is to extend decisions into the future by taking decisions on options that work for protecting against detrimental effects and that shape a development path which permits future decisions and revisions.

Recommended Readings

Dietz, T., Ostrom, E., & Stern, P. C. (2003). The struggle to govern the commons. *Science*, *302*,1907–1912. doi:10.1126/science.1091015.

Hammond, J. S., Keeney, R. L., & Raiffa, H. (1999). *Smart choices: A practical guide to making better decisions*. Boston: Harvard Business School Press.

Parson, E. A., & Karwat, D. (2011). Sequential climate change policy. *WIREs Climate Change*, *2*, 744–756. doi:10.1002/wcc.128.

Trigeorgis, L. (2001). Real options. An overview. In E. S. Schwartz & L. Trigeorgis (Eds.), *Real options and investment under uncertainty* (pp. 103–134). Cambridge, MA: The MIT Press.

Van Hoek, R. I. (2001). The rediscovery of postponement a literature review and directions for research. *Journal of Operations Managment*, *19*, 161–184.

References

Allenspach, U. (2013). *Sequences of choices with multiple criteria and thresholds. Implications for rational decisions in the context of sustainability*. Zurich: ETH. http://dx.doi.org/10.3929/ethz-a-009773097.

Andreou, C. (2012). Dynamic choice. In E. N. Zalta (Ed.), *The stanford encyclopedia of philosophy*. http://plato.stanford.edu/archives/fall2012/entries/dynamic-choice. Accessed 2 Jan 2015.

Beauchemin, K. A., Kreuzer, M., O'Mara, F., & McAllister, T. A. (2008). Nutritional management for enteric methane abatement: A review. *Australian Journal of Experimental Agriculture, 48*, 21–27.

Betz, G. (2016). Accounting for possibilities in decision-making. In S. O. Hansson & G. Hirsch Hadorn (Eds.), *The argumentative turn in policy analysis. Reasoning about uncertainty* (pp. 135–169). Cham: Springer. doi:10.1007/978-3-319-30549-3_6.

Bratman, M. E. (2012). Time, rationality, and self-governance. *Philosophical Issues, 22*, 73–88.

Broome, J. (2008). The ethics of climate change. *Scientific American*, June 2008: 69–73.

Brun, G., & Betz, G. (2016). Analysing practical argumentation. In S. O. Hansson & G. Hirsch Hadorn (Eds.), *The argumentative turn in policy analysis. Reasoning about uncertainty* (pp. 39–77). Cham: Springer. doi:10.1007/978-3-319-30549-3_3.

Brun, G., & Hirsch Hadorn, G. (2008). Ranking policy options for sustainable development. *Poiesis & Praxis, 5*, 15–30. doi:10.1007/s10202-007-0034-y.

Brunner, R. (2010). Adaptive governance as a reform strategy. *Policy Sciences, 43*, 301–341. doi:10.1007/s11077-010-9117-z.

Dietz, T., Ostrom, E., & Stern, P. C. (2003). The struggle to govern the commons. *Science, 302*,1907–1912. doi:10.1126/science.1091015.

Doorn, N. (2016). Reasoning about uncertainty in flood risk governance. In S. O. Hansson & G. Hirsch Hadorn (Eds.), *The argumentative turn in policy analysis. Reasoning about uncertainty* (pp. 245–263). Cham: Springer. doi:10.1007/978-3-319-30549-3_10.

Edvardsson, K. (2004). Using goals in environmental management: The Swedish system of environmental objectives. *Environmental Management, 34*, 170–180. doi:10.1007/s00267-004-3073-3.

Edvardsson Björnberg, K. (2008). Utopian goals. Four objections and a cautious defense. *Philosophy in the Contemporary World, 15*, 139–154.

Edvardsson Björnberg, K. (2016). Setting and revising goals. In S. O. Hansson & G. Hirsch Hadorn (Eds.), *The argumentative turn in policy analysis. Reasoning about uncertainty* (pp. 171–188). Cham: Springer. doi:10.1007/978-3-319-30549-3_7.

Elliott, K. C. (2016). Climate geoengineering. In S. O. Hansson & G. Hirsch Hadorn (Eds.), *The argumentative turn in policy analysis. Reasoning about uncertainty* (pp. 305–324). Cham: Springer. doi:10.1007/978-3-319-30549-3_13.

Folke, C., Hahn, T., Olsson, P., & Norberg, J. (2005). Adaptive governance of social-ecological systems. *Annual Review of Environment and Resources, 30*, 441–473. doi:10.1146/annurev.energy.30.050504.144511.

Forster, P., Ramaswamy, V., Artaxo, P., Berntsen, T., Betts, R., Fahey, D. W., Haywood, J., Lean, J., Lowe, D. C., Myhre, G., Nganga, J., Prinn, R., Raga, G., Schulz, M., & van Dorland, R. (2007). Changes in atmospheric constituents and in radiative forcing. In S. Solomon, D. Qin, M. Manning, Z. Chen, M. Marquis, K. Averyt, M. M. B. Tignor, & H. L. R. Miller (Eds.), *Climate change 2007: The physical science basis. Contribution of working group I to the fourth assessment report of the intergovernmental panel on climate change* (pp. 131–234). Cambridge/New York: Cambridge University Press.

Fox, C. R., Bardolet, D., & Lieb, D. (2005). Partition dependence in decision analysis, resource allocation, and consumer choice. In R. Zwick & A. Rapoport (Eds.), *Experimental business research* (Vol. III, pp. 229–251). Dordrecht: Springer.

Frederick, S., Loewenstein, G., & O'Donoghue, T. (2003). Time discounting and time preference: A critical review. In G. Loewenstein, D. Reid, & R. Baumeister (Eds.), *Time and decision.*

Economic and psychological perspectives on intertemporal choice (pp. 13–86). New York: Russell Sage Foundation.

Gregory, R., Ohlson, D., & Arvai, J. (2006). Deconstructing adaptive management: Criteria for applications to environmental management. *Ecological Applications, 16*, 2411–2425.

Gross, M., & Hoffmann-Riem, H. (2005). Ecological restoration as a real-world experiment: Designing robust implementation strategies in an urban environment. *Public Understanding Science, 14*, 269–284. doi:10.1177/0963662505050791.

Grüne-Yanoff, T. (2016). Framing. In S. O. Hansson & G. Hirsch Hadorn (Eds.), *The argumentative turn in policy analysis. Reasoning about uncertainty* (pp. 189–215). Cham: Springer. doi:10.1007/978-3-319-30549-3_8.

Hammitt, J. K., Lempert, R. J., & Schlesinger, M. E. (1992). A sequential decision stategy for abating climate change. *Nature, 357*, 315–318.

Hammond, J. S., Keeney, R. L., & Raiffa, H. (1999). *Smart choices: A practical guide to making better decisions*. Boston: Harvard Business School Press.

Hansson, S. O. (1996). Decision making under great uncertainty. *Philosophy of the Social Sciences, 26*, 369–386.

Hansson, S. O. (2016). Evaluating the uncertainties. In S. O. Hansson & G. Hirsch Hadorn (Eds.), *The argumentative turn in policy analysis. Reasoning about uncertainty* (pp. 79–104). Cham: Springer. doi:10.1007/978-3-319-30549-3_4.

Hansson, S. O., & Hirsch Hadorn, G. (2016). Introducing the argumentative turn in policy analysis. In S. O. Hansson & G. Hirsch Hadorn (Eds.), *The argumentative turn in policy analysis. Reasoning about uncertainty* (pp. 11–35). Cham: Springer. doi:10.1007/978-3-319-30549-3_2.

Hirsch Hadorn, G., Brun, G., Soliva, C., Stenke, A., & Peter, T. (2015). Decision strategies for policy decisions under uncertainties: The case of mitigation measures addressing methane emissions from ruminants. *Environmental Science & Policy, 52*, 110–119. http://dx.doi.org/10.1016/j.envsci.2015.05.011.

Holling, C. S. (1978). *Adaptive environmental assessment and management*. New York: Wiley.

Hulme, M. (2009). *Why we disagree about climate change: Understanding controversy, inaction and opportunity*. Cambridge: Cambridge University Press.

Kisperska-Moron, D., & Swierczek, A. (2011). The selected determinants of manufacturing postponement within supply chain context: An international study. *Internationl Journal of Production Economics, 133*, 192–200. doi:10.1016/j.ijpe.2010.09.018.

Levi, I. (1984). *Decisions and revisions. Philosophical essays on knowledge and value*. Cambridge: Cambridge University Press.

McClennen, E. F. (1990). *Rationality and dynamic choice. Foundational exporations*. Cambridge: Cambridge University Press.

Möller, N. (2016). Value uncertainty. In S. O. Hansson & G. Hirsch Hadorn (Eds.), *The argumentative turn in policy analysis. Reasoning about uncertainty* (pp. 105–133). Cham: Springer. doi:10.1007/978-3-319-30549-3_5.

Oenema, O., Wrage, N., Velthof, G. L., van Groenigen, J. W., Dolfing, J., & Kuikman, P. J. (2005). Trends in global nitrous oxide emissions from animal production systems. *Nutrient Cycling in Agroecosystems, 72*, 51–65. doi:10.1007/s10705-004-7354-2.

Oxford English Dictionary (OED). (2014). *strategy, n*. Oxford University Press. http://dictionary.oed.com/. Accessed 10 Sept 2014.

Pahl-Wostl, C. (2007). Transitions towards adaptive management of water facing climate and global change. *Water Resource Management, 21*, 49–62. doi:10.1007/s11269-006-9040-4.

Parson, E. A., & Karwat, D. (2011). Sequential climate change policy. *WIREs Climate Change, 2*, 744–756. doi:10.1002/wcc.128.

Schreiber, E. S. G., Berlin, A. R., Nicol, S. J., & Todd, C. R. (2004). Adaptive management: A synthesis of current understanding and effective application. *Ecological Management & Restauration, 5*, 117–182. doi:10.1111/j.1442-8903.2004.00206.x.

Singer, P., & Mason, J. (2006). *The way we eat. Why our food choices matter*. Emmaus: Rodale.

Smith, P., Martino, D., Cai, Z., Gwary, D., Janzen, H., Kumar, P., McCarl, B., Ogle, S., O'Mara, F., Rice, C., Scholes, B., & Sirotenko, O. (2007). Agriculture. In B. Metz, O. R. Davidson, P. R. Bosch, R. Dave, & L. A. Meyer (Eds.), *Climate change 2007: Mitigation. Contribution of working group III to the fourth assessment report of the intergovernmental panel on climate change* (pp. 498–540). Cambridge/New York: Cambridge University Press.

Steinfeld, H., Geber, P., Wassenaar, T., Castel, V., Rosales, M., & de Haan, C. (2006). *Livestock's long shadow: Environmental issues and options*. Rome: FAO, Food and Agriculture Organization of the United Nations. ftp://ftp.fao.org/docrep/fao/010/a0701e/a0701e00.pdf. Accessed 2 Jan 2015.

Swanson, D., Barg, S., Tyler, S., Venema, H., Tomar, S., Badwahl, S., Nair, S., Roy, D., & Drexhage, J. (2010). Seven tools for creative adaptive policies. *Technological Forecasting & Social Change, 11*, 924–939. doi:10.1016/j.techfore.2010.04.005.

Tigges, R. (2011). *Moratorium 2011 – Das Schicksalsjahr für deutsche Atomkraftwerke: Aufbruch zu einer neuen Energiestrategie für unser Land*? http://www.moratorium2011.de/. Accessed 10 Sept 2014.

Tol, R. S. (2005). Adaptation and mitigation: Trad-offs in substance and methods. *Environmental Science & Policy, 8*, 572–758. doi:10.1016/j.envsci.2005.06.011.

Trigeorgis, L. (2001). Real options. An overview. In E. S. Schwartz & L. Trigeorgis (Eds.), *Real options and investment under uncertainty* (pp. 103–134). Cambridge, MA: The MIT Press.

UNFCCC, United Nations Framework Convention on Climate Change. (2008). *Challenges and opportunities for mitigation in the agricultural sector* (Technical paper no 8). http://unfccc.int/resource/docs/2008/tp/08.pdf. Accessed 2 Jan 2015.

Van der Pas, J. W. G. M., Walker, W. E., Marchau, V. A. W. J., van Wee, B., & Kwakkel, J. H. (2013). Operationalizing adaptive policymaking. *Futures, 52*, 12–26. doi:10.1016/j.futures.2013.06.004.

Van Hoek, R. I. (2001). The rediscovery of postponement a literature review and directions for research. *Journal of Operations Managment, 19*, 161–184.

Van Reedt Dortland, M., Voordijk, H., & Dewulf, G. (2014). Making sense of future uncertainties using real options and scenario planning. *Futures, 55*, 15–31. doi:10.1016/j.futures.2013.12.004.

Walters, C. (1986). *Adaptive management of renewable resources*. New York: McMillan.

Webster, M., Jabobovits, L., & Norton, J. (2008). Learning about climate change and implications for near-term policy. *Climatic Change, 89*, 67–85. doi:10.1007/s10584-008-9406-0.

Part III
Case Studies

Chapter 10
Reasoning About Uncertainty in Flood Risk Governance

Neelke Doorn

Abstract The number and impact of catastrophic floods have increased significantly in the last decade, endangering both human lives and the environment. Although there is a broad consensus that the probability and potential impacts of flooding are increasing in many areas of the world, the conditions under which flooding occurs are still uncertain in several ways. In this chapter, I explore how argumentative strategies for framing, timing, goal setting, and dealing with value uncertainty are being employed or can be employed in flood risk governance to deal with these uncertainties. On the basis of a discussion of the different strategies, I sketch a tentative outlook for flood risk governance in the twenty-first century, for which I derive some important lessons concerning the distribution of responsibilities, the political dimension of flood risk governance, and the use of participatory approaches.

Keywords Uncertainty • Wicked problem • Flood risk management • Water governance • Building with nature • European Flood risk directive (2007/60/EC) • Flood safety • Flood risk • Water management • Water safety

1 Introduction

The number and impact of catastrophic floods have increased significantly in the last decade, endangering both human lives and the environment, and causing severe economic losses (Smith and Petley 2009). With climate change, the risk of flooding is likely to increase even further in the coming decades (EEA 2010; CRED 2009). Although there is a broad consensus that the probability and potential impact of flooding are increasing in many areas of the world, the conditions under which flooding occurs are still uncertain in several ways.

N. Doorn (✉)
Department of Values, Technology and Innovation, School of Technology, Policy and Management, Technical University Delft, Delft, The Netherlands
e-mail: N.Doorn@tudelft.nl

S.O. Hansson, G. Hirsch Hadorn (eds.), *The Argumentative Turn in Policy Analysis*, Logic, Argumentation & Reasoning 10, DOI 10.1007/978-3-319-30549-3_10

First, many of the data that are needed to base decisions on are still uncertain: What will the quantitative effect of climate change be on the probability of flooding? How will demographic conditions like urbanization and aging develop?

Second, two major policy developments take place in flood risk management affecting the way in which flood risks are currently "managed." The first development concerns the so-called "governance turn," which has taken place in European flood risk policy. Until the late twentieth century, safety against flooding was seen as a purely economic good, and the responsibility for managing flood risks was seen as the exclusive task of the state. In the past decades, this centralized approach is increasingly replaced by a more flexible and adaptive "governance" approach (Butler and Pidgeon 2011; Meijerink and Dicke 2008; McDaniels et al. 1999). The term governance stems from political science and it is used to refer to the way in which authority is exercised and shared between different actors in order to come to collectively binding decisions (Bell 2002; Wolf 2002). Applied to flood risks, governance refers to the interplay of public and private institutions involved in decision making on flood risk management (Asselt, Marjolein, and Renn 2011). The governance approach in flood risk management (in short: flood risk governance) puts less emphasis on the prevention of flooding and more on the minimization of negative consequences (Heintz et al. 2012). Additionally, it ascribes more responsibility to private actors and decentralized governmental bodies (Meijerink and Dicke 2008). The second policy development concerns the introduction of the European Flood risk directive (2007/60/EC). The Flood risk directive does not contain concrete standards nor does it prescribe specific measures, but it does require Member States of the European Union to review their systems of flood risk management.[1] Although the Flood risk directive itself is legally binding only to European member states, experiences with this directive will probably be transferred to non-European countries as well.

Taken together, the uncertainties with respect to the impact and severity of flooding and the developments in the flood policy domain prompt some urgent moral questions (Mostert and Doorn 2012; Doorn 2015): How should the money available for minimizing the risk of flooding be distributed? How should the responsibilities pertaining to flood risk management (both between private and public actors and between several governmental bodies or countries sharing a water course) be distributed? How should environmental impact be taken into

[1] The Flood risk directive prescribes Member States to assess the flood risks in their river basins and prepare flood hazard and flood risk maps for all areas with a significant flood risk (Art. 4–6 and 13). Moreover, they have to establish flood risk management plans for these areas, containing "appropriate objectives" for managing the risks and measures for achieving these objectives (Art. 7). These plans have to be coordinated at the river basin level (Art. 8) and may not include measures that increase flood risks in other countries, unless agreement on these measures has been reached (Art. 7.4, cf. preamble 15 and 23). Moreover, Member States have to encourage active involvement in the development of the plans (Art. 10.2, Art. 9.3). In doing all this, Member States have to consider human health and the effects on the environment and cultural heritage (Art. 2.2, 7.2 and 7.3).

account in the management of flood risks? Moreover, the uncertainties with regard to the risks of flooding and the developments in flood risk policy put limits to the applicability of traditional risk analysis. Decisions in risk governance cannot be based on probabilistic information alone (Doorn and Hansson 2011) and alternative strategies should be employed to base the decisions on.

In this chapter, I explore how argumentative strategies are being or can be employed in flood risk governance. The outline of this chapter is as follows. Following this introduction, I first describe the basic terminology and definitions (Sect. 2). In Sect. 3, I describe argumentative strategies. In the concluding Sect. 4, I summarize the findings and sketch a tentative outlook for flood risk governance in the twenty-first century. In the remainder of this text, I use the term flood risk *governance* to refer to the policy and decision making process on flood risks and the term flood risk *management* to the technical aspects of dealing with flood risks.

2 Basic Terminology and Definitions

Before discussing the argumentative strategies employed in the context of flood risk governance, it is important to clarify the terminology and to distinguish between different types of flooding.

To start with the notions of risk, it is important to distinguish between *risk* and *uncertainty*. This distinction dates back to work in the early twentieth century by the economists Keynes and Knight (Knight 1935 [1921]; Keynes 1921). Knight proposed to reserve the term "risk" for situations where one does not know for sure what will happen, but where the chance can be quantified (for example, rolling a dice). Uncertainty refers to situations where one does not know the chance that some undesirable event will happen (Knight 1935 [1921]:19–20). This terminological reform has spread to other disciplines, including engineering, and it is now commonly assumed in most scientific and engineering contexts that "risk" refers to something that can be assigned a probability, whereas "uncertainty" may be difficult or impossible to quantify.

The distinction between risk and uncertainty has been criticized by scholars working in risk governance (Asselt, Marjolein, and Renn 2011; Löfstedt 2005; Millstone et al. 2004). They argue that this framing of risks mistakenly suggests that risks can be captured by a simple cause-and-effect model with statistics available to assign probabilities. Most risks are not of this simple type but they are so-called "systemic risks"; that is, risks that are complex, multi-causal, and surrounded by uncertainty and ambiguity (Renn 2008; Klinke and Renn 2002). Although I agree with the observation that most risks are not of the simple type, it does not preclude the distinction between risk and uncertainty. I therefore propose to categorize systemic risks as uncertainty. I do agree with the observation, though, that, contrary to what is often assumed, we are far more often in a situation of uncertainty than one of risk (see Hansson and Hirsch Hadorn 2016 and Hansson 2009 for a similar observation).

If we define floods as the presence of water on land that is usually dry, we can distinguish between different types of floods. A first distinction to be made is that between seasonal flooding and extreme flood events. Seasonal flooding occurs on a recurrent basis and it is not necessarily harmful. It may provide agricultural land with nutrients. Usually, relatively reliable data is available to predict the occurrence of seasonal flooding and it is therefore meaningful to assess the risks in statistical terms. Van Asselt and Renn mention seasonal flooding as one of the paradigmatic examples of – what in risk governance is labeled – simple risks (Asselt, Marjolein, and Renn 2011). However, climate change may of course also have an impact on seasonal flooding, so the label "simple risk" is probably an oversimplification also for seasonal flooding.

Flood risk governance is less concerned with seasonal flooding than with extreme flood events that do not occur on a recurrent basis. The effects of these extreme flood events are significantly worse than the potential nuisance of seasonal flooding. They can, for example, be caused by extreme weather events or the collapse of existing (flood protection) structures. These extreme events are usually distinguished after their causes:

- Fluvial or riverine flooding: these floods are usually caused by rainfall over an extended period and an extended area. Downstream areas may be affected as well, even in the absence of heavy rainfall in these areas;
- Flash floods: these floods occur in areas where heavy rainfall or sudden melting of snow leads to rapid water flows downhill, which cause an almost instantaneous overflowing of the river banks; dam breaches can be seen as a type of flash flood;
- Coastal flooding: flooding of the land from the sea, usually a combination of high water level and severe wave conditions due to extreme weather events.

Although the impact of the consequences of extreme floods differs per area, they are in almost all situations potentially large.

The conditions under which these extreme flood events occur and their impact are uncertain in several ways.

First, there is uncertainty on the occurrence of these types of floods. Climate change may increase the probability that these events occur. Though it is by now widely accepted in the scientific community that our climate is subject to change, it is still difficult to quantify the effects of climate change. The sea will probably rise in the coming decades and centuries but predictions as to the exact rise in sea level range from approximately 30 cm (lower limit scenario RCP2.6) to 100 cm (upper limit scenario RCP8.5) at the end of the twenty-first century (IPCC 2014). Similarly, more extreme weather events are expected to occur (both in terms of heavy rainfall and in terms of drought), but these predictions are hard to quantify.

Second, demographic conditions may change, and so does the impact of extreme flooding. Urbanization, for example, may lead to more casualties in cases of coastal flooding. Since these demographic developments are hard to predict with accuracy the expected flood risk (in terms of probability times effect) is hard to quantify.

Third, the knowledge base for identifying possible solutions is insufficient and disputed (Driesssen and Van Rijswick 2011). Some engineers call for traditional (hard) flood protection measures, whereas others opt for "green solutions," where agricultural land is "given back to the river." Hence, the governance of flood risks involves value conflicts which may in turn lead to incomplete preference orderings (Espinoza and Peterson 2008). Together, these uncertainties and ambiguities may influence each other: policy choices are affected by societal and environmental developments and vice versa. This is often referred to as *deep uncertainty* (Hallegatte et al. 2012; Lempert et al. 2003) or *great uncertainty* (Hansson and Hirsch Hadorn 2016).

If we bring these two elements together (potentially large impact and uncertain conditions), we can see the main challenge for the governance of flood risks: to develop a response (both in technical and policy terms) to a hazard with potentially large impact under conditions of uncertainty (Haasnoot 2013). In the terminology of policy sciences, flood risk governance is a typical example of a wicked problem; that is, a problem that is difficult or impossible to solve because of incomplete, contradictory, and changing requirements that are often difficult to pin down (Brunner et al. 2005). Wicked problems are characterized by ambiguity with regard to the problem definition, uncertainty about the causal relations between the problem and potential solutions, and a wide variety of interests and associated values (Rittel and Webber 1973).

In the remainder of this paper, I will talk about the governance of extreme flood events rather than seasonal flooding. Although it is common to refer to flood *risk* governance, it should be clear by now that the term "uncertainty" is more in place.

3 Argumentative Strategies

If we look at the governance of flood risks, we see that a number of argumentative strategies are relevant for flood risk governance. In this section, I discuss the following five strategies: framing strategies, timing strategies, goal setting, dealing with value uncertainty, and participatory decision making.

3.1 *Framing*

Here by framing will be meant the way a problem is presented and, as a result of which, what solutions people see as being in their interest and, accordingly, what solutions they see as conflicting (Schön and Rein 1994). Framing is one of the most important strategies when reasoning about uncertainty in the governance of flood risks. As explained in Grüne-Yanoff (2016), framing in the policy domain can be used to justify certain policies but also instrumentally to steer certain behavior.

An interesting country to look at is the US and its way of framing flood risks. Characteristic for the American coastal flood risk policy is an emphasis on flood hazard mitigation (Wiegel and Saville 1996). Rather than trying to prevent flooding, the focus has always been on prediction of floods and on insurance, which suggests that the very fact of flooding is accepted (Bijker 2007). In this view, it is not the government's responsibility to provide safety against flooding, but rather to limit its consequences and (possibly) provide financial compensation or make insurance possible. Elements of the governance approach that are new for European flood risk policy have since long been present in the United States. This policy was broadly accepted until the New Orleans area was hit by Hurricane Katrina in summer 2005 and the governmental agencies failed to contain the flood effectively (Warner 2011). Congressional hearings pointed at the role of the Federal Emergency Management Agency (FEMA), the agency responsible for disaster management. Established in 1978, the FEMA was an independent agency until the beginning of the twenty-first century. After the 2001 terrorist attacks, the agency was subsumed under the newly established Department of Homeland Security (DHS). The focus of the FEMA shifted to terrorism, as a result of which preparedness for natural hazards (including flooding) was given low priority. After the country was caught unawares by Hurricane Katrina, it turned out that no federal funding had been awarded to disaster preparedness projects unless it was presented as a terrorism function (Davis et al. 2006). These two factors, the conception of flood risk as something to be accepted and FEMA's focus on terrorism prevention at the exclusion of natural disaster planning both strongly influenced the way the US shaped its flood risk policy in the past (Bijker 2007).

In the Netherlands, flood risks are framed quite differently compared to the United States. The Netherlands is a country below sea level and central in the Dutch history of flood risk management is the 1953 storm surge disaster. The combination of a long-lasting storm, unfavorable wind conditions, and high spring tide led to the flood disaster that still marks the Dutch view on coastal engineering (Bijker 1996). More than 1,800 people drowned and 200.000 ha of land was inundated. After the 1953 floods, the credo of Dutch engineering became "never again!" However, if we look at the Dutch history of flood risk management since 1950s more closely, we can distinguish between different periods with different policy frames and different ways to achieve this goal.

Immediately after the 1953 floods, there was ample room for technocratic solutions. Already drafted before the 1953 disaster, a "Deltaplan" was put in place, which included the norm that the coastal flood defense system should be able to withstand 1:10,000 year storm conditions. This criterion was laid down in the "Delta Law," which was unanimously approved by Parliament (Bijker 2007). Because Dutch engineers had already developed plans for improving the coastal defense system before the 1950s, the Dutch water agency *Rijkswaterstaat* was able to fall back on these plans and they could immediately start working on the large-scale Delta Works project that would allow the Netherlands to fight against the water (Lintsen 2002).

In the 1970s, opposition to the hegemonic position of *Rijkswaterstaat* grew. Water issues became framed in terms of safety versus ecology. *Rijkswaterstaat* was criticized for their shortsighted technocratic solutions that were supposedly harmful to the environment and landscape. One of the last elements of the Delta Plan, the closure of the Eastern Scheldt, met so much opposition that *Rijkswaterstaat* was forced to cancel the original plan of full closure of the Eastern Scheldt and to carry out an alternative plan that comprised the construction of a storm surge barrier storm; that is, a barrier that would normally be open and allow water to pass through but could be closed if a flood threatened the hinterland (Bijker 2002).

In two subsequent winters in the 1990s, the Netherlands again proved vulnerable to flooding, this time riverine flooding. In December 1993 and again in January 1995, a large part of the country was at risk of flooding and almost 200,000 people in areas along the rivers Rhine and Meuse were evacuated. With the disaster of 1953 still in mind, the warning was taken seriously and within a record time of only 6 weeks after the peak of the river discharge, a new law that would lower the acceptable risk of riverine flooding from 1:100 to 1:1.250 year was approved by Parliament (Borman 1995). However, this higher safety level was now to be achieved by giving more "Room for the River," as the new policy line was aptly called. In the official policy announcement, the lack of room for the river – due to, for example, embankments and the construction of buildings on floodplains – was mentioned as the primary cause of riverine flooding. Hence, flood prevention was from then on to be achieved by building *with* rather than *against* nature.

Interestingly, after the flooding caused by Hurricane Katrina, both policy makers and engineers in the US looked at the Netherlands to see how flood risks should be governed, but the policy makers and engineers in the Netherlands used Katrina to put the prevention of flooding back on the agenda (Disco 2006). Whereas the Room for the River policy of the 1990s considered hard preventive infrastructural measures as less desirable than soft spatial measures, the increased awareness of climate change and the disruptive effect of Katrina on the US society created room again for solutions aimed at flood control. In 2007, the Delta committee was installed with the task to advise the government on how to develop a long term vision on flood safety for the Netherlands, taking into account climate change. Additionally, the committee was asked to bring across a sense of urgency to the Dutch society. This latter task is striking: apparently, communicating the urgency of flood safety was presupposed in the committee's task description. In its advice in 2008, the committee presented flood safety as something too important to be left to regular short-term focused politics or decentralized governmental bodies. Flood safety is of national importance and the solidarity principle should therefore be guiding. The committee argued that the responsibility for flood safety should lie with the central government (Vink et al. 2013).

Since the committee's report in 2008, the financial situation has changed dramatically and the financial means for flood prevention are limited. With climate change and demographic developments increasingly framed as deep uncertainties,

Dutch policy in the 2010s shows a gradual shift from flood control to adaptation in Dutch flood risk policy (Haasnoot 2013).

To summarize, the framing of flood risks in the Netherlands has shifted from "fight against water" in the 1950, to "building with nature" in the 1990s; and from "centralized flood control" in the first decade of the twenty-first century to "adaptation" in the second decade of the twenty-first century.

3.2 Timing

The second argumentative strategy that is often used in flood risk policy is timing. Timing can be relevant both in the sense of *when* the decision is made and in the sense of the *time horizon* taken into account in the decision itself. The two elements cannot be fully distinguished, as Hirsch Hadorn (2016) shows.

Regarding the timing of the decision, natural disasters (like flooding) are often the starting point for considering or implementing new policy. In that sense, the implementation of flood protection policy is often reactive. However, such a reactive policy can only be considered rational if one can or is prepared to bear the consequences of the flood event. The more severe the consequences, the less likely it becomes that society is indeed willing to accept these consequences.[2]

Once flood protection has failed, there is usually wide public support for implementing policy and building new infrastructures. If we look at the Netherlands, for example, both after the 1953 flood and after the high waters in the 1990s, new policy was adopted within only a few weeks after the flood and high water respectively. In 1953, three weeks after the flood, a governmental committee was formed, which delivered an interim "Delta Plan" only one week later. The implementation of this plan started already before the political procedures had been completed and construction work started in 1955 (Bijker 2002). Similarly in the 1990s, it took only six weeks to complete the implementation of the new river law and in this case, the construction work started only two months later (Borman 1995). Strikingly, also the flooding resulting from Hurricane Katrina was used in the Netherlands as an opportunity to put flood prevention back on the agenda. These examples show that, in the Netherlands at least, natural disasters may be used to put flood protection on the agenda and to create support for implementing new policy.

In flood risk governance, the timing of the decision is less important than the time horizon to take into account. It makes a large difference on which time horizon flood risk policy is based. Given the deep uncertainty involved in climate policy, the challenge is to predict the relevant conditions for the time horizon chosen.

[2] For an example in which such an approach was indeed considered rational, see Schefczyk (2016). In this chapter, Schefczyk explains how Alan Greenspan, the chairman of the US Federal Reserve Bank of the United States, considered relying on insurance measures against unlikely but highly adverse events to be the rational approach, which means that he explicitly accepted the potential consequences.

A distinction is usually made between predictive (top-down) approaches and adaptive or resilience-based (bottom-up) approaches (Dessai and Van der Sluijs 2007).[3] Top-down approaches focus on scenarios to predict possible conditions for the time horizon chosen and to assess the impact. Top-down approaches are most widely used (Carter et al. 2007; Adger et al. 2007). However, given the deep uncertainty involved in climate policy, the predictions for the long term may vary significantly between different scenarios. In its advisory report on how to prepare for climate change, for example, the Dutch committee on flood safety (the Delta Committee) used more extreme scenarios than the ones used by the Intergovernmental Panel on Climate Change (IPCC). Whereas the IPCC scenarios indicated a sea level rise between 20 and 60 cm at the end of the twenty-first century (IPCC 2007), the Delta Committee argued that it would be better to be prepared for a sea level rise of 0.65–1.3 m in the year 2010 because they considered it "prudent to reckon with upper limits [of sea level rise, ND], so that decisions and measures will hold for a long time span" (Delta Committee press release, quoted in Vink et al. 2013). In other words, by including a large time horizon in combination with extreme scenarios, the Committee wanted to develop a robust policy that would suffice for the long term. Although the use of these worst case scenarios may be warranted from a safety point of view, for policy makers it is often problematic to rely too much on these projections. Not only is it difficult for policy makers to select scenarios, the predictions may also vary significantly when the scenarios are being updated (to illustrate this point, the IPCC report of 2013 gives a 40 cm higher upper limit for the sea level rise in 2,100 than the report of 2007). Robust rational decision making vis-à-vis deep uncertainty requires a shift from probabilistic to possibilistic knowledge dynamics (Betz 2016).

In order to avoid the disadvantages of traditional top-down approaches, much attention is now put in developing bottom-up approaches. Whereas the traditional probabilistic top-down approaches rely heavily on climate predictions, the bottom-up approaches focus on vulnerability and the adaptive capacity of a system (here: flood risk management system) and on the measures required to improve its resilience and robustness (Carter et al. 2007). This adaptive capacity is assessed by looking at the social factors that determine the ability to cope with climatic hazards; the outcomes are partly based on qualitative data (experiences of stakeholders, expert judgments, etc.). Although the majority of approaches are still top-down, some promising bottom-up approaches are currently being developed. In the context of water governance, an approach has been developed based on so-called adaptation tipping points (ATP). These tipping points indicate under what conditions current water management strategies stop being effective for clearly specified objectives. If a tipping point is reached, additional actions are needed (Kwadijk et al. 2010). Based on these tipping points, adaptation pathways may be developed which describe a sequence of water management strategies enabling

[3] It should be noted that different taxonomies exist. Some scholars talk about top-down approaches as hazard-based and bottom-up approaches as vulnerability-based (cf. Burton et al. 2005).

policy makers to explore options for adapting to changing environmental and societal conditions (Haasnoot 2013). The clear advantage of such an approach is that it does not rely on specific future predictions of, say, sea level rise. Given the uncertainties involved in these predictions, a bottom-up approach may therefore be considered more suitable for decision making. Bottom-up approaches, in turn, are sometimes criticized for relying too much on expert judgment and qualitative data (Füssel 2007). Despite these challenges for bottom-up approaches, adaptive management is increasingly considered the preferred approach to deal with timing issues in flood risk governance.

3.3 Goal Setting

The third argumentation strategy is about goal setting and revision of goals. As indicated in Edvardsson Björnberg (2016), goal revision can be both achievability-related and desirability-related. In flood risk governance, goal revision occurs on the basis of both considerations.

As stated in the introductory section, until the end of the twentieth century, flood risk management in Europe was primarily focused on the control and prevention of flooding. Since the late 1990s, the emphasis has shifted from a sole focus on the prevention of flood risks to mitigation of the negative consequences of flooding (Heintz et al. 2012). Not only was it considered unrealistic to prevent all flooding, it was also considered undesirable because a sole focus on prevention would result in environmental damage and damage to cultural heritage.

In line with this shift from sole prevention towards mitigation, the Dutch Delta Committee introduced the concept of multi-layer safety to strengthen flood protection in the Netherlands. The idea of "multi-layer safety" is that flood risk governance consists of three layers: prevention, spatial planning, and disaster management. Though coined differently, a similar shift in the goal of flood risk governance is taking place in other European countries, most notably in the UK (Cashman 2011; Scrase and Sheate 2005) and Germany (Merz and Emmermann 2006).[4]

Although the idea of multi-level safety is not unanimously supported – opponents argue that multi-layer safety is not cost-effective because in low-lying countries the most effective way to deal with floods is to prevent them (Vrijling 2009) – the concept itself clearly shows how the goal of flood risk policy has shifted from prevention *sec* to the mitigation of negative consequences. By discouraging the construction of buildings in flood-prone areas and by investing in evacuation

[4] For a cross-country comparison, see Bubeck et al. (2013). The authors notice convergence between flood risk policies in Europe, although Dutch flood risk policy is still more technocratic than the flood risk policy in Germany and the UK. Adaptation to climate change is still not considered in the US flood risk policy because, contrary to Europe, the potential negative effects of global warming are still topic of debate.

schemes, a higher probability of flooding may be considered acceptable in some regions.

Multi-layer safety will probably lead to more differentiation in safety levels between different regions, which is – from a moral point of view – not uncontroversial. When differentiating between safety levels, a pressing issue is how to balance equity with efficiency considerations (Doorn 2014a; Peterson 2003). The answer to this – as yet – open question should also be seen in the light of a changing view on the role of government in society, entailing a redistribution of responsibilities between public and private actors (Butler and Pidgeon 2011; Wolsink 2006).

3.4 *Conflicting Values*

The fourth reasoning strategy concerns dealing with value uncertainty (Möller 2016). Like in other environmental domains, flood risk management involves different values, with priorities varying over time.

In the last decades, new strategies have been proposed for improving the level of protection against flooding. Whereas flood protection in the beginning of the twentieth century was still limited to dyke construction or strengthening, with or without additional fixed structures, both urbanization and a growing awareness of ecological impact have prompted the design of alternative flood protection measures. This is partly related to the introduction of competing interests in the domain of flood protection. The value of safety has lost its monopoly and other values have become important as well.

The landmark example in hydraulic engineering in which new values were included in flood risk governance is the design of the Dutch Eastern Scheldt storm surge barrier in the 1970s and 1980s, already mentioned in Sect. 3.1. The original plan was to close off the Eastern Scheldt, but by the late 1960s, both environmentalists and fishermen opposed its full closure. As an alternative, a storm surge barrier was designed that would normally be open and allow water to pass through, but would close in case the water at the sea side exceeded a certain level. Although significantly more costly than the original design, the storm surge barrier was considered to be the optimal solution because it was able to include both the value of safety and the value of ecology. For a discussion of how these values translate into different design goals, see the work by Edvardsson Björnberg (2013) on goal setting in the design of the Venice Storm surge barrier.

In this particular example, the ecological value was not included at the expense of safety. Opponents of the more recent "Room for the River" projects warn that these projects *do* actually come at the expense of safety (Warner and Van Buuren 2011). If this is indeed the case, it will be difficult to evaluate different flood risk strategies in quantitative terms. The original technical question (how to make a flood defence structure as safe as possible or how to achieve a particular level of safety) then turns into a more abstract question of prioritisation of values, which are probably "operationalized" differently in the different flood protection strategies. If we want to compare the ecological damage of a traditional "hard" intervention (e.g., dyke strengthening) with that of a "soft" intervention (e.g., a retention basin), the ecological damage produced by the former may be so different from that produced by the latter that the most we can say is that one strategy is preferable from an ecological point of view. We cannot quantitatively express this preference (Doorn 2014b). How can we compare in quantitative terms, for example, the ecological damage caused by a lowered ground water level to the extinction of a unique species? This same impossibility of quantification probably holds for other values, such as social-cultural ones. On a smaller scale, this is a trade-off that regional water boards need to make when deciding about the ground water level. The level that is preferable from an agricultural perspective is not necessarily preferable from a safety or ecological perspective. This suggests that flood risk governance, apart from being a technological challenge, also is a political one. Ultimately, these political decisions should be made by democratically legitimate bodies. Moreover, the political nature of flood risk governance also warrants the call for participatory approaches.

3.5 Participation

Participation is increasingly seen as an indispensable element of flood risk policy. The right to participation in water-related issues is also partly laid down in international conventions and directives. Participatory methods can be used for two different reasons (Rowe and Frewer 2004). They derive either from the recognition of the very nature of democracy, or they are a means to enrich the assessment and decision making through involving citizens and stakeholders in the process. In the former case, participation is considered a way to empower citizens and stakeholders; hence, the participatory process is a goal in itself (Maasen and Weingart 2005; Perhac 1998; Dryzek 1997). In the latter case, participation is a way to improve the quality of the decisions (Raadgever et al. 2012; Pahl-Wostl 2007; Brunner et al. 2005).

Regarding the democratic right to participation in environmental issues, on June 25 1998, the United Nations Economic Commission for Europe (UNECE) adopted the *Convention on Access to Information, Public Participation in Decision-Making and Access to Justice in Environmental Matters*, often referred to as the Aarhus Convention after the Danish city of Aarhus (Århus) where the Convention gathered. The Aarhus Convention establishes a number of rights of citizens and organizations with regard to:

1. The access to environmental information that is held by public authorities, including policies and measures taken;

2. Participation in environmental decision-making, including the possibility to comment on proposals for policies and interventions affecting or relating to the environment; and
3. Review procedures to challenge public decisions that have been made without respecting the two aforementioned rights or environmental law in general.

Although the provisions made in the Aarhus Convention are only indirectly implemented in the European Flood directive (see Art. 9.3 of the Flood directive), the Aarhus Convention is mentioned explicitly on the official EU website on the Flood directive.[5] This suggests that public participation in flood risk policy is considered important by the EU.

In practice, participatory approaches are as yet not systematically included in flood risk governance, although considerable effort has been made to involve stakeholders in drafting flood risk policy on an ad hoc basis. In Europe, quite a number of projects have been initiated by water authorities to ensure the involvement of key stakeholders in the implementation of the Water directive and the Flood directive (both are relevant for flood risk policy).[6]

Although the idea of participation (or public engagement) is supported almost unanimously, it turns out difficult to put it in practice. It is therefore questionable whether the underlying motivations (democratisation and improved decision-making) are actually achieved. The following concerns or challenges are mentioned in the literature on participatory approaches in the context of flood risk management and water policy:

1. Water authorities tend to focus on major stakeholders (mostly organizations) rather than individual citizens (Woods 2008). At the same time, practitioners notice a lack of willingness by individual citizens to become involved, partly because they see flood risk management as the sole responsibility of the government (WMO 2006). The WMO points at the importance of education in this regard. Other reported obstacles in securing the involvement of citizens are limited financial resources and practical barriers like stakeholders' spatial distribution (Almoradie et al. 2015) and the large amount of technicalities involved (Howarth 2009). Not all decisions lend themselves to stakeholder consultation. The more local the level at which the decisions are made, the more useful is the stakeholders' input (Woods 2008). Citizens of countries with decentralized water authorities are therefore at an advantage for successful participation.
2. There are limits to what can be achieved through public participation. For example, public participation cannot remove deeply rooted and conflicting interests (Van Buuren et al. 2013). Some flood risk management decisions involve zero-sum games, making it impossible to have a "mutual gain" for all parties involved. In those situations, the final decision should be made by

[5] http://ec.europa.eu/environment/water/flood_risk/implem.htm (last accessed: February 22, 2016).

[6] E.g., the UK (Nye et al. 2011; Woods 2008), Germany (Heintz et al. 2012), Italy (Soncini-Sessa 2007). See also Warner et al. (2013) for a comprehensive discussion.

political bodies (Lubell et al. 2013). Additionally, when transboundary aspects are at stake, international and bilateral agreements may be more important than stakeholder participation at the community level (cf. Elliott 2016). There is a potential tension between the need for global arrangements and a meaningful mandate at the lower community levels (Doorn 2013). Lastly, Howarth argues that the emphasis on procedures to include stakeholders ("proceduralization") in environmental legislation may come at the expense of substantive content (Howarth 2009). If the implementation of the European directives only requires that stakeholders are consulted, important environmental concerns may remain unaddressed.

3. Analysis of European flood risk legislation shows a lack of possibilities for EU citizens to rely on substantive provisions before the administrative courts (Bakker et al. 2013). This means that the third element in the Aarhus Convention ("access to justice") is currently not adequately implemented.

The points mentioned above indicate that participation is not without effort. Effective involvement of local stakeholders requires context-specific approaches with a focus on content (Doorn 2016). These participatory approaches should be complemented with adequate legal provisions before administrative courts.

4 Conclusions

In this chapter, I have shown how argumentative strategies are currently being employed in flood risk policy. The use of these strategies cannot be seen isolated from the "governance-turn" in flood risk policy. Dealing with flood risks is no longer a strictly technological issue; neither is flood safety the sole responsibility of the central government.

The preamble of the European Flood directive states that "Floods are natural phenomena which cannot be prevented. However, some human activities (such as increasing human settlements and economic assets in floodplains and the reduction of the natural water retention by land use) and climate change contribute to an increase in the likelihood and adverse impacts of flood events" (second consideration in the preamble). In other words, flood risks are partly a natural hazard and partly a man-made one. In practice, there are limits to the prevention of flooding by technological means; flood risks can only be controlled to some extent. With the deep uncertainties involved (both in terms of climate change but also in terms of demographic developments), future strategies in flood risk management will probably focus on reducing vulnerability and improving resilience; that is, on the adaptive capacity of the system.

Some important lessons could be derived from the discussion of the different strategies. The first concerns the distribution of responsibilities. Especially the section on goal setting showed a redistribution of responsibilities. Safety against flooding is no longer the sole responsibility of the central government. If decentralized governmental bodies and private parties (including citizens) get more responsibility, they should also have capacity to fulfill this responsibility. This means that money should be made available for capacity-building and education. The second lesson concerns the political dimension of flood risk governance. If flood risk management is more than a technological issue (a claim which I hope is not controversial after having read this chapter), flood risk policy should conform to appropriate democratic procedures. The last lesson concerns the use of participatory approaches. Participation is necessary, also in the light of the previous remark. At the same time, participation does not suffice for achieving adequate flood risk policy. More insight is needed into the effects of participatory approaches and methodologies on the actual content of the policy measures. Simply saying that the general public will be included is probably not sufficient to reach this public, let alone, to actually have it engaged. At the same time, some issues cannot be solved by simply involving the public. A mixture of traditional top-down approaches and local arrangements is required for adequately addressing the flood risk challenges.

Acknowledgement This research is supported by the Netherlands Organisation for Scientific Research (NWO) under grant number 016-144-071.

Recommended Readings

Haasnoot, M. (2013). *Anticipating change: Sustainable water policy pathways for an uncertain future*. Enschede: University of Twente.

Lankford, B., Bakker, K., Zeitoun, M., & Conway, D. (Eds.). (2013). *Water security: Principles, perspectives and practices*. New York: Earthscan/Routledge.

Warner, J. F. (2011). *Flood planning: The politics of water security*. London: I.B. Taurus.

References

Adger, W. N., Agrawala, S., Monirul Qader Mirza, M., Conde, C., O'Brien, K., Pulhin, J., Pulwarty, R., Smit, B., & Takahashi, K. (2007). Assessment of adaptation practices, options, constraints and capacity. In M. L. Parry, O. F. Canziani, J. P. Palutikof, P. J. Van der Linden, & C. E. Hanson (Eds.), *Climate change 2007: Impacts, adaptation and vulnerability. Contribution of working group II to the fourth assessment report of the Intergovernmental Panel on Climate Change* (pp. 717–743). Cambridge: Cambridge University Press.

Almoradie, A., Cortes, V. J., & Jonoski, A. (2015). Web-based stakeholder collaboration in flood risk management. *Journal of Flood Risk Management, 8*, 19–38.

Asselt, V., Marjolein, B. A., & Renn, O. (2011). Risk governance. *Journal of Risk Research, 14*, 573.

Bakker, M. H., Green, C., Driessen, P., Hegger, D. L. T., Delvaux, B., Rijswick, M. V., Suykens, C., Beyers, J.-C., Deketelaere, K., Doorn-Hoekveld, W., & Dieperink, C. V. (2013). *Flood risk management in Europe: European flood regulation [Star-Flood Report Number D1.1.1]*. Utrecht: Utrecht University.

Bell, S. (2002). *Economic governance and institutional dynamics*. Oxford: Oxford University Press.

Betz, G. (2016). Accounting for possibilities in decision making. In S. O. Hansson & G. Hirsch Hadorn (Eds.), *The argumentative turn in policy analysis. Reasoning about uncertainty* (pp. 135–169). Cham: Springer. doi:10.1007/978-3-319-30549-3_6.

Bijker, E. W. (1996). History and heritage in coastal engineering in the Netherlands. In N. C. Kraus (Ed.), *History and heritage of coastal engineering* (pp. 390–412). New York: American Society of Civil Engineers.

Bijker, W. E. (2002). The Oosterschelde storm surge barrier: A test case for Dutch water technology, management, and politics. *Technology and Culture, 43*, 569–584.

Bijker, W. E. (2007). American and Dutch coastal engineering: Differences in risk conception and differences in technological culture. *Social Studies of Science, 37*, 143–151.

Borman, T. C. (1995). Deltawet grote rivieren. *Ars Aequi, 44*, 594–603.

Brunner, R. D., Steelman, T. A., Coe-Juell, L., Cromley, C. M., Edwards, C. M., & Tucker, D. W. (2005). *Adaptive governance: Integrating science, policy and decision-making*. New York: Columbia University Press.

Bubeck, P., Kreibich, H., Penning-Rowsell, E. C., Wouter Botzen, W. J., De Moel, H., & Klijn, F. (2013). Explaining differences in flood management approaches in Europe and the USA. In F. Klijn & T. Schweckendiek (Eds.), *Comprehensive flood risk management: Research for policy and practice* (pp. 1199–1209). London: Taylor & Francis Group.

Burton, I., Malone, E., Huq, S., Lim, B., & Spanger-Siegfried, E. (2005). *Adaptation policy frameworks for climate change: Developing strategies, policies and measures*. Cambridge: Cambridge University Press.

Butler, C., & Pidgeon, N. (2011). From 'flood defence' to 'flood risk management': Exploring governance, responsibility, and blame. *Environment and Planning C – Government & Policy, 29*, 533–547.

Carter, T. R., Jones, R. N., Lu, X., Bhadwal, S., Conde, C., Mearns, L. O., O'Neill, B. C., Rounsevell, M. D. A., & Zurek, M. B. (2007). New assessment methods and the characterisation of future conditions. In M. L. Parry, O. F. Canziani, J. P. Palutikof, P. J. Van der Linden, & C. E. Hanson (Eds.), *Climate change 2007: Impacts, adaptation and vulnerability. Contribution of working group II to the fourth assessment report of the Intergovernmental Panel on Climate Change* (pp. 133–171). Cambridge: Cambridge University Press.

Cashman, A. C. (2011). Case study of institutional and social responses to flooding: Reforming for resilience? *Journal of Flood Risk Management, 4*, 33–41.

CRED. (2009). *Annual disaster statistical review 2008: The numbers and trends*. Brussels: Centre for Research on the Epidemiology of Disasters (CRED).

Davis, T., Rogers, H., Shays, C., Bonilla, H., Buyer, S., Myrick, S., Thornberry, M., Granger, K., Pickering, C. W., Shuster, B., & Miller, J. (2006). *A failure of initiative. The final report of the select bipartisan committee to investigate the preparation for and response to Hurricane Katrina*. Washington, DC: U.S. Government Printing Office.

Dessai, S., & Van der Sluijs, J. P. (2007). *Uncertainty and climate change adaptation – a scoping study [report NWS-E-2007-198]*. Utrecht: Copernicus Institute, Utrecht University.

Disco, C. (2006). Delta blues. *Technology and Culture, 47*, 341–348.

Doorn, N. (2013). Water and justice: Towards an ethics for water governance. *Public Reason, 5*, 95–111.

Doorn, N. (2014a). Equity and the ethics of water governance. In A. Gheorghe, M. Masera, & P. F. Katina (Eds.), *Infranomics – sustainability, engineering design and governance* (pp. 155–164). Dordrecht: Springer.

Doorn, N. (2014b). Rationality in flood risk management: The limitations of probabilistic risk assessment (PRA) in the design and selection of flood protection strategies. *Journal of Flood Risk Management, 7*, 230–238. doi:10.1111/jfr3.12044.

Doorn, N. (2015). The blind spot in risk ethics: Managing natural hazards. *Risk Analysis, 35*, 354–360. doi:10.1111/risa.12293.

Doorn, N., & Hansson, S. O. (2011). Should probabilistic design replace safety factors? *Philosophy & Technology, 24*, 151–168. doi:10.1007/s13347-010-0003-6.
Doorn, N. (2016). Governance experiments in water management: From interests to building blocks. *Science and Engineering Ethics*. doi:10.1007/s11948-015-9627-3.
Driesssen, P. J., & Van Rijswick, H. F. M. W. (2011). Normative aspects of climate adaptation policies. *Climate Law, 2*, 559–581.
Dryzek, J. S. (1997). *The politics of the earth: Environmental discourses*. Oxford: Oxford University Press.
Edvardsson Björnberg, K. (2013). Rational goals in engineering design: The Venice dams. In M. J. De Vries, S. O. Hansson, & A. W. M. Meijers (Eds.), *Norms in technology* (pp. 83–99). Dordrecht: Springer.
Edvardsson Björnberg, K. (2016). Setting and revising goals. In S. O. Hansson & G. Hirsch Hadorn (Eds.), *The argumentative turn in policy analysis. Reasoning about uncertainty* (pp. 171–188). Cham: Springer. doi:10.1007/978-3-319-30549-3_7.
EEA. (2010). *Mapping the impacts of natural hazards and technological accidents in Europe: An overview of the last decade (European Environment Agency)*. Luxembourg: Publications Office of the European Union.
Elliott, K. C. (2016). Climate geoengineering. In S. O. Hansson & G. Hirsch Hadorn (Eds.), *The argumentative turn in policy analysis. Reasoning about uncertainty* (pp. 305–324). Cham: Springer. doi:10.1007/978-3-319-30549-3_13.
Espinoza, N., & Peterson, M. (2008). Incomplete preferences in disaster risk management. *International Journal of Technology, Policy and Management, 8*, 341–358.
Füssel, H.-M. (2007). Adaptation planning for climate change: Concepts, assessment approaches, and key lessons. *Sustainability Science, 2*, 265–275.
Grüne-Yanoff, T. (2016). Framing. In S. O. Hansson & G. Hirsch Hadorn (Eds.), *The argumentative turn in policy analysis. Reasoning about uncertainty* (pp. 189–215). Cham: Springer. doi:10.1007/978-3-319-30549-3_8.
Haasnoot, M. (2013). *Anticipating change: Sustainable water policy pathways for an uncertain future*. Enschede: University of Twente.
Hallegatte, S., Shah, A., Lempert, R.J., Brown, C., & Gill, S. (2012). *Investment decision making under deep uncertainty application to climate change*. Tech. Rep. Policy research working paper 6193. http://elibrary.worldbank.org/doi/pdf/10.1596/1813-9450-6193. Accessed 5 May 2015.
Hansson, S. O. (2009). From the casino to the jungle. *Synthese, 168*, 423–432. doi:10.1007/s11229-008-9444-1.
Hansson, S. O., & Hirsch Hadorn, G. (2016). Introducing the argumentative turn in policy analysis. In S. O. Hansson & G. Hirsch Hadorn (Eds.), *The argumentative turn in policy analysis. Reasoning about uncertainty* (pp. 11–35). Cham: Springer. doi:10.1007/978-3-319-30549-3_2.
Heintz, M. D., Hagemeier-Klose, M., & Wagner, K. (2012). Towards a risk governance culture in flood policy: Findings from the implementation of the "Floods Directive" in Germany. *Water, 4*, 135–156.
Hirsch Hadorn, G. (2016). Temporal strategies for decision making. In S. O. Hansson & G. Hirsch Hadorn (Eds.), *The argumentative turn in policy analysis. Reasoning about uncertainty* (pp. 217–242). Cham: Springer. doi:10.1007/978-3-319-30549-3_9.
Howarth, W. (2009). Aspirations and realities under the water framework directive: Proceduralisation, participation and practicalities. *Journal of Environmental Law, 21*, 391–417.
IPCC. (2007). *Climate change 2007: The physical science basis. Working group 1 contribution to the fourth assessment report of the IPCC*. Cambridge: Cambridge University Press.
IPCC. (2014). *Climate change 2013: The physical science basis. Working group 1 contribution to the fifth assessment report of the IPCC (draft)*. Cambridge: Cambridge University Press.
Keynes, J. M. (1921). *A treatise on probability*. London: Macmillan.
Klinke, A., & Renn, O. (2002). A new approach to risk evaluation and management: Risk-based, precaution-based and discourse-based management. *Risk Analysis, 22*, 1071–1094.
Knight, F. H. (1935[1921]). *Risk, uncertainty and profit*. Boston: Houghton Mifflin.

Kwadijk, J. C. J., Haasnoot, M., Mulder, J., Hoogvliet, M., Jeuken, A., Van der Krogt, R., Van Oostrom, N., Schelfhout, H., Van Velzen, E., Van Waveren, H., & De Wit, M. (2010). Using adaptation tipping points to prepare for climate change and sea level rise: A case study in the Netherlands. *Wiley Interdisciplinary Reviews: Climate Change, 1*, 729–740.

Lempert, R. J., Popper, S., & Bankes, S. (2003). *Shaping the next one hundred years: New methods for quantitative, long term policy analysis* (Technical Report MR-1626-RPC). Santa Monica: RAND Corporation.

Lintsen, H. (2002). Two centuries of central water management in the Netherlands. *Technology and Culture, 43*, 549–568.

Löfstedt, R. E. (2005). *Risk management in post-trust societies*. Hampshire: Palgrave.

Lubell, M., Gerlak, A., & Heikkila, T. (2013). CalFed and collaborative watershed management: Success despite failure? In J. F. Warner, A. Van Buuren, & J. Edelenbos (Eds.), *Making space for the river: Governance experiences with multifunctional river flood management in the US and Europe* (pp. 63–78). London: IWA Publishing.

Maasen, S., & Weingart, P. (2005). *Democratization of expertise? Exploring novel forms of scientific advice in political decision-making*. Dordrecht: Springer.

McDaniels, T. L., Gregory, R. S., & Fields, D. (1999). Democratizing risk management: Successful public involvement in local water management decisions. *Risk Analysis, 19*, 497–510.

Meijerink, S., & Dicke, W. (2008). Shifts in the public-private divide in flood management. *International Journal of Water Resources Development, 24*, 499–512. doi:10.1080/07900620801921363.

Merz, B., & Emmermann, R. (2006). Zum Umgang mit Naturgefahren in Deutschland. Vom Reagieren zum Risikomanagement. *GAIA, 15*, 265–274.

Millstone, E., Van Zwanenberg, P., Marris, C., Levidow, L., & Torgesen, H. (2004). *Science in trade disputes related to potential risks: Comparative case studies*. Seville: Institute for Prospective Technological Studies (JRC-IPTS).

Möller, N. (2016). Value uncertainty. In S. O. Hansson & G. Hirsch Hadorn (Eds.), *The argumentative turn in policy analysis. Reasoning about uncertainty* (pp. 105–133). Cham: Springer. doi:10.1007/978-3-319-30549-3_5.

Mostert, E., & Doorn, N. (2012). The European flood risk directive and ethics. *Water Governance, 2*, 10–14.

Nye, M., Tapsell, S., & Twigger-Ross, C. (2011). New social directions in UK flood risk management: Moving towards flood risk citizenship? *Journal of Flood Risk Management, 4*, 288–297.

Pahl-Wostl, C. (2007). Transitions towards adaptive management of water facing climate and global change. *Water Resources Management, 21*, 49–62.

Perhac, R. M. (1998). Comparative risk assessment: Where does the public fit in? *Science, Technology & Human Values, 23*, 221–241.

Peterson, M. (2003). Risk, equality, and the priority view. *Risk Decision and Policy, 8*, 17–23.

Raadgever, G. T., Mostert, E., & Van de Giesen, N. C. (2012). Learning from collaborative research in water management practice. *Water Resources Management, 26*, 3251–3266.

Renn, O. (2008). *Risk governance: Coping with uncertainty in a complex world*. London: Earthscan.

Rittel, H. W. J., & Webber, M. M. (1973). Dilemmas in a general theory of planning. *Policy Sciences, 4*, 155–169.

Rowe, G., & Frewer, L. J. (2004). Evaluating public-participation exercises: A research agenda. *Science, Technology & Human Values, 29*, 512–557.

Schefczyk, M. (2016). Financial markets: The stabilisation task. In S. O. Hansson & G. Hirsch Hadorn (Eds.), *The argumentative turn in policy analysis. Reasoning about uncertainty* (pp. 265–290). Dordrecht: Springer. doi:10.1007/978-3-319-30549-3_11.

Schön, D. A., & Rein, M. (1994). *Frame reflection: Towards the resolution of intractable policy controversies*. New York: Basic Books.

Scrase, J. I., & Sheate, W. R. (2005). Re-framing flood control in England and Wales. *Environmental Values, 14*, 113–137.

Smith, K., & Petley, D. N. (2009). *Environmental hazards: Assessing risk and reducing disaster*. London: Routledge.

Soncini-Sessa, R. (Ed.). (2007). *Integrated and participatory water resources management: Practice [volume 1, part B]*. Amsterdam: Elsevier.

Van Buuren, A., Edelenbos, J., & Warner, J. F.(2013). Space for the river: Governance challenges and lessons. In J. F. Warner, A. Van Buuren, & J. Edelenbos (Eds.), *Making space for the river: Governance experiences with multifunctional river flood management in the US and Europe* (pp. 187–201). London: IWA Publishing.

Vink, M. J., Boezeman, D., Dewulf, A., & Termeer, C. J. A. M. (2013). Changing climate, changing frames Dutch water policy frame developments in the context of a rise and fall of attention to climate change. *Environmental Science & Policy, 30*, 90–101. doi:10.1016/j.envsci.2012.10.010.

Vrijling, J. K. (2009). *The lessons from New Orleans, risk and decision analysis in maintenance optimization and flood management*. Delft: IOS Press.

Warner, J. F. (2011). *Flood planning: The politics of water security*. London/New York: I.B. Tauris.

Warner, J. F., & Van Buuren, A. (2011). Implementing room for the river: Narratives of success and failure in Kampen, the Netherlands. *International Review of Administrative Sciences, 77*, 779–801. doi:10.1177/0020852311419387.

Warner, J. F., Van Buuren, A., & Edelenbos, J. (Eds.). (2013). *Making space for the river: Governance experiences with multifunctional river flood management in the US and Europe*. London: IWA Publishing.

Wiegel, R. L., & Saville, T. (1996). History of coastal engineering in the USA. In N. C. Kraus (Ed.), *History and heritage of coastal engineering* (pp. 513–600). Washington, DC: American Society of Civil Engineers.

WMO. (2006). *Social aspects and stakeholder involvement in integrated flood management*. APFM technical document No. 4. http://www.adpc.net/v2007/Resource/downloads/socialaspect13oct_2.pdf. Accessed 5 May 2015.

Wolf, K. D. (2002). Contextualizing normative standards for legitimate governance beyond the state. In J. R. Grote & B. Gbikpi (Eds.), *Participatory governance: Political and societal implications* (pp. 35–50). Opladen: Leske + Budrich Verlag.

Wolsink, M. (2006). River basin approach and integrated water management: Governance pitfalls for the Dutch space-water-adjustment management principle. *Geoforum, 37*, 473–487.

Woods, D. (2008). Stakeholder involvement and public participation: A critique of water framework directive arrangements in the United Kingdom. *Water and Environment Journal, 22*, 258–264.

Chapter 11
Financial Markets: Applying Argument Analysis to the Stabilisation Task

Michael Schefczyk

> Reality is immensely more complex than models, *with millions of potential weak links. Ex post, it is easy to highlight the one that blew up, but ex ante is a different matter.*
>
> (Caballero and Kurlat 2009: 20)

Abstract This article applies argument analysis techniques in order to identify shortcomings in Alan Greenspan's justification for the Federal Reserve's inactivity regarding the housing price boom between 2002 and 2005. The application of argument analysis techniques does not only help to detect fallacies in the argumentative underpinning of a policy. Such techniques also help to raise awareness for dubious premises and make it more likely that the need to adjust confidence will be recognized. I thus conclude that their use has the potential to improve stabilisation policy in the future.

Keywords Great recession • Risk management approach • Alan Greenspan • Federal reserve • Housing price bubble • Argument analysis

1 Introduction

Among other things, central banks have the task of maintaining the stability of the financial system and containing systemic risk (*stabilisation task*). Modern financial systems are vulnerable to banking crises, and it is a core task of central banks to

The paper profited very much from comments by the editors, Gregor Betz, Georg Brun and the participants in a workshop on uncertainty at the ETH Zürich.

M. Schefczyk (✉)
Karlsruhe Institute of Technology, Karlsruhe, Germany
e-mail: michael.schefczyk@kit.edu

S.O. Hansson, G. Hirsch Hadorn (eds.), *The Argumentative Turn in Policy Analysis*, Logic, Argumentation & Reasoning 10, DOI 10.1007/978-3-319-30549-3_11

prevent them. A typical sequence of events leading to a banking crisis is the following (see Cooper 2008; Galbraith 1990/1993; Mackay 1841/1995; Minsky 1986/2008): A large expansion in credit, for instance due to low interest rates or increased market optimism, causes an increased demand for assets in fixed supply. As a consequence, the prices of these assets rise. Rising asset prices attract investors, who speculate that the price trend will continue. Price rises due to increased demand by speculative investors attract more speculative investors. Finally, the price level exceeds market fundamentals and is then driven by so-called "speculative debt"— that is, debt which can only be serviced if the price of the asset does not fall. The perception of constant price rises eventually causes growing concern among market participants about a possible trend reversal. More and more investors are ready to sell. "Small events" (Allen and Gale 2007: 126ff.) are often interpreted as indicators a reversal of the price trend is about to take place and investors start selling. The falling prices due to this selling make speculative debt incurred increasingly unserviceable. Customers and business partners of financial institutions which financed the speculative purchases of these assets begin to be concerned about their possible insolvency. As a precautionary measure, they withdraw deposits and stop making transactions. As the affected financial institutions become insolvent, a banking crisis is created.

This is, roughly, the pattern of events in September 2008 which caused the most severe financial crisis in a century. The years before had been marked by a strong increase in US property prices. This increase, in turn, had resulted from a remarkable growth in credit and speculative debt. Lehmann Brothers, a huge financial institution, had a significant share of mortgage-related securities on its balance sheet and was thus heavily exposed to the danger of a reversal in housing prices (Kindleberger and Aliber 1978/2011: 257). When it went bankrupt in 2008, a panic ensued.

Ben Bernanke, then chairman of the Federal Reserve Board, claimed that the regulators could not have foreseen the danger (Angelides et al. 2011: 3). The official report of the US Financial Crisis Inquiry Commission, however, concludes that the collapse was neither unforeseeable nor completely unforeseen and that "profound lapses in regulatory oversight" (Angelides et al. 2011: xxviii) contributed to the instability of the financial system.

In retrospect, to be sure, the mechanisms which produced the global economic and financial crisis seem straightforward (Sinn 2010/2011; Stiglitz 2010; Krugman 1999/2009; Posner 2009; Wolf 2009; Soros 2008/2009; Shiller 2008). But, at the time, they were neither obvious to the policymakers at the Federal Reserve nor to the vast majority of economic experts. Why did so few anticipate the imminent danger? One answer blames ideological blinders. According to Paul Krugman (2009) and others, pre-crisis mainstream economics was strongly biased towards the view that financial markets are inherently stable (*stability view*). An extreme version of the stability view, the efficient market hypothesis (EMH), even denies the existence of economic bubbles; in this view, EMH might have led regulators to ignore the potential dangers for the financial system from a drastic decline of inflated house prices. Some have thus argued that the crisis was a kind of false negative (Stiglitz 2014; Bezemer 2009).[1] Although available at the time, theoretical alternatives which would have enabled policymakers to assess the risks more realistically were not considered. Besides ideological blinders, the economist Robert Shiller argues that the "social contagion of boom thinking" (Shiller 2000/2015, 2008) was a reason why regulators and economists failed to identify the danger. Boom thinking neutralises worries about rapidly rising asset prices with what Shiller calls "new era stories". Such stories purport to provide reasons to believe that past experience is misleading for the understanding of current economic affairs in general and price booms in particular. Regulators are not immune to social contagion by new era thinking (Shiller 2008: 51–52). Furthermore, Shiller established "from the Lexis-Nexis database that in the English language the term *new era economy* did not have any currency until a *Business Week* cover story in July 1997 attributed this term to Alan Greenspan, marking an alleged turning point in his thinking since the 'irrational exuberance' speech some months earlier" (Shiller 2000/2015: 124). According to Shiller, the social contagion of new era thinking, which destabilised financial markets, originated from an announcement by no less a figure than the then chairman of the Federal Reserve.

In this article, I examine various public announcements of Alan Greenspan in order to do three things: First, I analyse how Greenspan conceived the role of uncertainty for central bank policy (Sect. 2). Second, Greenspan's arguments for inactivity with regard to the housing market are reconstructed in detail (Sect. 3). Third, I show that Greenspan's position was open to serious objections at the time (Sect. 4).[2]

The argument analysis of this article reveals that neither the stability thesis nor uncritical new era thinking loomed large in Greenspan's view of the stabilisation task. His decision to stay inactive was mainly based on considerations concerning uncertain causes of price developments and the relative costs of intervention.[3] The flaws of Greenspan's position are obvious in retrospect. This article contends that the application of argument analysis techniques makes the discovery of unreasonable policy positions easier and thus more likely; in particular, it might thereby contribute to the improvement of stabilisation policy.

[1] "No one would, or at least should, say that macroeconomics has done well in recent years. The standard models not only didn't predict the Great Recession, they also said it couldn't happen—bubbles don't exist in well-functioning economies of the kind assumed in the standard model." (Stiglitz 2014: 1).

[2] For an introduction to reconstructing and assessing arguments see Brun and Betz (2016).

[3] For an overview on rules for the evaluation and prioritization of uncertainties see Hansson (2016).

2 Uncertainty and the Risk-Management Approach

According to the traditional approach in policy analysis, central banks should choose the path of action which best advances the bank's objectives in view of the most likely development of the economy. This presumes that either the policymakers can be certain of outcomes or that any lack of certainty is irrelevant for the plan of action. If central bank policy is based on economic models which only allow for uncertainty from "random shocks", the optimal policy plan is "certainty-equivalent" (Jenkins and Longworth 2002: 4–5; Batini et al. 1999: 183–184). Uncertainty is practically irrelevant in these models.

Greenspan's risk-management approach, by contrast, is based on the view that uncertainty "is not just a pervasive feature of the monetary policy landscape; it is the defining feature of that landscape" (Greenspan 2003: 1, 2004: 36).

The economic literature distinguishes between three types of uncertainty facing central banks (Dennis 2005). First, they can be uncertain about the data, as sometimes measurements are difficult, or there is no or no sufficiently complete data set (*data uncertainty*).[4] Second, policymakers face uncertainty regarding particular parameters within a given model, such as price elasticity or the turnover rate (*parameter uncertainty*). Third, sometimes the available models do not fully capture crucial structural aspects of the economy (*model uncertainty*). Data, parameter, and model uncertainty may affect both the policymakers' knowledge about the state of the economy and the effects of a policy action on the economy.

In the relevant literature, the notion of "uncertainty" signifies both situations *with* and situations *without* known probability distributions. In practice, Alan Greenspan emphasised, the distinction between "risk" and "Knightian uncertainty" is often difficult to make, so that "one is never quite sure what type of uncertainty one is dealing with" (Greenspan 2004: 36). Greenspan seems to differentiate between uncertainty *within* a model and uncertainty *about* a model. A policymaker may use a model which employs a particular probability distribution (uncertainty as risk) but may be uncertain about the adequacy of the model itself; if the model were wrong, the policymakers would be dealing with Knightian uncertainty instead of risk. Which situation obtains, "one is never quite sure".[5]

In contrast to the traditional approach to central bank policy, the risk-management approach considers the outcomes of different scenarios and assesses their respective probabilities.[6] These probability assessments often cannot be based

---

[4] Shiller reports that in 2004 there were no data on long-term performance for home prices in the US or other countries (Shiller 2008: 31).

[5] For an overview on different notions of uncertainty and risk see Hansson and Hirsch Hadorn (2016).

[6] "For example, policy A might be judged as best advancing the policymakers' objectives, conditional on a particular model of the economy, but might also be seen as having relatively severe adverse consequences if the true structure of the economy turns out to be other than the one assumed. On the other hand, policy B might be somewhat less effective in advancing the policy objectives under the assumed baseline model but might be relatively benign in the event that the structure of the economy turns out to differ from the baseline" (Greenspan 2004: 37).

on established macroeconomic models or past experience, but have to rely on judgement; such "judgments, by their nature, are based on bits and pieces of history that cannot formally be associated with an analysis of variance" (Greenspan 2004: 39).

Martin Feldstein has argued that Greenspan's risk-management approach amounts to the Bayesian theory of decision-making (Feldstein 2004: 42). The central bank assigns subjective probabilities to states of the world and to the correctness of theories; the optimal policy is then the one with the highest expected utility in terms of the bank's targets. However, Greenspan stresses that it is often impossible to quantify risks with any confidence. Central banks are permitted to act on the basis of uncertain judgments under two conditions: (a) The action insures the economy against very adverse outcomes within the scope of the bank's responsibility; (b) the costs of the action are low in terms of the bank's objectives, namely maximum long-term economic growth and price stability. Greenspan refers to such actions as insurance measures.

Insurance measures may contain some information about implicit subjective probability and cost assessment. But this information is too unspecific to produce a concrete figure which could be employed in maximizing expected utility (see also Blinder and Reis 2005: 18–24).[7]

3 Reasoning About Bubbles: Greenspan's Arguments for Inactivity

This section analyses 14 of Chairman Greenspan's arguments for not taking countervailing measures against surging prices in the housing market.[8] Before I turn to Greenspan's announcements, I shall briefly discuss a conjecture of Paul Krugman (2009), Joseph Stiglitz (2010: 269–270) and others. They claim that Greenspan failed to take measures against the possible build-up of a housing bubble because he accepted the efficient market hypothesis (EMH).

The reasoning behind EMH, which once passed for the best corroborated theory in economics (Jensen 1978: 95), is the following (Shleifer 2000): There are three types of investors: Rational types who value investments adequately on the basis of information (A-types); irrational types who trade randomly (B-types); and irrational types who imitate the trades of other investors (C-types). (a) If the market is populated by A-types, the price adequately represents all available information. (b) If the market is populated by A-types and B-types, the trades of the B-types cancel each other out; A-types will not be influenced by B-types; thus, the price adequately represents all available information. (c) If the market is populated by A,

[7] This is the fallacy of treating uncertain probability estimates as certain (Hansson 2016).

[8] For an overview on core arguments for inactivity and counter arguments in this debate see Fig. 11.1 at the end of Sect. 4.

B, and C-types, the potentially distorting effect of C-types on prices (when imitating B-types) will be neutralised by A-types who use arbitrage opportunities,[9] i.e. they sell (short) overpriced and buy (long) underpriced items until the price adequately represents all available information. Thus, as long as there is a critical number of A-types "the price must always be right"; bubbles are impossible and changes in market prices in t_1 must be described as random movements in t_0 (because the information which the price change in t_1 responds to is not known in t_0 – if it were known, it would have already been included in the price).

If we follow Krugman, Stiglitz, and others, Greenspan's argument for inactivity has roughly this form:

Reconstruction 1 (Inactivity Argument)[10]
Sub-Argument 1 (R1SA1) (theoretical)

Premise 1	If EMH applies to market *M* in period *t*, economic bubbles cannot occur in *M* in *t*.
Premise 2	EMH applies to *M* in *t*.
Conclusion	Economic bubbles cannot occur in *M* in *t*.

Sub-Argument 2 (R1SA2) (practical)

Premise 1	It is irrational to take measures against events which cannot occur.
Premise 2	Economic bubbles cannot occur in *M* in *t*.
Conclusion	It is irrational to take measures against economic bubbles in *M* in *t*.

Both sub-arguments in R1 are valid. Premise 1 in R1SA1 is uncontroversial. EMH implies the impossibility of bubbles.[11] In order to judge the truth of premise 2 in R1SA1, one has, first, to interpret the meaning of "EMH applies to *M* in *t*". I shall assume that "EMH applies to *M* in *t*" roughly means that "EMH is a well corroborated model of *M* in *t*". Second, one has to specify the market *M* in *t*; in this case, *M* refers to the US housing market after 2000. Thus, the truth of premise 2 depends on whether EMH is a well corroborated model of the US housing market after 2000. There are very good reasons to be sceptical about this claim (cf. Shiller 2000/2015). But we can bypass a further discussion of R1 as Greenspan did not

[9] Arbitrage is the "purchase of one security and simultaneous sale of another to give a risk-free profit" (Brealey and Myers 1981/1991: G1).

[10] Strictly speaking, the following is an informal argument scheme. Thus, the term "reconstruction" as used here has to be taken with a pinch of salt.

[11] Its main author, Eugene Fama, even went so far as to remark: "The word 'bubble' drives me nuts. For example, people say 'the Internet bubble'. Well, if you go back to that time, most people were saying the Internet was going to revolutionize business, so companies that had a leg up on the Internet were going to become very successful" (https://www.minneapolisfed.org/publications/the-region/interview-with-eugene-fama).

accept the conclusion of R1SA1. He repeatedly considered the possibility of a bubble on the US housing market after 2000 (pars pro toto Greenspan 2002a, b) and thus rejected premise 2 in R1SA2.[12]

I shall now examine documents in which Greenspan addresses "concerns about the possible emergence of a bubble in home prices" (Greenspan 2002a). The first document under scrutiny is a testimony before the Joint Economic Committee of the US Congress on 17 April 2002; it contains the following passage:

> The ongoing strength in the housing market has raised concerns about the possible emergence of a bubble in home prices. However, the analogy often made to the building and bursting of a stock price bubble is imperfect. . . . [U]nlike in the stock market, sales in the real estate market incur substantial transactions costs and, when most homes are sold, the seller must physically move out. Doing so often entails significant financial and emotional costs and is an obvious impediment to stimulating a bubble through speculative trading in homes. Thus, while stock market turnover is more than 100 percent annually, the turnover of home ownership is less than 10 percent annually—scarcely tinder for speculative conflagration. (Greenspan 2002a)

I propose to call the argument in this passage the *turnover argument*. The turnover argument justifies the view that under normal circumstances there are no bubbles in the real estate market.

Reconstruction 2 (Turnover Argument)

Premise 1	Sales in the real estate market incur substantial transaction costs.
Premise 2	If transaction costs are substantial, the market turnover is low.
Premise 3	If market turnover is low, no "speculative conflagration" develops.
Premise 4	If no "speculative conflagration" develops, prices do not rise significantly above their fundamentals.
Premise 5	If prices do not rise significantly above their fundamentals, bubbles do not occur.
Conclusion	Thus, bubbles do not occur in the real estate market.

In the following passage, Greenspan addresses a further difference between the stock market and the real estate market.

> A home in Portland, Oregon is not a close substitute for a home in Portland, Maine, and the "national" housing market is better understood as a collection of small local housing markets. Even if a bubble were to develop in a local market, it would not necessarily have implications for the nation as a whole. (Greenspan 2002a)

I propose to call the argument in this passage the *spatial fragmentation argument*. The spatial fragmentation argument justifies the view that under normal circumstances there are no bubbles in the real estate market.

---

[12] With reference to the stock market in the summer of 2000, Greenspan remarks that prices "had risen to levels in excess of any economically supportable base" (Greenspan 2002b: 3).

Reconstruction 3 (Spatial Fragmentation Argument)

Premise 1	The US housing market is a collection of local markets.
Premise 2	It is unlikely that bubbles in local markets have strong detrimental effects on the economy of the whole nation.
Conclusion	Thus, it is unlikely that bubbles on the US housing market have strong detrimental effects on the economy of the whole nation.

The spatial fragmentation argument hedges the turnover argument. If, against expectations, bubbles were to develop on the US real estate market, they would be, in all likelihood, a limited number of local phenomena which would not pose a threat for the economy as a whole.

In a testimony before the Joint Committee on 9 June 2005, Greenspan retreated from the turnover argument, which had proved untenable in the light of new developments.

> [I]n recent years, the pace of turnover of existing homes has quickened. It appears that a substantial part of the acceleration in turnover reflects the purchase of second homes—either for investment or vacation purposes. Transactions in second homes, of course, are not restrained by the same forces that restrict the purchases or sales of primary residences—an individual can sell without having to move. This suggests that speculative activity may have had a greater role in generating the recent price increases than it has customarily had in the past. (Greenspan 2005a)

Surging home turnover and a steep climb in home prices contradict premise 1 and 2 of the turnover argument of June 2002. Greenspan responded to this contradiction (a) by distinguishing between two types of transaction on the housing market, namely transactions in primary residences and transactions in second homes, and (b) by limiting the scope of the turnover argument. In its limited form, the turnover argument claims that bubbles cannot occur in markets for primary residences. But, since the transaction costs of the sale of second homes are low enough to allow for high turnover and speculative activity, bubbles can develop.

Reconstruction 4 (Speculative Conflagration Argument)

Premise 1	Sales of second homes do not incur substantial transaction costs.
Premise 2	If transaction costs are not substantial, the market turnover can be high.
Premise 3	If market turnover can be high, "speculative conflagration" can develop.
Premise 4	If speculative conflagration can develop, prices can rise significantly above their fundamentals.
Premise 5	If prices can rise significantly above their fundamentals, bubbles can occur.
Conclusion	Thus, bubbles can occur in the real estate market.

In essence, Greenspan stood fast to the spatial fragmentation argument of 2002. In one passage, he added considerations which were meant to endorse premise 2 in view of possible objections:

> Although we certainly cannot rule out home price declines, especially in some local markets, these declines, were they to occur, likely would not have substantial macroeconomic implications. Nationwide banking and widespread securitization of mortgages make it less likely that financial intermediation would be impaired than was the case in prior episodes of regional house price corrections. (Greenspan 2005a)

Can a correction of regional house prices impair financial intermediation and thus have substantial macroeconomic implications for the US? The question arises because Greenspan takes note of the possibility that some households make use of "exotic forms of mortgages" in order to "purchase a home that would otherwise be unaffordable" (Greenspan 2005a). In the case of a price reversal, mortgage lenders face the danger of massive losses and bankruptcy. As a consequence, house price corrections have impaired financial intermediation in the past. Why is Greenspan not overly concerned?

Reconstruction 5 (Diversification Argument)

Premise 1	Nationwide banking and widespread securitization diversify the risk of mortgage lenders (in the case of home price declines in some local markets).
Premise 2	The diversification of risk makes the impairment of financial intermediation (in the case of home price declines in some local markets) unlikely.
Conclusion	Thus, nationwide banking and widespread securitization make the impairment of financial intermediation (in the case of home price declines in some local markets) unlikely.

A financial crisis is unlikely, but not impossible. Greenspan added the following reflection in order to strengthen his point that the situation in the housing market did not pose a substantial threat to the US economy:

> Moreover, a substantial rise in bankruptcies would require a quite-significant overall reduction in the national housing price level because the vast majority of homeowners have built up substantial equity in their homes despite large home equity withdrawals in recent years financed by the mortgage market. (Greenspan 2005a)

Reconstruction 6 (Financial Intermediation Argument)

Premise 1	If the national housing price level were to drop significantly, financial intermediation would be impaired.
Premise 2	If financial intermediation were to be impaired, this would have detrimental macroeconomic effects.
Premise 3	A significant drop in the national housing price level is unlikely in period *t*.
Conclusion	It is unlikely in *t* that detrimental macroeconomic effects occur as a consequence of a significant drop in the national housing price level.

Why did Greenspan think that a significant reduction in the national housing price level was unlikely (premise 3)? The implicit assumption in the quoted passage seems to be that a significant reduction in housing prices can only occur as the result of widespread foreclosures. However, widespread foreclosures were unlikely since the "vast majority of homeowners have built up substantial equity in their homes" (Greenspan 2005a).

Greenspan gave another argument to the effect that a significant reduction in the national housing price level was unlikely:

> [P]roductivity gains in residential construction have lagged behind the average productivity increases in the United States for many decades. This shortfall has been one of the reasons that house prices have consistently outpaced the general price level for many decades. (Greenspan 2005a)

Reconstruction 7 (Productivity Shortfalls Argument)

Premise 1	Price rises in *M* in *t* above the general price trend are either due to speculative activity or to productivity shortfalls.
Premise 2	Speculative activity cannot go on for decades.
Premise 3	Productivity shortfalls can go on for decades.
Premise 4	Price rises in *M* in *t* above the general price trend have gone on for decades.
Conclusion	Thus, price rises in *M* in *t* are not due to speculative activity, but to productivity shortfalls.

In a nutshell, between April 2002 and June 2005 Alan Greenspan developed a series of arguments to the effect that speculative bubbles in local housing markets were possible, albeit unlikely. In any case, they posed no danger for the US economy. In the documents under scrutiny, Greenspan never explicitly discussed the possibility of a credit crisis or a bank panic as a consequence of a sharp decline in house prices. The closest he came to the topic of possible repercussions in the financial sector was in his testimony on 26 September 2005. It is "encouraging", Greenspan said, that the majority of homeowners have enough equity "to absorb a potential decline in house prices"; he also adds that "the situation clearly will require our ongoing scrutiny in the period ahead, lest more adverse trends emerge" (Greenspan 2005c).

Generally, Greenspan assumed that the eventual bursting of the property bubble would consist of a number of uncorrelated events. The harm for the local economy would be limited. The risks are well diversified, and the most likely cause of (the lion's share of) recent price increases is a productivity shortfall in home construction.

*

I shall now analyse a second line of Greenspan's reasoning which was occasioned by criticism concerning the Fed's response to the tech bubble. In this second line of reasoning Greenspan argued that the Federal Reserve should not use monetary policy to prevent the development of bubbles. The following passages from an introductory talk at the annual Federal Reserve Bank of Kansas City's Jackson Hole Economic Symposium contain the first of a series of arguments that constitute the second line of reasoning:

> We at the Federal Reserve considered a number of issues related to asset bubbles—that is, surges in prices of assets to unsustainable levels. As events evolved, we recognized that, despite our suspicions, it was very difficult to definitively identify a bubble until after the fact—that is, when its bursting confirmed its existence. (Greenspan 2002b: 4)

Reconstruction 8 (Identification Argument)
Part 1:

Premise 1	The Federal Reserve would be able to take appropriate measures against the development of an asset price bubble if, and only if, it could identify a bubble with certainty before it bursts.
Premise 2	As a rule, the Federal Reserve cannot identify a bubble with certainty before it bursts.
Conclusion	As a rule, the Federal Reserve is not able to take appropriate measures against the development of a bubble.

But why should one think that it is difficult for the Federal Reserve to identify a bubble with certainty? Greenspan offers an interesting justification.

> [I]f the central bank had access to this information [evidence of a developing bubble], so would private agents, rendering the development of bubbles highly unlikely. (Greenspan 2002b: 7)[13]

Part 2:

Premise 1	If the central bank had evidence of developing bubbles, private agents would also have access to this evidence.
Premise 2	If private agents were to have access to evidence of developing bubbles, the development of bubbles would be highly unlikely.
Premise 3	If the development of bubbles were highly unlikely, there would be no need for the central bank to take appropriate measures.
Conclusion	If the central bank had evidence for a developing bubble, there would be no need for the central bank to take appropriate measures.

[13] See also: "A large number of analysts have judged the level of equity prices to be excessive, even taking into account the rise in 'fair value' resulting from the acceleration of productivity and the associated long-term corporate earnings outlook. But bubbles generally are perceptible only after the fact. To spot a bubble in advance requires a judgment that hundreds of thousands of informed investors have it all wrong. Betting against markets is usually precarious at best" (Greenspan 1999a).

The identification problem is relevant as it makes pre-emptive interventions risky. The central bank might fight a fire which is not there. Interventions triggered by a false positive would yield two adverse outcomes. First, they would interfere with the rational investment decisions of market participants, distort the price mechanism, and thereby distort the market process. Second, the direct costs of intervention would constitute sheer waste.

However, Greenspan did not oppose risky interventions as a matter of principle. Quite the reverse, he repeatedly underlined the need to deal with risk and uncertainty in monetary policy (Greenspan 2003); as pointed out in Sect. 2, he advocated a so-called risk management approach which acknowledges "the need to reach a judgment about the probabilities, costs, and benefits of the various possible outcomes under alternative choices for policy" (Greenspan 2003: 3).[14] Greenspan mentions the Russian debt default in 1988 as a case in which the risk management approach led policymakers to intervene in order to avoid a severely adverse low-probability outcome. Such interventions have the character of insurance "that might prove unnecessary" (Greenspan 2003: 4).[15]

Nonetheless, he was opposed to act pre-emptively with regard to possible bubbles, as the following passage indicates:

> In fact, our experience over the past fifteen years suggests that monetary tightening that deflates stock prices without depressing economic activity has often been associated with subsequent *increases* in the level of stock prices. . . . It seems reasonable to generalize from our recent experience that no low-risk, low-cost, incremental monetary tightening exists that can reliably deflate a bubble. (Greenspan 2003: 5)[16]

For the sake of simplicity, I shall call monetary tightening which does not depress economic activity "soft monetary tightening".

Reconstruction 9 (Ineffectiveness of Low-Cost Intervention Argument)

Premise 1	If monetary tightening is soft, it is often associated with a subsequent increase in the level of stock prices.
Premise 2	If monetary tightening is often associated with a subsequent increase in the level of stock prices, it cannot deflate a bubble.
Conclusion	If monetary tightening is soft, it cannot deflate a bubble.

[14] He did not subscribe to Brainard's (1967) proposition that policymakers can, under a restrictive set of assumptions, ignore uncertainty and proceed as if they knew the structure of the economy (see Greenspan 2003: 3).

[15] "The product of a low-probability event and a potentially severe outcome was judged a more serious threat to economic performance than the higher inflation that might ensue in the more probable scenario . . . Given the potentially severe consequences of deflation, the expected benefits of the unusual policy action were judged to outweigh its expected costs" (Greenspan 2005b: 5).

[16] Greenspan repeated parts of his opening remarks at the 2002 Jackson Hole conference word by word in an article for the *American Economic Review* which appeared in 2004.

For the sake of simplicity, I shall call monetary tightening that is associated with a subsequent increase in the price level "counter-productive".

Reconstruction 10 (Counter-Productivity of Low-Cost Interventions Argument)

Premise 1	Soft monetary tightening frequently has counter-productive effects.
Premise 2	Policymakers ought to abstain from using monetary policy with counter-productive effects.
Conclusion	Policymakers ought to abstain from soft monetary tightening.

According to Greenspan, policymakers faced the problem that (a) the identification of a bubble entails model uncertainty and that (b) soft monetary tightening, which does not depress economic activity and is thus a form of low-cost intervention, is not only ineffective, but counter-productive.

Apart from the arguments in R9 and R10, Greenspan presented two further considerations in support of his view that the central bank should not try to prevent the development of bubbles by monetary tightening. One of the considerations can be found in a testimony on 17 June 1999:

> While bubbles that burst are scarcely benign, the consequences need not be catastrophic for the economy. The bursting of the Japanese bubble a decade ago did not lead immediately to sharp contractions in output or a significant rise in unemployment. Arguably, it was the subsequent failure to address the damage to the financial system in a timely manner that caused Japan's current economic problems. . . . And certainly the crash of October 1987 left little lasting imprint on the American economy. (Greenspan 1999a)

Greenspan conceived the bursting of a bubble on a par with other forms of economic shocks,[17] like a war or a sudden rise in oil prices, and he recommended the same cure, namely monetary easing—according to his critics the very cause of the disease he wanted to cure. Yet during his long term as chairman, the Fed repeatedly and very successfully responded to economic shocks with massive injections of liquidity through a lowering of interest rates.

Reconstruction 11 (Timely Response Argument)
Part 1 (consequences not always catastrophic):

Premise 1	If the consequences of bursting bubbles were always catastrophic, there would be no example of a crash without some lasting impact on the economy.
Premise 2	There is an example of a crash without some lasting impact on the economy (crash of October 1987).
Conclusion	Thus, the consequences of bursting bubbles need not be catastrophic for the economy.

[17] Economic shocks are unexpected events with a depressing effect on economic performance.

The upshot of this argument is that there is no need to prevent the development of bubbles because they do not (necessarily) cause dramatic economic problems. Greenspan offered an alternative explanation of Japan's predicament. For the sake of simplicity, I shall call the failure of policymakers to address the damage to the financial system in a timely manner "lack of timely response".

Part 2 (lack of timely response):

Premise 1	If an external shock occurs and there is no timely policy response, an economic crisis follows.
Premise 2	An external shock occurred in Japan in 1990.
Premise 3	There was no timely policy response in Japan in 1990.
Conclusion	An economic crisis followed in Japan in 1990.

Part 2 gives an explanation of Japan's economic crisis by applying the deductive-nomological model (also known as the Hempel-Oppenheim model). The interesting part of the argument is the covering law in premise 1 which explains in conjunction with premise 2 and 3 Japan's crisis, if all premises are true. This successful explanation of Japan's crisis lends at the same time inductive support to premise 1 and hence Greenspan's theory. According to Greenspan, dramatic economic problems result as the conjunction of two necessary conditions: (a) an external shock and (b) the lack of a timely response by economic policy.

The idea of monetary easing was to facilitate what Greenspan termed a "soft landing" after a shock, after which money supply would be tightened when the economy had recovered. The crisis management after the stock market crash in October 1987 was exemplary for the monetary policy of the Fed under Greenspan. In an article for the *American Economic Review* Greenspan explains:

> Instead of trying to contain a putative bubble by drastic actions with largely unpredictable consequences, we chose, as we noted in our mid-1999 congressional testimony, to focus on policies "to mitigate the fallout when it occurs and, hopefully, ease the transition to the next expansion". (Greenspan 2004: 36)

He gave a similar description in an earlier speech:

> The broad success of that paradigm seemed to be most evident in the United States over the past two and one-half years. Despite the draining impact of a loss of $8 trillion of stock market wealth, a sharp contraction in capital investment and, of course, the tragic events of September 11, 2001, our economy is still growing. Importantly, despite significant losses, no major U.S. financial institution has been driven to default. Similar observations pertain to much of the rest of the world but to a somewhat lesser extent than to the United States. These episodes suggest a marked increase over the past two or three decades in the ability of modern economies to absorb unanticipated shocks. (Greenspan 2002c)

Greenspan's examples aim at warranting the thesis that a timely intervention of monetary policymakers can ward off the otherwise harmful effects of external shocks at relatively low costs.

Reconstruction 12 (Benign Neglect Argument)

Premise 1	The net costs of mitigation (by monetary easing after the bursting of a bubble) are lower than the net costs of pre-emptive tightening (in order to neutralize a bubble).
Premise 2	Policymakers should prefer policies with lower net cost to policies with higher net costs.
Conclusion	Policymakers should prefer mitigation to pre-emptive tightening.

Greenspan repeatedly emphasised that the mitigation approach has to confront unquantifiable risks and thus "involves significant judgement on the part of the policymakers" (Greenspan 2003: 5). Basing policy not only on quantitative models but on "broader, though less mathematically precise, hypotheses of how the world works" (Greenspan 2003: 5) seemed to be a new and superior paradigm of policymaking.

In general, low interest rates encourage investment and consumption and thereby stimulate economic performance. In the 1990s, the US experienced a period of low inflation and strong growth, combined with a bullish stock market; the perception of risk on the part of investors was low. Greenspan was well aware that the reduced sense of risk might breed unrealistic expectations about future profits, asset price trends, and other economic parameters.[18] In combination with low interest rates, it was a distinct possibility that the proverbial "irrational exuberance" would fuel the development of bubbles. But according to the identification argument, one cannot be certain about the causes of price developments. Rising household wealth, in form of share packages or homes, spurs consumption and thus economic output. Fighting a bubble which is not there would have had high opportunity costs in terms of economic growth.

The other consideration in favour of premise 1 in R12 is brought to bear in the following lengthy passage in which Greenspan reflects on the reasons for the "ability of modern economies to absorb unanticipated shocks":

> The wide-ranging development of markets in securitized bank loans, credit card receivables, and commercial and residential mortgages has been a major contributor to the dispersion of risk in recent decades both domestically and internationally. These markets have tailored the risks associated with such assets to the preferences of a broader spectrum of investors. *Especially important in the United States have been the flexibility and the size of the secondary mortgage market. Since early 2000, this market has facilitated the large debt-financed extraction of home equity that, in turn, has been so critical in supporting consumer outlays in the United States throughout the recent period of cyclical stress. This market's flexibility has been particularly enhanced by extensive use of interest rate swaps and options*

[18] "As recent experience attests, a prolonged period of price stability does help to foster economic prosperity. But, as we have also observed over recent years, as have others in times past, such a benign economic environment can induce investors to take on more risk and drive asset prices to unsustainable levels. This can occur when investors implicitly project rising prosperity further into the future than can reasonably be supported. By 1997, for example, measures of risk had fallen to historic lows as businesspeople, having experienced years of continuous good times, assumed, not unreasonably, that the most likely forecast was more of the same" (Greenspan 1999a).

to hedge maturity mismatches and prepayment risk. Financial derivatives, more generally, have grown at a phenomenal pace over the past fifteen years ... These increasingly complex financial instruments have especially contributed, particularly over the past couple of stressful years, to the development of a far more flexible, efficient, and resilient financial system than existed just a quarter-century ago. (Greenspan 2002c, emphasis added)

Reconstruction 13 (Dispersion Argument: Similar to R5)

Premise 1	Securitisation of mortgages improves the dispersion of risks.
Premise 2	Improved dispersion of risks makes the financial system more resilient.
Conclusion	Securitisation of mortgages makes the financial system more resilient.

Reconstruction 14 (Resilience Argument)

Premise 1	Securitisation of mortgages makes the financial system more resilient.
Premise 2	A more resilient financial system reduces the economic costs of bursting bubbles.
Conclusion	Securitisation of mortgages reduces the economic costs of bursting bubbles.

Traditionally, mortgages were offered by local lenders. When a local housing bubble burst, or when an economic shock, such as the dislocation of a major employer, hit the community, then the mortgage lender faced the danger of bankruptcy as many borrowers became unable to service their debts. Since Greenspan conceived the US housing market to be a collection of local markets, he assumed that the respective risks were not correlated. However, securitisation created a national, even a global, market for mortgages and thus reduced the likelihood that a local bank might go out of business after a massive decline in local prices or some other shock to the community's economy.

4 How Strong Were Greenspan's Arguments?

The reconstruction of Greenspan's case for inaction has not revealed any invalid arguments. This section makes a cursory check of the reasonableness of his position. One condition for reasonableness is defensibility. A position is defensible if a critical number of experts hold it to be relevant and possibly true. These experts do not need to accept the position themselves. It suffices that they agree the position is informative and not in conflict with well corroborated claims. So, even if one assumes that some of Greenspan's arguments were not sound (because their premises were wrong), it does not follow that his approach was indefensible. Another condition for reasonableness is confidence adjustment. This means that a proponent adjusts her confidence in a position in response to objections to it. For instance, EMH was once considered to be among the best validated theories in economics. Since the 1970s, though, the hypothesis was confronted with various empirical anomalies. As a result, a growing number of economists reasonably adjusted their confidence in EMH (Shleifer 2000: 175).

Greenspan was hailed in his day by a critical number of experts as the "greatest central banker who has ever lived" (Blinder and Reis 2005: 13). All the same, objections to Greenspan's premises were made by highly respectable academics and were known to him. I shall argue in this section that these objections were certainly robust enough to justify confidence adjustments. A confidence adjustment should be accompanied by a hedging strategy if (a) the effect of the position being wrong was highly adverse and (b) a cost-effective hedging strategy was available (insurance principle). Greenspan did accept the insurance principle, as pointed out in Sect. 3; moreover, (a) and (b) were indeed the case. Therefore, I conclude that he failed to adjust his confidence in his position.

In the following, I shall peruse Greenspan's thinking in light of supporting arguments and objections.

1. **First objection to R2 (turnover argument): price rises above fundamentals very likely**
   In 2002, Greenspan concluded that the probability of a bubble in the housing market was very low. He based his conclusion mainly on a transaction cost argument. Speculation requires a high turnover, which is unlikely when transaction costs are high. On first inspection, the reasoning appears to be plausible because moving house is burdensome in financial and emotional respects. The home price boom may thus reflect low interest rates and higher incomes. However, in 2001 home prices in many US cities *began* to rise by 10 % even though it was a recession year (Shiller 2007: 90). Price rises in the US real estate market since the late 1990s were extraordinarily high by historical standards; they far outpaced productivity growth, inflation, GDP growth, or the growth of real incomes of average Americans (Stiglitz 2010: 86; Shiller 2008: 29–41). This development was not easy to square with the view that home prices reflect strong fundamentals.
2. **Second objection to R2 (turnover argument): argument inconsistent with acknowledged facts**
   Besides, Greenspan did not apply the turnover argument consistently. In his testimony on 17 June 1999 he referred to the Japanese real estate bubble between 1985 and 1991. According to the turnover argument, the development of such a bubble is very unlikely. The obvious challenge for Greenspan would have been to explain why the argument does not apply to Japan; or, more generally, why real estate bubbles occur more often than one would expect on the basis of the turnover argument.

3. **Support for R3 (spatial fragmentation argument): resilience of US economy over the past decades**

   The spatial fragmentation argument of 2002 concluded that local housing bubbles, in all likelihood, would not have a strong detrimental effect on the US economy as a whole.

   The spirit of the spatial fragmentation argument has to be seen in the context of the popular "great moderation" narrative. Between the mid-1980s and 2007, inflation in the US was low and relatively stable while economic growth was unusually strong. Why precisely volatility of output and of inflation decreased is not entirely clear (Bernanke 2004). At any rate, the fact that the US economy weathered a number of severe shocks during that period without dramatic effects on economic growth and inflation led many academic economists to believe that improved monetary policy, namely inflation targeting,[19] in combination with structural improvements due to technology, neutralized the danger of volatile asset prices. Along these lines, Greenspan's successor Ben Bernanke and Mark Gertler concluded in 2001 that "inflation-targeting central banks need not respond to asset prices, except insofar as they affect the inflation forecast" (Bernanke and Gertler 2001: 253).
4. **Objection to R4 (speculative conflagration argument): misidentification of cause**

   In September 2005, when the speculative character of the boom became increasingly difficult to deny, Greenspan finally conceded the existence of local real estate bubbles in the US.[20] He conjectured that the higher turnover was due to the transaction of second homes, thereby missing the core problem of subprime clients.[21] The subprime market was in Greenspan's estimation just "adding to the pressure in the marketplace".[22]

[19] "In an inflation-targeting framework, publicly announced medium-term inflation targets provide a nominal anchor for monetary policy, while allowing the central bank some flexibility to help stabilize the real economy in the short run" (Bernanke and Gertler 2001: 253).

[20] "In the United States, signs of froth have clearly emerged in some local markets where home prices seem to have risen to unsustainable levels. It is still too early to judge whether the froth will become evident on a widening geographic scale, or whether recent indications of some easing of speculative pressures signal the onset of a moderating trend" (Greenspan 2005a).

[21] "According to data collected under the Home Mortgage Disclosure Act (HMDA), mortgage originations for second-home purchases rose from 7 % of total purchase originations in 2000 to twice that at the end of last year. Anecdotal evidence suggests that the share may currently be even higher" (Greenspan 2005a).

[22] "The apparent froth in housing markets may have spilled over into mortgage markets. The dramatic increase in the prevalence of interest-only loans, as well as the introduction of other, more-exotic forms of adjustable-rate mortgages, are developments that bear close scrutiny. To be sure, these financing vehicles have their appropriate uses. But to the extent that some households may be employing these instruments to purchase a home that would otherwise be unaffordable, their use is adding to the pressures in the marketplace" (Greenspan 2005a).

5. **First objection to R8 (identification argument): bubbles can be identified**
   Greenspan was not the only central banker who was convinced that it is impossible to identify developing asset-price bubbles with certainty (see King 2004: 44; see also Bernanke and Gertler 2000: 8). Among academic economists, Ben Bernanke and Mark Gertler were the most prominent advocates of this view.[23] It was not undisputed, though. Claudio Borio and William White of the Bank for International Settlements argued at the Jackson Hole conference in 2003 that excessive growth in asset prices leading to financial crises can be fairly well predicted on the basis of two indicators, "namely the ratio of (private sector) credit to GDP and inflation-adjusted equity prices" (Borio and Drehmann 2009; Borio and White 2003: 153; see also Borio and Lowe 2002).
6. **Support for R8 (identification argument): objection to Borio and White**
   In his comment on the Borio and White paper, Mark Gertler responded to this point by insisting that strong credit growth could also be indicative of "efficient financial development" (Gertler 2003: 214) without giving further explanations about his views on the relation between efficient financial development and credit growth.[24]
7. **Second objection to R8 (identification argument): argument irrelevant and practically misleading**
   Another important contribution to the debate was made by Michael Bordo and Olivier Jeanne. They argued that the identification problem is irrelevant for the question of pre-emptive monetary policy since a credit crunch could also occur in a world without bubbles. Even an asset price reversal which is responsive to a change in fundamentals could result in a credit crisis. "Hence, the debate about proactive versus reactive monetary policies should not be reduced to a debate over the central bank's ability to assess deviations in asset prices from fundamental values" (Bordo and Jeanne 2002: 160).

[23] In practice, Ben Bernanke was well prepared to do what he declared to be impossible in theory. In October 2005, Bernanke, then chairman of the President's Council of Economic Advisers, identified the causes of the house price rises as follows: "House prices have risen by nearly 25 % over the past 2 years. Although speculative activity has increased in some areas, at a national level these price increases largely reflect strong economic fundamentals, including robust growth in jobs and incomes, low mortgage rates, steady rates of household formation, and factors that limit the expansion of housing supply in some areas. House prices are unlikely to continue rising at current rates. However, as reflected in many private-sector forecasts such as the Blue Chip forecast mentioned earlier, a moderate cooling in the housing market, should one occur, would not be inconsistent with the economy continuing to grow at or near its potential next year" (Bernanke 2005).

[24] In a speech before the New York Chapter of the National Association for Business Economics on 15 October 2002, Ben Bernanke addressed an earlier paper by Borio and Lowe, arguing that rapid growth of credit may "reflect simply the tendency of both credit and asset prices to rise during economic booms" (Bernanke 2002).

8. **First support for R12 (benign neglect argument): general agreement about successful application**
   The conclusion of the benign neglect argument to the effect that policymakers should prefer mitigation to pre-emptive tightening was widely accepted among central bankers since the 1990s (Bordo and Jeanne 2002: 141). The approach appeared to have passed several empirical tests with good results. In 2004, it seemed not unreasonable to conclude that the "strategy of addressing the bubble's consequences rather than the bubble itself has been successful" (Greenspan 2004: 36).
9. **Second support for R12 (benign neglect argument): optimism about possibility of timely monetary easing justified**
   Similar to Greenspan, Gertler and Bernanke emphasized that it is unnecessary to solve the identification problem as their "reading of history is that asset price crashes have done sustained damage to the economy only in cases when monetary policy remained unresponsive or actively reinforced deflationary pressures" (Bernanke and Gertler 2000: 3).
10. **Objection to R13 (dispersion argument), R14 (resilience argument): incentive problems in the housing market**
    With regard to housing, Greenspan argued that the securitisation of mortgages reduces the economic costs of bursting bubbles. Arguably, this was his single most important misjudgement.
    On the surface, the case for the conclusions in R14 (and R13) looked plausible enough. But securitisation changed the incentives for lenders (Stiglitz 2010: 77–108). In the old days, local lenders had a strong motive to assess diligently the creditworthiness of individual borrowers as they had to bear the potential losses. Mortgages were mostly fixed rate and long term, and lenders did not offer to finance more than 80 % of the house price. With the opportunity to sell the mortgages to third parties, lenders were less inclined to check the borrowers' ability to shoulder the debt. As long as one could successfully pass on the default risk to others, it became lucrative simply to generate mortgages. Since banks and mortgage originators receive fees, they also earn with refinancing. This explains the trend to short-term, adjustable-rate mortgages. Lenders encouraged customers to take advantage of low interest rates and seemingly ever rising house prices, thereby producing the high turnover which reinforced the price trend and generated fees. With short-term interest rates at 1 % in 2003, it was clear that many borrowers would face unsustainable debt in the near future. It was also clear that house prices would drop due to a growing number of sales and foreclosures. Falling prices triggered more sales from speculators and from borrowers who became aware that their mortgages were worth more than their houses.[25] An increasing number of foreclosures in

[25] In the US, borrowers are not obliged to service a mortgage which is higher than the house price. All they have to do is to hand over the house to the creditor.

combination with dropping house prices amounted to significant losses for financial intermediaries with mortgages and mortgage-backed securities on their balance sheets. During the bubble, highly leveraged banks like Lehmann Brothers were extraordinarily profitable. The exceptional profitability of highly leveraged intermediaries puts competitive pressure on all suppliers in the market as the more profitable establishments are able to offer customers better conditions. But high leverage comes with a substantial risk of bankruptcy when the price trend reverses. That happened in the summer of 2008. The proximate cause of Lehman Brothers' bankruptcy in September 2008 was the loss of confidence of creditors and investors concerned about the rapidly declining value of its mortgage-related securities (Kindleberger and Aliber 1978/2011: 257).

The incentive problem was well known and widely discussed in the academic literature. Joseph Stiglitz warned as early as 1992 "that the securitization of mortgages would end in disaster, as buyers and sellers alike underestimated the likelihood of a price decline and the extent of correlation" between seemingly independent risks (Stiglitz 2010: 19).

At a meeting in Jackson Hole in 2003, Gertler blamed ill-conceived liberalisation for the increased volatility of financial markets in the 1990s and thus stressed the importance of regulation and supervision. At another Jackson Hole conference 2 years later, Raghuram Rajan expressed concerns that deregulation created competitive pressures in finance which force financial institutions "to flirt continuously with the limits of illiquidity" (Rajan 2005: 314). Managers had the incentive to take on risks as their performance was measured on the basis of returns relative to their peers. Rajan mentioned two kinds of "perverse behaviour" which flourished in this environment. First, managers would gain from concealing "tail risks", severely adverse consequences with small probability, from investors; second, they had an incentive to imitate the behaviour of other managers in order not to underperform.

> Both behaviors can reinforce each other during an asset price boom, when investment managers are willing to bear the low-probability tail risk that asset prices will revert to fundamentals abruptly, and the knowledge that many of their peers are herding on the risk gives them comfort that they will not underperform significantly if boom turns to bust. (Rajan 2005: 317)

Rajan raised the question whether banks would be able to survive when the tail risk finally materialised. In a nutshell, the resilience argument, according to which securitisation of mortgages reduces the economic costs of bursting bubbles, was unconvincing in view of the market's incentive structure.

For an overview on the support and attack relations between core arguments of in the debate about bubbles see Fig. 11.1.

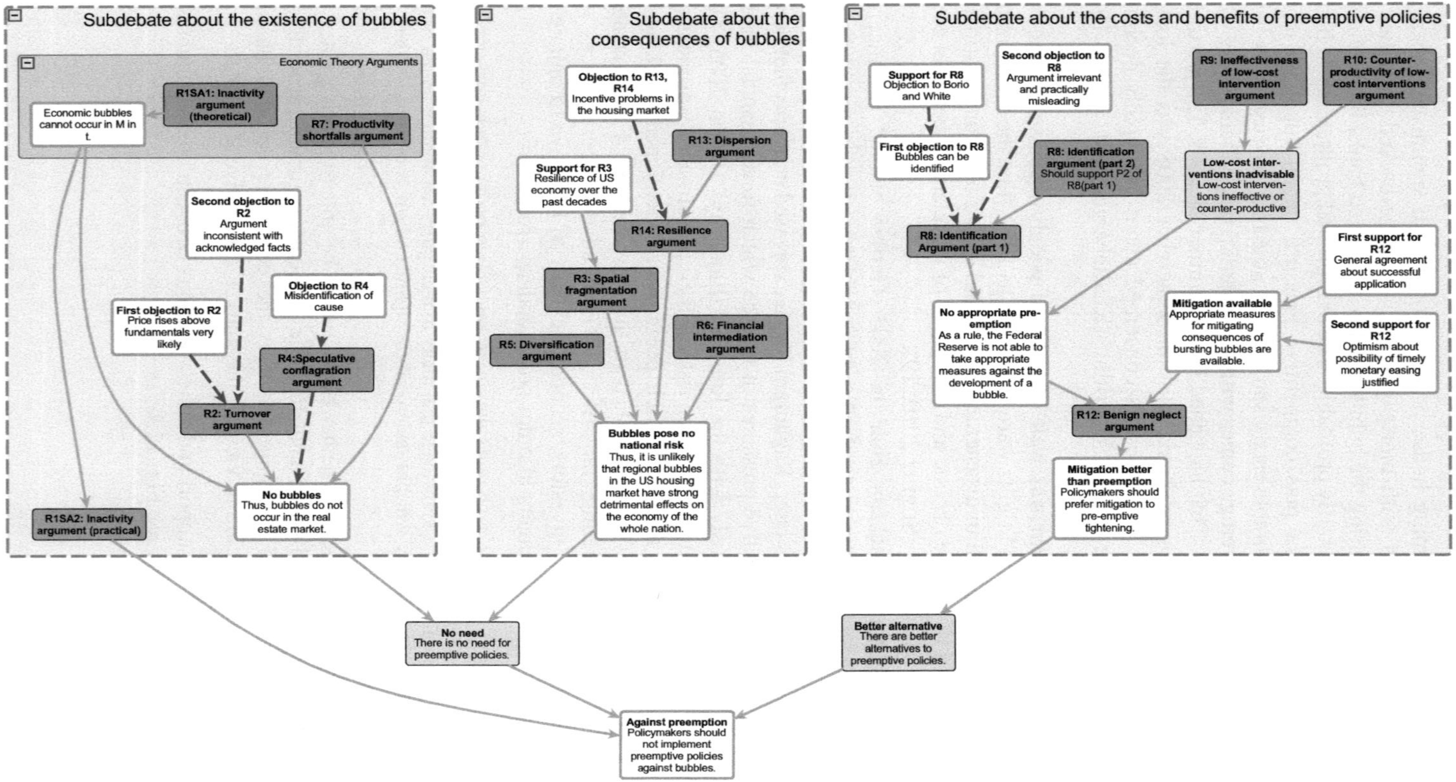

Fig. 11.1 Argument map illustrating support (*solid arrows*) and attack relations (*dashed arrows*) between core arguments of the debate about bubbles

5 Conclusion

Greenspan's arguments for inactivity are to a large degree congruent with the position of Gertler and Bernanke. However, in contrast to Greenspan, Bernanke and Gertler emphasized that benign neglect is plausible only if an adequate regulatory structure is in place. "Financial imbalances", writes Gertler in response to Borio and White, are "largely the product of ill-designed financial deregulation" (Gertler 2003: 215). With appropriate regulatory and supervisory machinery operating, monetary policy need not concern itself with the possibility of bubbles. Whereas Gertler argued that monetary policy can ignore asset price developments as long as prudential policy is used to "prevent undesired financial risk exposure from building up" (Gertler 2003: 221), there is no mention of the importance of regulation and supervision in Greenspan's discussion of benign neglect. Gertler's qualified defence of the identification argument points to a grave shortcoming in Greenspan's position. An adjustment in confidence would have been appropriate.

Thus one, maybe *the*, central problem of risk management in the Greenspan era was the undue reliance on the stabilising effects of innovative financial instruments (Wolf 2009: 194). What surprised Greenspan was not that bubbles are possible but that the effects of the housing bust could not be contained and that the costs of "mitigation" became astronomical as a consequence (Caballero and Kurlat 2009: 20). In comparison, the costs of maintaining a regulatory structure would have been minuscule. It would have insured the global economy against the possibility of the harmful effects of a housing price reversal.

The application of argument analysis techniques does not only help to detect fallacies in the argumentative underpinning of a policy. Such techniques also help to raise awareness for dubious premises. They make it more likely that a need to adjust confidence will become conspicuous. I thus conclude that their use has the potential to improve stabilisation policy in the future.

Recommended Readings

Allen, F., & Gale, D. (2007/2009). *Understanding financial crises*. Oxford: Oxford University Press.

Cooper, G. (2008). *The origin of financial crises: Central banks, Credit bubbles and the efficient market fallacy*, New York: Vintage Books.

Kindleberger, C., & Aliber, R. (1978/2011). *Manias, panics, and crashes: A history of financial crises* (6th edn.). New York: Palgrave Macmillan.

Stiglitz, J. (2010). *Freefall: Free markets and the sinking of the global economy*. London: Allen Lane.

References

Allen, F., & Gale, D. (2007/2009). *Understanding financial crises*. Oxford: Oxford University Press.

Angelides, P. et al. (2011). Final report of the National Commission on the Causes of the Financial and Economic Crisis in the United States, submitted by the Financial Crisis Inquiry Commission, Pursuant to Public Law (pp. 111–121). http://www.gpo.gov/fdsys/pkg/GPO-FCIC/pdf/GPO-FCIC.pdf. Accessed 9 June 2015.

Batini, N., Martin, B., & Salmon, C. (1999). Monetary policy and uncertainty. http://www.bankofengland.co.uk/archive/Documents/historicpubs/qb/1999/qb990205.pdf. Accessed 9 June 2015.

Bernanke, B. S. (2002). Remarks by Governor Ben S. Bernanke. Before the New York Chapter of the National Association for Business Economics, New York, New York October 15, 2002. http://www.federalreserve.gov/Boarddocs/Speeches/2002/20021015/default.htm. Accessed 24 Nov 2015.

Bernanke, B. S. (2004, February 20). The great moderation: Remarks at the meetings of the Eastern Economic Association. Washington, DC. https://fraser.stlouisfed.org/title/?id=453#!8893. Accessed 9 June 2015.

Bernanke, B. S. (2005). Testimony before the Joint Economic Committee, October 20, 2005. *The Economic Outlook*. http://georgewbushwhitehouse.archives.gov/cea/econ-outlook20051020.html. Accessed 9 June 2015.

Bernanke, B. S., & Gertler, M. (2000). *Monetary policy and asset price volatility* (NBER Working Paper 7559).

Bernanke, B. S., & Gertler, M. (2001). Should central banks respond to movements in asset prices? *American Economic Review, 91*, 253–257.

Bezemer, D. (2009). "No One Saw This Coming": Understanding financial crisis through accounting models. MPRA Paper No. 15892, posted 24. June 2009. http://mpra.ub.uni-muenchen.de/id/eprint/15892. Accessed 9 June 2015.

Blinder, A., & Reis, R. (2005). Understanding the Greenspan standard. Economic policy symposium "The Greenspan Era: Lessons for the Future", 22–24 August in Jackson Hole, Wyoming: 11–96. http://www.kc.frb.org/publicat/sympos/2005/pdf/Blinder-Reis2005.pdf. Accessed 9 June 2015.

Bordo, M., & Jeanne, O. (2002). Monetary policy and asset prices: Does 'Benign Neglect' make sense. *International Finance, 5*, 139–164.

Borio, C., & Lowe, P. (2002). Assessing the risk of banking crises. *BIS Quarterly Review* (December Issue), 43–54.

Borio, C., & White, W. R. (2003). Whither monetary and financial stability? The implications of evolving policy regimes. Economic policy symposium "Monetary Policy and Uncertainty: Adapting to a Changing Economy", 22–24 August in Jackson Hole, Wyoming: 131–211. https://www.kansascityfed.org/publicat/sympos/2003/pdf/Boriowhite2003.pdf. Accessed 9 June 2015.

Borio, C., & Drehmann, M. (2009). Assessing the risk of banking crises – revisited. *BIS Quarterly Review* (March Issue), 29–46.

Brainard, W. (1967). Uncertainty and the effectiveness of policy. *Cowles Foundation Paper, 257*, 411–425.

Brealey, R. A., & Myers, S. C. (1981/1991). *Principles of corporate finance*. New York: McGraw-Hill.

Brun, G., & Betz, G. (2016). Analysing practical argumentation. In S. O. Hansson & G. Hirsch Hadorn (Eds.), *The argumentative turn in policy analysis. Reasoning about uncertainty* (pp. 39–77). Cham: Springer. doi:10.1007/978-3-319-30549-3_3.

Caballero, R. J., & Kurlat, P. (2009). The 'Surprising' nature of financial crisis: A macroeconomic policy proposal. Economic policy symposium "Financial Stability and Macroeconomic Policy", 20–22 August in Jackson Hole, Wyoming: 19–68. https://www.kansascityfed.org/~/media/files/publicat/sympos/2009/papers/caballerokurlat082409.pdf?la=en. Accessed 9 June 2015.

Cooper, G. (2008). *The origin of financial crises: Central banks, credit bubbles and the efficient market fallacy*. New York: Vintage Books.
Dennis, R. (2005). Uncertainty and monetary policy. *Federal Reserve Bank of San Francisco, 33*, 1–3.
Feldstein, M. (2004). Innovations and issues in monetary policy: Panel discussion. *American Economic Review, 94*, 41–43.
Galbraith, K. (1990/1993). *A short history of financial Euphoria*. New York: Penguin.
Gertler, M. (2003). Comment on Whither monetary and financial stability? The implications of evolving policy regimes. Economic policy symposium "Monetary Policy and Uncertainty: Adapting to a Changing Economy", 22–24 August in Jackson Hole, Wyoming: 213–223. https://www.kansascityfed.org/publicat/sympos/2003/pdf/Gertler2003.pdf. Accessed 9 June 2015.
Greenspan, A. (1999a). *Monetary policy and the economic outlook before the Joint Economic Committee, U.S. Congress*. http://www.federalreserve.gov/boarddocs/testimony/1999/19990617.htm. Accessed 9 June 2015.
Greenspan, A. (1999b). Opening remarks. Economic policy symposium "New Challenges for Monetary Policy". 22–24 August in Jackson Hole, Wyoming, 1–9. http://www.kc.frb.org/publicat/sympos/1999/S99gren.pdf. Accessed 9 June 2015.
Greenspan, A. (2002a). Monetary policy and the economic outlook. Testimony of chairman Greenspan before the Joint Economic Committee, U.S. Congress. http://www.federalreserve.gov/boarddocs/testimony/2002/20020417/default.htm. Accessed 9 June 2015.
Greenspan, A. (2002b). Opening remarks. In *Economic policy symposium "Rethinking Stabilization Policy", 22–24 August in Jackson Hole, Wyoming* (pp. 1–10). http://www.kc.frb.org/publicat/sympos/2002/pdf/S02Greenspan.pdf . Accessed 9 June 2015.
Greenspan, A. (2002c). International financial risk management. Remarks by Chairman Alan Greenspan before the council on foreign relations, Washington, DC. http://www.federalreserve.gov/boarddocs/Speeches/2002/20021119/default.htm. Accessed 9 June 2015.
Greenspan, A. (2003). Opening remarks. Economic policy symposium "Monetary Policy and Uncertainty: Adapting to a Changing Economy", 22–24 August in Jackson Hole, Wyoming, 1–7. http://www.kc.frb.org/publicat/sympos/2003/pdf/Greenspan2003.pdf. Accessed 9 June 2015.
Greenspan, A. (2004). Risk and uncertainty in monetary policy. *American Economic Review, 94*, 33–40.
Greenspan, A. (2005a). The economic outlook. Testimony of Chairman Greenspan before the Joint Economic Committee, U.S. Congress. http://www.federalreserve.gov/boarddocs/testimony/2005/200506092/default.htm. Accessed 9 June 2015.
Greenspan, A. (2005b). Opening remarks. Economic policy symposium "The Greenspan Era: Lessons for the Future", 25–27 August in Jackson Hole, Wyoming: 1–10. https://www.kansascityfed.org/publicat/sympos/2005/pdf/Green-opening2005.pdf. Accessed 9 June 2015.
Greenspan, A. (2005c). Mortgage banking. Remarks by Chairman Alan Greenspan to the American Bankers Association Annual Convention, Palm Desert, California (via satellite). http://www.federalreserve.gov/boardDocs/Speeches/2005/200509262/default.htm. Accessed 9 June 2015.
Hansson, S. O. (2016). Evaluating the uncertainties. In S. O. Hansson & G. Hirsch Hadorn (Eds.), *The argumentative turn in policy analysis. Reasoning about uncertainty* (pp. 79–104). Cham: Springer.
Hansson, S. O., & Hirsch Hadorn, G. (2016). Introducing the argumentative turn in policy analysis. In S. O. Hansson & G. Hirsch Hadorn (Eds.), *The argumentative turn in policy analysis. Reasoning about uncertainty* (pp. 11–35). Cham: Springer. doi:10.1007/978-3-319-30549-3_2.
Jenkins, P, & Longworth, D. (2002). Monetary policy and uncertainty. Bank of Canada Review, 3–10. http://www.bankofcanada.ca/wp-content/uploads/2010/06/longworth_e.pdf. Accessed 9 June 2015.

Jensen, M. (1978). Some anomalous evidence regarding market efficiency. *Journal of Financial Economics, 6*, 95–101.
Kindleberger, C. & Aliber, R. (1978/2011). *Manias, panics, and crashes: A history of financial crises* (6th edn.) New York: Palgrave Macmillan.
King, M. (2004). The institutions of monetary policy. *American Economic Review, 94*, 1–13.
Krugman, P. (1999/2009). *The return of depression economics and the crisis of 2008*. New York: W. W. Norton.
Krugman, P. (2009). How did economists get it so wrong? *The New York Times*. http://www.nytimes.com/2009/09/06/magazine/06Economic-t.html?pagewanted=all&_r=0. Accessed 9 June 2015.
Mackay, C. (1841/1995). *Extraordinary popular delusions and the madness of the crowds*. Herfordshire: Wordsworth.
Minski, M. (1986/2008). *Stabilizing an unstable economy*. New York: McGraw-Hill.
Posner, R. A. (2009). *A failure of capitalism. The crisis of '08 and the descent into depression*. Cambridge, MA: Harward University Press.
Rajan, R. G. (2005). Has financial development made the world riskier? Economic policy symposium "The Greenspan Era: Lessons for the Future", 25–27 August in Jackson Hole, Wyoming: 313–369. https://www.kansascityfed.org/publicat/sympos/2005/pdf/Rajan2005.pdf. Accessed 9 June 2015.
Shiller, R. J. (2007). Understanding recent trends in house prices and homeownership. In *Economic policy symposium "Housing, Housing Finance, and Monetary". Economic policy symposium "Housing, Housing Finance, and Monetary Policy". August 30 to September 1 in Jackson Hole, Wyoming* (pp. 89-123). https://www.kansascityfed.org/publicat/sympos/2007/PDF/Shiller_0415.pdf. Accessed 9 June 2015.
Shiller, R. J. (2008). *The subprime solution: How today's global financial crisis happened, and what to do about it*. Princeton: Princeton University Press.
Shiller, R. J. (2000/2015). *Irrational exuberance: Revised and expanded third edition*. Princeton: Princeton University Press.
Shleifer, A. (2000). *Inefficient markets: An introduction to behavioral finance*. Oxford: Oxford University Press.
Sinn, H-W. (2010/2011). *Kasino-Kapitalismus. Wie es zur Finanzkrise kam, und was jetzt zu tun ist*. Munich: Ullstein.
Soros, G. (2008/2009). *The crash of 2008 and what it means: The new paradigm for financial markets*. New York: Public Affairs.
Stiglitz, J. (2010). *Freefall: Free markets and sinking of the global economy*. London: Allen Lane.
Stiglitz, J. (2014). Reconstructing macroeconomic theory to manage economic policy (NBER Working Paper 20517).
Wolf, M. (2009). *Fixing global finance: How to curb financial crisis in the 21st century*. New Haven: Yale University Press.

Chapter 12
Uncertainty Analysis, Nuclear Waste, and Million-Year Predictions

Kristin Shrader-Frechette

Abstract What should government do with a former nuclear-reprocessing site, contaminated with hundreds of thousands of curies of shallow-buried radioactive waste, including high-level waste, some in only plastic bags and cardboard boxes, all sitting on a rapidly eroding plateau? Some of the waste will remain lethal for millions of years, and a contaminated underground plume has already reached drinking-water supplies. If cleanup costs are billions of dollars, government may unscientifically and unethically do what the US Department of Energy (DOE) is doing at West Valley, New York. This chapter argues that DOE is (1) avoiding doing any classic uncertainty analysis to assist in site decisionmaking, and (2) arbitrarily redefining "uncertainty analysis," so that it can claim to have shown that by the time lethal amounts of nuclear waste migrate, they will cause only minor harm. Therefore DOE (3) practicing special-interest science, using flawed analytic methods to arrive at questionable, predetermined conclusions.

Keywords Best estimate • Nuclear waste • Prediction • Probabilistic analysis • Reprocessing • West Valley

1 Introduction

Thirty miles from Buffalo, New York, the West Valley nuclear-waste site sits on a plateau that is eroding away – slowly collapsing into the Lake Erie watershed at the rate of roughly a meter per year. In the 1960s, Nuclear Fuel Services promised local economic prosperity when it began reprocessing spent-nuclear fuels at the New York site. After only 6 years of too-expensive and polluting reprocessing, the company abandoned the venture and left a regional health-and-safety threat, one that will continue for tens of thousands to millions of years. "Packaged in canisters, drums, cardboard boxes, and plastic bags, the [West Valley] list of contaminated wastes reads like a laundry list of dangerous elements: strontium 90, cesium-137,

K. Shrader-Frechette (✉)
Department of Philosophy and Biological Sciences, University of Notre Dame, 100 Malloy Hall, Notre Dame, IN 46556, USA
e-mail: Kristin.Shrader-Frechette.1@nd.edu

S.O. Hansson, G. Hirsch Hadorn (eds.), *The Argumentative Turn in Policy Analysis*, Logic, Argumentation & Reasoning 10, DOI 10.1007/978-3-319-30549-3_12

plutonium-238, -239, -240, and -241, uranium-238, curium-244, cobalt-60, americium-241, iodine-129, tritium. . .thorium-234," and others (Napoleon et al. 2008). Because radiation has no safe dose (National Research Council/National Academy of Sciences (NRC/NAS) 2006), anyone who ingests or inhales it, even at very low doses, can have it lodge in tissues, fat, or bone and cause leukemias and cancers. Ever since the 1970s, when West Valley stopped nuclear-fuel reprocessing, state governments and national and local citizens groups have been fighting over how to solve the West Valley dilemma: Site radioactive wastes are not safe where they are, even for a short time, but site remediation – for which the US Department of Energy (DOE) and the state of New York are responsible – will be expensive (Napoleon et al. 2008).

2 Overview

In 2010, DOE "solved" the West Valley, radioactive-waste dilemma. Responsible for costly cleanup, DOE instead did trivial remediation, ignored the huge plume of radioactive waste moving toward Lake Erie and drinking-water supplies, then issued an environmental impact statement (EIS). The 2010 EIS, more than 1000 pages, declared that the lethal waste was safe where it was – even for 10,000 to a million years.

How could EIS science support such a safety claim when the site began leaking radioactive wastes within a decade? The answer is simple. If the cost of site cleanup is in the billions of dollars, government may do what the US DOE recently did at West Valley, New York. This chapter argues that DOE (1) avoided doing any classic uncertainty analysis to assist in site decisionmaking, (2) arbitrarily redefined "uncertainty analysis," so that it could claim to have shown that by the time lethal amounts of waste migrated, they would cause only minor harm, and (3) appears to have fallen victim to special-interest science, science whose predetermined conclusions typically dictate flawed analytic methods. That is, rather than admitting West Valley threats, doing a legitimate uncertainty analysis, and choosing the cheapest long-term cleanup strategy, waste removal, DOE has done something else. It has used flawed science to defend the cheapest short-term strategy: leaving the dangerous waste where it is, in one of the most erosion-prone areas of the country.

3 Background on West Valley, New York

West Valley, New York is located in an area so hydrologically and geologically unstable and erosion-ridden (ground movement in meters per year) that government today would never allow anyone to site a nuclear-waste facility there (US Department of Energy 2010; Napoleon et al. 2008). Yet DOE did so, nearly a half-century ago. Given the need to protect the public from this New York radioactive contamination for the next 10,000–1,000,000 years (US Department of Energy 2010; Napoleon et al. 2008), one would expect DOE to perform a scientifically defensible analysis of site risks and how to manage them. After all, since reprocessing ended at West Valley, DOE has had more than 25 years to study the site.

Instead of performing a scientifically defensible EIS, one responsive to the many critical scientific peer reviews of earlier drafts, in 2010 DOE took an economically expedient course, one that requires only small current costs but imposes massive costs and risks on future people. Although DOE has not technically chosen a site clean-up solution, its accepted 2010 EIS claims that minor cleanup, plus leaving much of the waste onsite, will be safe for the next tens-to-hundreds of thousands of years (US Department of Energy 2010). This is a surprising conclusion, given that economists have shown that the least-expensive long-term strategy is to move West Valley wastes to a safer, drier location (Napoleon et al. 2008). This chapter argues that DOE's 2010 EIS reached its expedient, rather than an economical and safe, strategy for West Valley mainly because it has relied on a scientifically indefensible treatment of uncertainty.

As a result, the 2010 EIS concludes that even if DOE merely leaves much of the long-lived nuclear waste onsite at West Valley, without any government-institutional management such as fences, monitoring, and erosion controls, the maximum annual future dose to any person offsite will be only 0.2 mrem – about 1/2000 of normal background radiation. Even with completely uncontrolled erosion, DOE also says the future yearly maximum offsite radiation dose would be only 4 mrem – about 1 % of normal background radiation (US Department of Energy 2010).

How can DOE predict such tiny exposures 10,000 to a million years into the future? And if the future exposures are really so low, why would the government today not allow siting a nuclear-waste facility at West Valley? Scientific peer reviewers consider such low DOE predictions for West Valley highly unlikely (US Department of Energy 2010; Napoleon et al. 2008). After all, they concern one of the most radiologically contaminated, poorly contained, long-lived hazards on the planet – a site where radioactive contamination is already offsite, in nearby creeks that lead to Lake Erie (US Department of Energy 2010).

This chapter argues that to justify its questionable, long-term, low-radiation-dose predictions about West Valley, DOE did a scientifically indefensible EIS. This EIS (1) avoided all uncertainty analysis, except for a couple of invalidly done assessments. However, to cover-up EIS failure to do standard uncertainty analyses, and to make the EIS appear as if it had reliably drawn its conclusions, the EIS (2) arbitrarily changed the meaning of a number of classic scientific terms, including "uncertainty analysis." These redefinitions and flawed scientific and mathematical analyses mislead readers about EIS scientific validity. They suggest that the UIS authors pursued special-interest science, science used to "justify" pre-determined conclusions, conclusions that happen to endorse the cheapest short-term solution but to impose massive long-term costs on future generations. Consider these flaws.

4 Avoiding Uncertainty Analyses

Scientists recognize that West Valley is a scientifically complex site, with large uncertainties of many kinds (Garrick et al. 2009). As the US National Academy of Sciences has pointed out, scientists also recognize that it is impossible to make precise predictions about site historical, hydrological, geological, and meteorological events tens of thousands of years in the future (Garrick et al. 2009; National Research Council 1995). Yet, despite the dominance of uncertainty in long-term hydrogeological prediction, and despite DOE's providing EIS uncertainty analyses only for a few of the hundreds of relevant parameters, it concludes the site will be safe for the long-term future. DOE also does no uncertainty analysis of its model predictions (US Department of Energy 2010), and it ignores uncertainties that arise from factors such as spatial variability at the site. Instead of emphasizing uncertainty and sensitivity analyses – that would help reveal the scientific reliability of its findings – the EIS employs a largely subjective, "best estimate" set of mostly deterministic predictions about future site safety. It uses single values for model inputs and parameters and then, without documentation, asserts that these values are conservative (US Department of Energy 2010).

Regarding parameter uncertainty, the EIS provides analyses for only a few selected cases, and it ignores uncertainty analyses for nearly all of the hundreds of site-relevant parameters. For example, although the EIS admits that erosion is the main way that site radionuclides are likely to be transported, it gives neither error estimates, nor confidence intervals, nor uncertainty analyses for the parameters involved in erosion prediction. Yet, it admits that these parameters have a large potential range (US Department of Energy 2010; Garrick et al. 2009), and that they depend on precipitation and topography – which change over time (US Department of Energy 2010). Nevertheless, except for one modeling scenario, the EIS reflects arbitrary parameter-input values, especially for gully erosion and landscape evolution, that are "unjustifiable and unsupported by scientific evidence" (US Department of Energy 2010). Hence, it is no surprise that the EIS simulation results show no gully erosion in the South Plateau over the next 10,000 years. This conclusion is "wholly inconsistent" with the observed topography and observed, long-term, continuing, severe erosion and stream-downcutting at the site. These are some of the reasons that the long-term, site-parameter predictions of the EIS are not reliable (US Department of Energy 2010).

Regarding model uncertainty, of course there is no known way to quantitatively assess the uncertainty in a conceptual model (Bredehoeft 2005; Bredehoeft and Konikow 1992). If one knew the relevant empirical values for different parameters, one would not need to use models in the first place. Hence there is no precise, quantitative check on models. Nevertheless the EIS could have done uncertainty analysis of its model predictions, and it did not (US Department of Energy 2010). It also could have qualitatively assessed the uncertainties in its main computer model – a landscape evolution model – by listing all major assumptions, questionable predictions, idealizations, and application problems. However, again the EIS did not do even this qualitative analysis. Instead the EIS used a crude landscape-evolution model for long-term site prediction – although scientists agree they are crude and unsuitable for long-term prediction. Because the EIS admits such site models cannot predict locations of streams, gullies, and landslides; cannot address stream wandering, over time; and cannot predict knickpoint erosion that is causing rapid downcutting erosion of stream channels and increased gullying (US Department of Energy 2010). It is puzzling that the EIS used the models for precise long-term predictions. Similarly, the EIS used a crude, one dimensional model of groundwater transport at the site to predict future radiation doses to humans, for 10,000–1,000,000 years (US Department of Energy 2010), although such models cannot be validated or verified (Bredehoeft 2005; Bredehoeft and Konikow 1992), and although a three-dimensional model likely would have been more reliable than the one-dimensional model (US Department of Energy 2010). Yet, never did the EIS present a compelling argument for why it chose to use simplified one-dimensional flow-and-transport models for the purposes of calculating something as important as long-term radiation dose (US Department of Energy 2010).

Given the crudeness of all such hydrogeological and landscape-evolution models, there is no way to credibly use them in order to conclude the West Valley site will be safe for 10,000–1,000,000 years. If not, the EIS should have admitted this fact, done uncertainty analyses, and avoided generating nearly worthless computer-model predictions whose reliability has never been assessed. In fact, even the EIS short-term computer models of the site are nearly worthless, because none of them is able to predict gully erosion. Yet gully erosion is the principal surface threat to the radioactive wastes. Never did the EIS do model verification or validation by comparing model output with actual field data (US Department of Energy 2010).

5 Changing the Meaning of Normal Scientific Words So as to Mislead

Rather than admit all these sources of uncertainty, and rather than do an uncertainty analysis, however, the DOE West Valley EIS simply redefines various scientific terms in ways that cover up the flaws in the document and the failure of the authors to do standard uncertainty analysis. For instance, while its use of the term "best estimate" suggests a reasoned, empirical assessment, the EIS uses the term in a way that is contrary to standard scientific use. Scientists usually employ the term to mean the average or mean of a distribution or some other optimum, such as a median. However, the DOE EIS uses "best estimate" to mean merely some estimate, subjectively considered by the authors (without any justification provided) to be conservative (US Department of Energy 2010). Yet nowhere does the EIS explain or argue why its supposed analyses are conservative (US Department of

Energy 2010). Indeed, given all the gratuitous guestimates and undocumented claims, it is impossible to check the alleged conservatism of the EIS scenarios (US Department of Energy 2010).

One alleged conservative best estimate, for instance, is that "the probable maximum flood floodplain is very similar to the 100-year floodplain" (US Department of Energy 2010). Yet by definition, the maximum flood over 100 years has more extreme, therefore more dangerous values over 50,000 years, for example, than over 1/500 of that time span, namely 100 years. Likewise, the EIS says it provides a conservative estimate of West Valley radiation in drinking water because its "drinking-water dose analysis conservatively assumes no radionuclide removal in the water treatment system (US Department of Energy 2010). Yet such an assumption is not conservative, but typical. US water-treatment facilities typically do nothing except add chlorine to the water to kill bacterials. They are not equipped to remove radionuclides or any other contaminants. Hence to assume no such removal is not conservative but typical. Similarly the EIS says it presents a conservative "best estimate" for West Valley accidents because it presents the estimated worker "accidents and fatalities that could occur from actions planned for each of the proposed alternatives. These estimates were projected using data from DOE's historical database for worker injuries and fatalities at its facilities" (US Department of Energy 2010). Yet employer databases typically underestimate health problems and accidents, both because they include no long-term follow-up of workers and because workers are reluctant to admit their radiation accidents, exceed dose levels, and thus lose their jobs. Given the EIS misnomer of "best estimates," the pro-nuclear scientific peer reviewers of the EIS warned: "it appears to us that a more apt description" of many of these alleged EIS "best estimate" cases would be "nominal and non-conservative" (Bredehoeft et al. 2006).

Likewise the EIS repeatedly claims to have presented an uncertainty analysis of its conclusions. Yet according to standard scientific usage, an uncertainty analysis assesses the degree to which any particular conclusions and input parameters are reliable. The West Valley EIS, however, does not employ the term "uncertainty analysis" in this way. Instead, as the pro-nuclear scientific peer reviewers point out, the West Valley EIS uses this term to mean simply presenting several different deterministic cases. The reviewers warn that although the EIS "considers presenting three sets of cases to constitute an analysis of uncertainty," it "cannot substitute for a comprehensive uncertainty analysis" (Bredehoeft et al. 2006).

In general, the EIS seems to assume that it has done an uncertainty analysis because it considers several different deterministic cases or uses some supposedly conservative assumptions. For instance, the DOE says in the EIS that "the uncertainty about the reliability of institutional controls" of the West Valley site, to limit radioactive contamination, "has been addressed by conducting the long-term analyses under two different sets of assumptions" (US Department of Energy 2010). Thus DOE redefines "uncertainty analysis" to mean examining two different cases, among the thousands of scenarios that might take place in the next 10,000–1,000,000 years. Moreover, nothing in the EIS justifies choosing these two sets of deterministic, non-probabilistic assumptions rather than others. Because there is no probabilistic analysis that could provide the basis for quantifying uncertainty, the EIS provides no basis for confidence in the quality of its conclusions and no basis for precisely or reliably understanding the contributors to uncertainty (US Department of Energy 2010).

In response to criticisms of its arbitrary redefinition of "uncertainty analysis" and other scientific terms and their associated methods, DOE simply responds that

> Chapter 4, Section 4.3.5, of the EIS contains a comprehensive list of uncertainties that affect the results.... DOE's analyses account for these uncertainties using state-of-the-art models, generally accepted technical approaches, existing credible scientific methodology, and the best available data in such a way that the predictions of peak radiological and hazardous chemical risks are expected to be conservative.... DOE believes the information in the EIS is adequate to support agency decisionmaking for all the reasonable alternatives (US Department of Energy 2010).

In short, DOE says that it can use whatever models or assumptions that it wants, call them conservative, and have no measure of their uncertainty, verification, validation, or sensitivity, and yet claim to do reliable science. Note that the quoted material from DOE merely begs the question that its analyses are conservative and adequate to support agency decisionmaking. It gives no reasons whatsoever for its opinions.

6 An Objection

Reinforcing this question-begging DOE response to criticisms that it has arbitrarily redefined "uncertainty analysis," probabilistic risk assessors (PRA) frequently say it is acceptable not to do standard uncertainty analysis but instead to include uncertainty in the performance or risk measure. That is, a common practice of PRAs is to represent uncertainty by a probability distribution of the supposed "frequency" of occurrence for each scenario under consideration (Garrick 2008; Hoffman and Kaplan 1999; Kaplan 1981). At West Valley, different scenarios might be full exhumation of the waste, partial exhumation of the waste, or no exhumation of the waste at all. Probabilistic risk assessors thus might assign probability distributions to different possible outcomes in each scenario, as a possible way to handle uncertainty.

However, this PRA way of dealing with uncertainty is scientifically incomplete, most obviously because there is no real "frequency" data about different aspects of various scenarios that occur thousands of years into the future. In predicting hydrogeological or other events tens or hundreds of thousands of years in the future, for instance, obviously data are inadequate to enable the estimation of probabilities as limiting relative frequencies. Yet the most important requirement of scientific methods is empirical control. Hence many mathematicians, scientists, and economists, such as Sir Nicholas Stern (2008) and Kahneman et al. (1982) believe that where we don't know the probability distribution, deep uncertainty prevails, and it cannot be characterized by probabilities see (Hansson and Hirsch Hadorn 2016). I think Stern, Kahneman, and others are right (Shrader-Frechette

1991, 1996), mainly because without empirical control, such probabilities are not reliable, and psychologists repeatedly have demonstrated this fact (Stern 2008; Kahneman et al. 1982).

In short, doing performance assessment instead of uncertainty analysis faces given three well-known scientific problems. These include (a) the well-known problem of expert overconfidence and the poor calibration of many experts who estimate uncertainty (Lin and Bier 2008; Shrader-Frechette 2014); (b) the lack of empirical validation for expert opinions about probabilities that may not be able to be estimated as limiting relative frequencies, and (c) the difficulty with using a Bayesian inference mechanism because it requires the prior distribution to be elicited without any knowledge of the data upon which the prior assessment will be later updated (Shrader-Frechette 1991). Nevertheless, where there is objective, empirical validation of expert subjective probabilities, it sometimes is possible to have science-based uncertainty quantification. This quantification is needed because psychometric studies show most experts are overconfident, even in their own fields (Lin and Bier 2008). They often badly underestimate the long tails in the distributions of normalized deviations from the true values. The goal of empirical validation of expert subjective probabilities is to detect the experts who are not overconfident and to differentially weight expert opinions, based on the goal of avoiding overconfidence and underconfidence. For an overview of fallacies in the evaluation and prioritization of uncertainties see Hansson (2016).

To help reduce typical problems (a)–(c), uncertainty analyses are obvious correctives, especially if they include two main components. One component corrective is (1) guarding against common errors when developing prior distributions. One can guard against these errors by using techniques such as those outlined in Quigley and Revie (2011), Hammitt and Shlyakhter (1999), and Shlyakhter (1994).

A second corrective is (2) empirically validating expert subjective probabilities, by using well-known EU-US Nuclear Regulatory Commission (NRC) strategies (Cooke and Goossens 2000; Cooke and Kelly 2010). The EU and US NRC used empirical validation of expert probability assessors, dependence modeling, and differential weighting for combining expert judgments to provide a route to more reliable expert advice (Cooke and Goossens 2000; Cooke and Kelly 2010), as illustrated in many EU-US NRC studies (Goossens et al. 1997, 1998a, b; Brown et al. 1997; Haskin et al. 1997; Little et al. 1997; Cooke et al. 1995; Harper et al. 1995). The heart of this strategy is to calibrate the reliability of each expert probability-estimator, based on assessing the person's probability estimates for events for which frequency data exist. This strategy works because assessors tend to be overconfident or underconfident, regardless of the areas in which they are working. As a result, one can assesses the reliability of expert subjective probabilities by means of checking the expert's performance in areas where frequency data are available.

If used correctly, both correctives (1) and (2) provide for more reliable forms of uncertainty analysis, to be done in addition to traditional uncertainty analysis. Regarding (2), validation methods can be scientifically superior to typical performance analysis – because they employ empirical validation of different experts' judgments about uncertainty. Each expert hypothesis can be empirically tested on the basis of calibration variables (from the experts' field) whose values are known post hoc. The statistical hypothesis being tested is that the realizations are independent samples from a distribution complying with the expert's state percentiles. An expert's statistical accuracy is the p-value of falsely rejecting this hypothesis, based on the values of the calibration variable. In this sort of validation, independence is not an assumption about the experts' joint distribution but a desideratum of the decisionmaker. The expert's informativeness is measured as Shannon-relative information so as to be tail-insensitive and independent of the scales of the underlying variables. The product of the statistical accuracy and informativeness scores yields a combined score that satisfied a scoring-rule constraint: in the long run, an expert maximizes her expected score by and only by stating percentiles corresponding to her true beliefs. This performance-based combination of expert judgments serves rational consensus and can provide more reliable quantification of uncertainty (Cooke 1991; Cooke and Goossens 2000; Aspinall et al. 2002; Aspinall and Cooke 2013; Aspinall 2010).

Other correctives for the preceding DOE errors would have been (3) to do full uncertainty analysis and full sensitivity analysis for at least the 100 most sensitive parameters used by DOE, for all its models, and for all its conclusions, and (4) to use three-dimensional hydrogeological models. Most important, DOE should have taken care (5) to avoid all subjective probabilities, best estimates, and deterministic analyses, and (6) to avoid arbitrary redefinitions of classic terms such as "best estimate," "uncertainty analysis," and "conservative." Once all these correctives had been applied, DOE would have been forced to admit the deep uncertainty surrounding future behavior at the West Valley site, uncertainty that unavoidably requires admission of this uncertainty and of the fact that any purely scientific value judgments about the site would be dominated by value judgments.

7 Significance of the EIS Treatment of Uncertainty

As analysis of the preceding problems indicates, the DOE West Valley EIS is so question-begging, arbitrary, and unempirical – especially in its treatment of million-year uncertainty about the West Valley site – that one wonders why the government spent millions of dollars and more than a decade performing this EIS. Indeed, pro-nuclear scientific peer reviewers claimed, about the EIS, that "a less sophisticated but more credible alternative [to the EIS] would be to judiciously extrapolate observed short and long-term patterns and rates of erosion at the site and the surrounding region into the future, considering such patterns and rates recorded in similar terrains elsewhere, and quantifying the associated predictive uncertainties (which we expect to be very large)" (Bredehoeft et al. 2006). Thus, the DOE has merely avoided full site clean-up, for which it is responsible, and instead used decades of expensive and invalid scientific mumbo-jumbo that redefines "uncertainty analysis" in a wholly arbitrary way. DOE has pushed this redefinition in an attempt to claim that the West Valley site will be safe for 10,000–1,000,000 years into the future. It should have said it could not predict over such a time period, as already mentioned, or it should have based its conclusions on standard uncertainty analysis, especially with the two added correctives, already discussed. But such admissions would leave the government responsible for full and expensive clean up of the West Valley site. Hence the flawed West Valley treatment of uncertainty may well be an artifact of its economic conflicts of interest. DOE analyzes a dangerous site in invalid ways so that DOE is responsible, at least at present, for spending less money to clean up the site for which it is responsible.

The flawed DOE treatment of uncertainty also may be a product of special-interest science – biased science, funded by special interests, whose conclusions are predetermined, not by truth but by how to save money or enhance the profits of special interests (Shrader-Frechette 2007). Special interests fund scientists to give them the answers that they want, including incomplete, biased "science" affirming that the funders' pollution or products are safe or beneficial. This fact has been repeatedly confirmed for pharmaceutical and medical-devices research (Krimsky 2003), energy-related research (Shrader-Frechette 2011), and pollution-related research (Michaels 2008; McGarity and Wagner 2008).

After all, special-interest "science" helped US cigarette manufacturers avoid regulations for more than 50 years. It also explains why fossil-fuel industry "science" denies anthropogenic climate change.

8 Conclusion

As this DOE case shows, special-interest science can be used not only by corporations but by allegedly democratic governments, as has occurred at West Valley. They can redefine "uncertainty analysis," so that they can claim to have reliable million-year predictions about information about which they have no adequate empirical data. Such misuse of science and redefinition of "uncertainty analysis" may be even more deadly and unethical because often citizens cannot sue government, the way they can sue corporations or citizens who harm them. If democratic governments claim "sovereign immunity," in cases like the West Valley EIS, they are able to avoid citizens' complaints and lawsuits. They also force the citizens to pay for the obviously flawed science that betrays both democracy and scientific truth.

Recommended Readings

Kahneman, D., Slovic, P., & Tversky, A. (1982). *Judgment under uncertainty: Heuristics and biases*. New York: Cambridge University Press.

Shrader-Frechette, K. (2014). *Tainted: How philosophy of science can expose bad science*. New York: Oxford University Press. Available at Oxford Scholarship Online at www.oup.com/uk/oso

References

Aspinall, W. P. (2010). A route to more tractable expert advice. *Nature, 463*, 294–295.

Aspinall, W. P., & Cooke, R. M. (2013). Quantifying scientific uncertainty from expert judgment elicitation. In L. Hill, J. C. Rougier, & R. S. J. Sparks (Eds.), *Risk and uncertainty assessment for natural hazards* (pp. 64–99). New York: Cambridge University Press.

Aspinall, W. P., Loughlin, S. C., Michael, F. V., Miller, A. D., Norton, G. E., Rowley, K. C., Sparks, R. S. J., & Young, S. R. (2002). The Montserrat volcano observatory: Its evolution, organisation, role and activities. In T. H. Druitt & B. P. Kokelaar (Eds.), *The eruption of Soufrière Hills volcano, Montserrat, from 1995 to 1999* (pp. 71–92). London: Geological Society.

Bredehoeft, J. (2005). The conceptualization model problem. *Hydrogeology Journal, 13*(1), 37–46.

Bredehoeft, J., & Konikow, F. (1992). Ground-water models cannot be validated. *Advances in Water Resources, 15*(1), 75–83.

Bredehoeft, J. D., Fakundiny, R. H., Neuman, S. P., Poston, J. W., & Whipple, C. G. (2006). *Peer review of draft environmental impact statement for decommissioning and/or long-term stewardship at the West Valley demonstration project and Western New York Nuclear Service Center*. West Valley: DOE.

Brown, J., Goossens, L. H. J., Harper, F. T., Kraan, B. C. P., Haskin, F. E., Abbott, M. L., Cooke, R. M., Young, M. L., Jones, J. A., Hora, S. C., Rood, A., & Randall, J. (1997). *Probabilistic accident consequence uncertainty analysis: Food chain uncertainty assessment* (Report NUREG/CR-6523, EUR 16771). Washington, DC: USNRC.

Cooke, R. M. (1991). *Experts in uncertainty; opinion and subjective probability in science*. New York: Oxford University Press.

Cooke, R. M. (2013). Uncertainty analysis comes to integrated assessment models for climate change…and conversely. *Climatic Change, 117*(3), 467–479. doi:10.1007/s10584-012-0634-y.

Cooke, R. M., & Goossens, L. H. J. (2000). *Procedures guide for structured expert judgment*. Brussels: European Commission.

Cooke, R. M., & Kelly, G. N. (2010). *Climate change uncertainty quantification: Lessons learned from the joint EU-USNRC project on uncertainty analysis of probabilistic accident consequence codes*. Washington, DC: Resources for the Future.

Cooke, R. M., Goossens, L. H. J., & Kraan, B. C. P. (1995). Methods for CEC/USNRC accident consequence uncertainty analysis of dispersion and deposition: Performance based aggregating of expert judgments and PARFUM method for capturing modeling uncertainty. Prepared for the Commission of European Communities, EUR 15856, Brussels.

Garrick, B. J. (2008). *Quantifying and controlling catastrophic risk*. Amsterdam: Elsevier.

Garrick, B. J., Bennett, S. J., Neuman, S. P., Whipple, C. G., & Potter, T. E. (2009). *Review of the U.S. Department of Energy Responses to the U.S. Nuclear Regulatory Commission Re the West Valley demonstration project phase 1 decommissioning plan*. Albany: New York State Energy Research and Development Authority.

Goossens, L. H. J., Boardman, J., Harper, F. T., Kraan, B. C. P., Cooke, R. M., Young, M. L., Jones, J. A., & Hora, S. C. (1997). *Probabilistic accident consequence uncertainty analysis: External exposure from deposited material uncertainty assessment* (Report NUREG/CR-6526, EUR 16772). Washington, DC: USNRC.

Goossens, L. H. J., Cooke, R. M., Kraan, B. C. P. (1998a). *Evaluation of weighting schemes for expert judgement studies*. Prepared for the Commission of European Communities, Directorate-General for Science, Reserach and Development, XII-F-6, Delft University of Technology, Delft.

Goossens, L. H. J., Harrison, J. D., Harper, F. T., Kraan, B. C. P., Cooke, R. M., & Hora, S. C. (1998b). *Probabilistic accident consequence uncertainty analysis: Internal dosimetry uncertainty assessment* (Report NUREG/CR-6571, EUR 16773). Washington, DC: USNRC.

Hammitt, J. K., & Shlyakhter, A. I. (1999). The expected value of information and the probability of surprise. *Risk Analysis, 19*(1), 135–152.

Hansson, S. O. (2016). Evaluating the uncertainties. In the argumentative turn in policy analysis. In S. O. Hansson & G. Hirsch Hadorn (Eds.), *Reasoning about uncertainty* (pp. 79–104). Cham: Springer. doi:10.1007/978-3-319-30549-3_4.

Hansson, S. O., & Hirsch Hadorn, G. (2016). Introducing the argumentative turn in policy analysis. In The argumentative turn in policy analysis. In S. O. Hansson & G. Hirsch Hadorn (Eds.), *Reasoning about uncertainty* (pp. 11–35). Cham: Springer. doi:10.1007/978-3-319-30549-3_2.

Harper, F. T., Goossens, L. H. J., Cooke, R. M., Hora, S. C., Young, M. L., Päsler-Sauer, J., Miller, L. A., Kraan, B. C. P., Lui, C., McKay, M. D., Helton, J. C., & Jones, J. A. (1995). *Probabilistic accident consequence uncertainty analysis: Dispersion and deposition uncertainty assessment* (Report NUREG/CR-6244, EUR 15855). Washington, DC: USNRC.

Haskin, F. E., Harper, F. T., Goossens, L. H. J., Kraan, B. C. P., Grupa, J. B., & Randall, J. (1997). *Probabilistic accident consequence uncertainty analysis: Early health effects uncertainty assessment* (Report NUREG/CR-6545, EUR 16775). Washington, DC: USNRC.

Hoffman, F. O., & Kaplan, S. (1999). Beyond the domain of direct observation: How to specify a probability distribution that represents the "State of Knowledge" about uncertain inputs. *Risk Analysis, 19*(1), 131–134.

Kahneman, D., Slovic, P., & Tversky, A. (1982). *Judgment under uncertainty: Heuristics and biases*. New York: Cambridge University Press.

Kaplan, S. (1981). On the method of discrete probability distributions in risk and reliability calculations – application to seismic risk assessment. *Risk Analysis, 1*(3), 189–196.

Krimsky, S. (2003). *Science in the private interest – Has the lure of profits corrupted biomedical research*. Lanham: Rowman & Littlefield.

Lin, S.-W., & Bier, V. M. (2008). A study of expert overconfidence. *Reliability Engineering and System Safety, 93*, 711–721.

Little, M., Muirhead, C., Goossens, L. H. J., Harper, F. T., Kraan, B. C. P., Cooke, R. M., Hora, S. C. (1997). *Probabilistic accident consequence uncertainty analysis: Late (somatic) health effects uncertainty assessment* (Report NUREG/CR-6555, EUR 16774). Washington, DC: USNRC.

McGarity, T., & Wagner, W. (2008). *Bending science*. Cambridge: Harvard University Press.

Michaels, D. (2008). *Doubt is their product*. Cambridge: Harvard University Press.

Napoleon, A., Fisher, J., Steinhurst, W., Wilson, M., Ackerman, F., Resnikoff, M., & Brown, E. (2008). *The real costs of cleaning up nuclear waste: A full cost accounting of cleanup options for the west valley nuclear waste site*. Cambridge: Synapse Energy Economics.

National Research Council. (1995). *Technical bases for Yucca Mountain Standards*. Washington, DC: National Academy Press. *Peer review of draft environmental impact statement for decommissioning and/or long-term stewardship at the West Valley Demonstration Project and Western New York Nuclear Service Center*. West Valley: New York State Energy Research and Development Authority.

National Research Council/National Academy of Sciences (NRC/NAS). (2006). *Health risks from exposure to low levels of ionizing radiation: BEIR VII, phase 2*. Washington, DC: National Academy Press.

Quigley, J., & Revie, M. (2011). Estimating the probability of rare events: Addressing zero failure data. *Risk Analysis, 31*(7), 1120–1132.

Shlyakhter, A. I. (1994). An improved framework for uncertainty analysis: Accounting for unsuspected errors. *Risk Analysis, 14*(4), 441–447.

Shrader-Frechette, K. (1991). *Risk and rationality*. Berkeley: University of California Press.

Shrader-Frechette, K. (1996). Science versus educated guessing. *BioScience, 46*(7), 488–489.

Shrader-Frechette, K. (2007). *Taking action, saving lives: Our duties to prevent environmental and public-health harms*. New York: Oxford University Press.

Shrader-Frechette, K. (2011). *What will work: Fighting climate change with renewable energy, not nuclear power*. New York: Oxford University Press.

Shrader-Frechette, K. (2014). Tainted: How philosophy of science can expose bad science. New York: Oxford University Press; available at Oxford Scholarship Online at www.oup.com/uk/oso

Stern, N. (2008). The economics of climate change. *American Economic Review, 98*(2), 1–37.

US Department of Energy (DOE). (2010). *Final environmental impact statement for decommissioning and/or long-term stewardship at the West Valley demonstration project and Western New York nuclear service center* (DOE/EIS-0226 vols. 1–2). West Valley: DOE.

Chapter 13
Climate Geoengineering

Kevin C. Elliott

Abstract Climate geoengineering is in many ways a "poster child" for the value of the argumentative approach to decision analysis. It is fraught with so many different kinds of uncertainty that the reductive approach described in the first chapter of this volume is seriously inadequate on its own. Instead, debates about climate geoengineering incorporate a wide variety of issues that can be fruitfully addressed using argumentative analysis. These include conceptual questions about how to characterize and frame the decision problem; ethical questions about the values and principles that should guide decision makers; and procedural questions about how to make decisions about climate geoengineering in a fair, legitimate manner.

Keywords Geoengineering • Climate change • Ethics • Uncertainty • Framing • Governance • Risk management • Precautionary principle • Argumentative analysis • Solar radiation management

1 Introduction

Climate geoengineering refers to the deliberate manipulation of earth systems, specifically in response to climate change (see e.g., Royal Society 2009: ix; Schneider 2001: 47). Commonly discussed strategies for climate geoengineering include the emission of sulfur aerosols to mimic the cooling effects of volcanic eruptions, seeding the oceans with iron to stimulate the growth of plankton that absorbs carbon dioxide, or spraying sea water into the air in order to create whiter clouds that reflect more solar radiation (Royal Society 2009). Climate geoengineering is in many ways a "poster child" for the value of argumentative analysis. It is fraught with so many different kinds of uncertainty that traditional, reductive approaches to decision analysis are of very limited use in addressing it (see Hansson and Hirsch Hadorn 2016). As a result, debates about climate

---

K.C. Elliott (✉)
Lyman Briggs College, Department of Fisheries & Wildlife, and Department of Philosophy, Michigan State University, East Lansing, MI, USA
e-mail: kce@msu.edu

S.O. Hansson, G. Hirsch Hadorn (eds.), *The Argumentative Turn in Policy Analysis*, Logic, Argumentation & Reasoning 10, DOI 10.1007/978-3-319-30549-3_13

geoengineering incorporate a wide variety of conceptual, ethical, and procedural questions that can be fruitfully addressed using argumentative analysis.

The next section provides an introduction to climate geoengineering and the major forms of uncertainty that make it exceedingly difficult to analyze using traditional forms of decision analysis. The following sections show how argumentative analysis can help to address three different kinds of questions that arise in debates about climate geoengineering: (1) questions about how to characterize and frame the decision problem; (2) questions about ethical values and principles such as distributive justice, moral hazards, and the precautionary principle; and (3) procedural questions about how to make decisions about climate geoengineering in a fair, legitimate manner. While the primary focus is on clarifying these debates rather than developing a specific normative position on the acceptability of climate geoengineering, this contribution suggests some important lessons. For example, it highlights the weaknesses of framing climate geoengineering as an insurance policy or a form of compensation; instead, the "technical fix" frame may be more fruitful. Another important lesson is that research projects on climate geoengineering raise many of the same ethical and political issues as efforts to implement it, and so serious effort should be put into developing governance schemes that can address the growing calls for research in this area. Third, efforts to justify climate geoengineering via "lesser of two evil" arguments should be evaluated with great care. Finally, it remains unclear how to develop fair, legitimate procedures for governing climate geoengineering.

2 Climate Geoengineering and Uncertainty

The British Royal Society (2009) coined the terms "carbon dioxide removal" (CDR) and "solar radiation management" (SRM) for two broad categories of climate geoengineering strategies. CDR strategies operate by removing carbon dioxide from the atmosphere, whereas SRM strategies lessen the amount of solar radiation absorbed by the earth. As the next section will discuss, there are strengths and weaknesses of dividing climate geoengineering approaches into these two broad categories. Nevertheless, these categories are commonly used, at least in part because their risk-benefit profiles tend to have different characteristics. For example, SRM strategies tend to be associated with more significant risks and uncertainties, whereas CDR approaches are often slower and more costly.

One of the most widely discussed SRM strategies involves emitting sulfur aerosols into the atmosphere. These aerosols have been found to cool the earth after volcanic eruptions and are frequently mentioned as one of the quickest and cheapest potential geoengineering strategies (Royal Society 2009). Other frequently discussed SRM strategies include painting urban structures white to increase reflection of solar radiation, deploying mirrors into space, or spraying sea water into the air to create more reflective clouds (Elliott 2010a: 241). A commonly discussed approach to CDR is to fertilize the oceans with iron in order to stimulate the growth of phytoplankton that absorb carbon dioxide (Cullen and Boyd 2008). Other examples of CDR include using new technologies to capture carbon dioxide from the air or from power plants, promoting the reactions of silicate rocks with atmospheric carbon dioxide, or promoting the growth of forests (Royal Society 2009).

Deciding whether to study or to employ various climate geoengineering techniques is an extremely complicated matter that illustrates many of this book's important themes. In particular, there are numerous forms of uncertainty in this case that make it very difficult to employ traditional forms of cost-benefit analysis. Many of these forms of uncertainty fall under the category of "great uncertainty" (Hansson and Hirsch Hadorn 2016). These include uncertainties about the range of possible outcomes, difficulties deciding how to frame the decision problem, contested ethical values, and challenges predicting how multiple agents will act in the future.

Before even turning to the uncertainties associated with predicting the various positive and negative consequences of climate geoengineering, there are numerous uncertainties associated with climate science that need to be taken into account. Without clear-cut information about the likely effects of climate change, it becomes very questionable to perform a complete cost-benefit analysis that compares the risks associated with performing climate geoengineering to the risks of going without it. For example, there are obvious uncertainties associated with calculating the plausible climate trajectories associated with particular emission scenarios for greenhouse gases, or the likelihood of particular emission scenarios, or the likelihood of specific harmful effects associated with particular climate trajectories (e.g., floods, droughts, or sea-level rise), or the details of how those effects might be distributed across time and space, or the climatic tipping points that would result in particularly catastrophic results (Tuana et al. 2012: 149–151). Even when experts claim to be able to provide fairly precise quantitative probabilities and estimates of uncertainty for some of these outcomes, their estimates can be influenced by problematic modeling assumptions or cognitive biases (Elliott and Resnik 2015; Elliott and Dickson 2011; Parker 2011; Jamieson 1996).

Turning to the uncertainties associated with climate geoengineering, even the earliest discussions of it emphasized the possibility of unexpected side effects and the importance of finding strategies for dealing with them (e.g., Kellogg and Schneider 1974). Some of the potential side effects of climate geoengineering strategies include changes to regional precipitation patterns, depletion of the ozone hole (especially from stratospheric aerosol emissions), altered ecosystems, and various sorts of environmental damage (Royal Society 2009; Robock 2008). It is also difficult to predict the effectiveness of various climate geoengineering strategies, including how the effects of the strategies will be distributed across time and space. Adding to the complexity is the fact that it could be ethically questionable to engage in the sorts of large-scale field trials that would be necessary to alleviate some of these uncertainties (NRC 2015).

These uncertainties about the effects of climate geoengineering could be addressed at least partially through further scientific research, and they could be evaluated to determine their relevance for decision making (Hansson 2016). However, there are social and political uncertainties that are much less amenable to scientific investigation and much more difficult to evaluate. For example, Dale Jamieson points out that the potential effects of climate change are so varied and pervasive that "it is extremely difficult to make an informed judgment between intentional or inadvertent climate change [i.e., engaging in climate geoengineering or not] on grounds of socio-economic preferability" (Jamieson 1996: 328). Moreover, deliberations about climate geoengineering need to take into account the possibility that "rogue" states, corporations, or individuals would attempt to implement it unilaterally, thereby creating serious political conflicts. They also have to consider whether it is even feasible to create fair and widely accepted international governance procedures for making decisions about climate geoengineering. Furthermore, given that there could be catastrophic consequences if SRM strategies were suddenly halted and the climate shifted dramatically, it would also be important to evaluate the likelihood that stable political entities could be maintained for as long as these strategies were needed. However, the probability of these social and political events cannot be predicted reliably (see Royal Society 2009; Robock 2008; Jamieson 1996).

Given all these uncertainties associated with climate geoengineering, it becomes all the more important to reflect on the general moral principles that should guide decision making in this context. However, these ethical principles and values represent yet another crucial category of uncertainty and ambiguity. For example, some scholars have argued that climate geoengineering could pose a *moral hazard*, in the sense that it could encourage risky behaviors by providing a sort of insurance policy against catastrophic climate change (NRC 2015: 8; Betz 2012: 479; Royal Society 2009: 39). However, there is confusion about the nature of moral hazards and the extent to which they should be avoided (Hale 2012). It is also unclear precisely how to obtain adequate consent from those who will be affected by climate change and climate geoengineering (both now and in the future) (Betz 2012: 478). There is also moral confusion about whether it would be inherently ethically problematic to manipulate the entire climate system intentionally (Preston 2012a; Katz 1992). Finally, a number of ethical principles that are relevant to the decision to geoengineer remain deeply contested. These include principles of distributive justice and procedural justice, the doctrine of doing and allowing, and the precautionary principle (Elliott 2010a).

Various forms of argumentative analysis can play a valuable role in addressing complicated decision problems like this one, which do not fulfill the preconditions for applying formal approaches of policy analysis (Hansson and Hirsch Hadorn 2016). Given the wide array of scientific, social, and moral uncertainties associated with climate geoengineering, it would be foolhardy to rely primarily on formal cost-benefit analyses for making decisions about implementing it. Instead, any formal analyses need to be embedded in a broader discussion about the moral and political principles that should govern the decision and the most appropriate ways of framing it (Grüne-Yanoff 2016). The following sections explore three ways in which argumentative analysis can be helpful in this case: (1) reflecting on how to frame and characterize the decision problem; (2) clarifying the key moral concepts and principles at stake; and (3) identifying the issues that need to be addressed in order to formulate an adequate governance scheme. Argumentative analysis can also play a valuable role in uncovering implicit assumptions and values inherent in efforts to characterize and alleviate scientific uncertainties (Hansson 2016; Tuana et al. 2012; Elliott 2011; NRC 1996). Nevertheless, this chapter focuses on the ethical and political uncertainties associated with climate geoengineering and explores the scientific uncertainties mainly as they arise in the ethical and political debates.

3 Framing and Characterizing the Decision Problem

When decision makers face particularly thorny decision problems that are not amenable to formal analysis, they are often forced to draw analogies and comparisons to other decisions in an effort to obtain insight and guidance. They may also attempt to break complex decisions into simpler pieces so that they are more tractable (Hirsch Hadorn 2016). With this in mind, an important role for argumentative analysis is to evaluate the ways in which a complex decision has been framed and characterized so that decision makers can understand whether they are implicitly introducing important assumptions or values to the decision problem (Grüne-Yanoff 2016). While this section cannot provide a comprehensive analysis of the major ways that climate geoengineering has been characterized, it provides three examples of the sorts of issues that deserve further consideration. First, the terminology used for describing environmental issues, including climate geoengineering, can influence the ways people think about the decision problem, and thus it merits scrutiny (Elliott 2009). Second, climate geoengineering is sometimes compared to other social phenomena, such as insurance policies or technical fixes, which means that it is very important to evaluate these comparisons. Third, some ethicists have attempted to simplify decisions about climate geoengineering by dividing the decision problem into separate categories, and these efforts also deserve close analysis.

Turning first to the choice of terminology, an initial question is whether it is even wise to use the term "geoengineering." One worry is that the reference to engineering could be misleading, given that many forms of climate geoengineering do not literally involve work by engineers. This might not seem significant, except that people may associate engineering projects with particular characteristics – for example, a relatively high degree of control and predictability – that are not present in the case of climate geoengineering (NRC 2015: 1; Elliott 2010a: 240). An additional problem with the reference to geoengineering is that it confuses efforts to manipulate the climate with other engineering efforts that take place in a geological context, such as water resources management, resource extraction, and ecological restoration (NRC 2015: 1; Bipartisan Policy Center 2011: 33). In part for these reasons, recent reports have chosen to use terms like "climate remediation technologies" or "climate intervention" rather than "geoengineering" (NRC 2015; Bipartisan Policy Center 2011).

Another worry is that the term "geoengineering" might inappropriately lump a number of very different activities together into a single category, which could make it more difficult for those deliberating about climate geoengineering to make valuable distinctions. For example, engaging in massive tree-planting efforts to remove carbon dioxide from the atmosphere seems exceedingly different from an ethical and social perspective than emitting sulfur aerosols into the atmosphere, but they could both be labeled as forms of climate geoengineering. The distinction between solar radiation management (SRM) and carbon dioxide removal (CDR) is intended to help alleviate this problem by providing distinctions between climate geoengineering techniques with significantly different characteristics, but it too faces significant difficulties. For one thing, it is not clear that the SRM/CDR distinction actually captures the important ethical distinctions that need to be made. SRM technologies tend to be regarded with extra suspicion, in part because they are often riskier and in part because CDR technologies often seem closer to natural processes that already remove carbon from the atmosphere (NRC 2015; Preston 2013: 24). Nevertheless, these intuitions can be misguided. For example, the CDR technology of seeding the oceans with iron as a strategy for stimulating phytoplankton growth appears to have serious ecological risks (Cullen and Boyd 2008). An additional problem with the CDR category is that climate geoengineering is typically regarded as an *alternative* to mitigation and adaptation strategies for responding to climate change, but some CDR approaches (such as reforestation or altered agricultural practices) can be regarded as *forms* of mitigation (Bipartisan Policy Council 2011: 7).

Even if one were to accept the distinction between carbon dioxide removal (CDR) and solar radiation management (SRM) as unproblematic, the term "solar radiation management" has been challenged for some of the same reasons as the term "geoengineering." Specifically, referring to "management" could give the false impression that humans can effectively alter solar radiation in fairly precise ways. Thus, some reports have tried to use terms like "sunlight reflection methods" or "albedo modification" in an effort to be more neutral about the effectiveness of these techniques (Pierrehumbert 2015; NRC 2015). One scientist has even suggested that a term like "albedo hacking" would better express the dangerous and experimental nature of these techniques (Pierrehumbert 2015).

In sum, the major terms and categories used for describing climate geoengineering continue to be a matter for debate. These terminological issues cannot be settled here, but they do illustrate the importance of argumentative analysis. Given that climate geoengineering encompasses so many different kinds of technologies, some of which may have severe and unpredictable consequences, the terms and categories used for describing it can have a significant influence on how people respond to it. The following sections will employ the traditional term "geoengineering" as well as the categories of "SRM" and "CDR" technologies, but they do so while acknowledging that the terms deserve ongoing scrutiny and analysis. Moreover, to avoid confusion between geoengineering of the climate and other sorts of geoengineering activities like mining or ecosystem restoration, the sections will consistently refer to "climate geoengineering."

In addition to these questions about what terminology to use for describing climate geoengineering, a second important issue is whether particular frames or analogies can assist in guiding people's responses to the phenomenon. Consider, for example, two particularly common frames: treating climate goengineering as an insurance policy or as a technological fix. At first glance, it does indeed seem like climate geoengineering could act like an insurance policy, in the sense that it would provide a valuable resource in case a climatic catastrophe were to occur (see e.g., Royal Society 2009: 45; Caldeira 2007; Hoffert et al. 2002: 986). But Dane Scott (2012) has analyzed this frame and argued that it faces severe limitations. He notes: "People act prudently when they buy insurance policies that are trustworthy because they are legally binding agreements and they are confident in the currency of compensation" (Scott 2012: 157). He points out that climate geoengineering is not at all like a typical insurance policy, given that it is so risky and unpredictable. He suggests that it would make more sense to compare climate geoengineering to an emergency medical technology like dialysis, which can keep a person alive but which has significant risks of its own.

Given that frames invariably focus people's attention on some considerations or arguments rather than others, Scott (2012) argues that a "technological fix" frame is likely to generate more fruitful social discussions about climate geoengineering than the insurance frame. He distinguishes, however, between two closely related versions of this frame: the "techno-fix" versus the pragmatic "technical fix" (Scott 2012: 158). He notes that the term "techno-fix" has become a polarizing catch-phrase that critics of technology use to disparage ill-conceived technological solutions to social problems. In contrast, he suggests that the pragmatic "technical fix" frame may strike a more appropriate tone that highlights the important issues in need of discussion. As a society, we have found that technical fixes can be helpful. Nevertheless, we have also found that technical fixes often fail to address underlying problems, and they tend to have unintended consequences of their own. Moreover, technical fixes can lead to "revenge effects," also known as risk homeostasis, in which people engage in inappropriately risky behavior because they feel protected by the technical fixes (Wilde 1998). Thus, Scott argues that this frame may prove to be particularly appropriate for describing climate geoengineering, because it focuses social discussions on some of the most important issues that need to be considered (Scott 2012: 167).

A third issue to examine when characterizing climate geoengineering is how to structure the decision problem in a manageable way. Hansson and Hirsch Hadorn (2016) emphasize that this is a valuable role for argumentative analysis. I have previously suggested that it can be helpful to break down decisions about climate geoengineering into three sub-decisions: (1) choices about discussing geoengineering; (2) choices about researching geoengineering; and (3) choices about implementing geoengineering (Elliott 2010a). Christopher Preston (2013: 24) has argued that a fourth sub-decision is also valuable to consider: (4) choices that occur after implementation. The distinction between the second and third sub-decisions (researching versus implementing geoengineering) is highly significant, insofar as numerous articles and reports have suggested that preliminary forms of *research* on geoengineering are to be recommended, whereas most forms of geoengineering should not be *implemented* without a great deal of further discussion (e.g., NRC 2015; Bipartisan Policy Center 2011; Keith et al. 2010; Royal Society 2009; Cicerone 2006). This approach of making an initial decision to engage in preliminary research that can inform future decisions about more intensive research or implementation exemplifies the sequential approach to decision making (Hirsch Hadorn 2016).

While these divisions of the decision problem seem reasonable at first glance, it turns out that they merit further discussion and analysis. Perhaps the most questionable aspect of this four-part decision structure is the distinction between researching climate geoengineering versus implementing it. On further inspection, this distinction becomes more complicated than it first appears. For one thing, a number of authors have pointed out that performing research on climate geoengineering could increase the likelihood that it will actually be implemented (Betz 2012: 476–477). For example, Dale Jamieson argues that "in many cases research leads unreflectively to development" (1996: 333). In a similar vein, Kyle Powys Whyte (2012b) worries that research efforts could "crystallize" paths for developing the technology, and Stephen Gardiner (2010) argues that research creates "institutional momentum." This connection between research and implementation becomes highly significant when evaluating the ethical implications of pursuing climate geoengineering research. While some scholars have argued that the risks of not doing research on climate geoengineering are greater than the risks of doing so (Keith et al. 2010), this conclusion depends on the potentially misleading assumption that the risks associated with researching it can be distinguished from the risks associated with implementing it.

A second reason for questioning the distinction between researching and implementing climate geoengineering is that some forms of research cannot be successfully performed without doing field tests, and it is not clear that some of these field tests are truly distinct from the actual implementation of climate geoengineering (Betz 2012: 480). In fact, Alan Robock and his coauthors (2010) argue that effective testing of at least some climate geoengineering approaches cannot occur without actually implementing them. For example, they note that one cannot distinguish natural weather and climate variability from the effects of climate geoengineering without a "large, decade-long forcing," which would require a large enough implementation to "disrupt food production on a large scale" (Robock et al. 2010: 530). Finally, even if one were able to distinguish limited field tests from full-scale attempts at implementation, many of the same ethical and political concerns would still apply to both. These concerns include the worry that national or corporate interests could hijack the technology for selfish purposes, the potential for side effects to cross national boundaries and create international tension, and the need to assign liability for potential harms (Blackstock and Long 2010). Given that climate geoengineering research and implementation raise many of the same ethical issues and that they may in some cases be indistinguishable, efforts to structure decisions about climate geoengineering using this distinction should arguably be scrutinized with great care.

4 Ethical Issues

Since there are such pervasive uncertainties associated with climate geoengineering, it is a fool's errand to try to quantify the likely costs and benefits associated with various climate geoengineering schemes with precision in an effort to determine a rational choice. Therefore, various ethical principles become particularly important for determining how to handle this uncertainty and deciding where to place the burden of proof when dealing with potential impacts that are difficult to predict. Many of the foundational documents discussing climate geoengineering highlight the necessity of thinking through its ethical ramifications (e.g., Royal Society 2009; Crutzen 2006). This section shows how argumentative analysis can help to clarify five important ethical principles and concepts that are relevant to the climate geoengineering case: appeals to the natural order, the precautionary principle, the concept of moral hazard, "lesser evil" arguments, and the concept of distributive justice (see Brun and Betz 2016 for an integrated assessment, with the help of argument maps, of several of these ethical principles in the case of climate geoengineering).

An initial and highly significant ethical issue posed by climate geoengineering is whether it is problematic to engage in intentional actions to alter the earth's "natural" climate system. At first glance, this may sound like an appeal to naturalness, which Hansson (2016) identifies as a fallacious form of argumentation. In other words, it does not follow from the fact that something is natural that it is morally good, and one cannot conclude that something is morally bad based on the fact that it is unnatural. Nevertheless, ethicists have pointed both to "extrinsic" and "intrinsic" reasons for thinking that it is ethically questionable to manipulate the climate system intentionally (Preston 2012b: 4). The extrinsic argument is relatively obvious; we have already seen that the global climate system is so complex that there are significant dangers of causing unanticipated harms when attempting to manipulate it. Moreover, efforts to alter the global climate are almost certain to generate significant international political disputes. Thus, this form of the argument does not really rest on an appeal to naturalness but rather on the practical and political difficulties of trying to control an exceedingly large and complex system.

The intrinsic argument against intentionally altering the earth's natural climate system is based on the notion that it is ethically problematic to violate earth's naturalness by turning it into a human artifact. As Christopher Preston (2012a) clarifies, the intentional manipulation of the earth's climate would represent a momentous shift in humanity's relationship with nature. Many environmental ethicists have argued that there is something valuable about maintaining elements of nature that are relatively free of human influence, and climate geoengineering seems to violate this principle by turning the entire climate system into an intentionally manipulated artifact (Preston 2012a). But further analysis is needed to determine what is meant by turning the earth into an artifact and whether this is indeed ethically problematic. Proponents of an ancient account, going back to Aristotle, argue that once an object has been manipulated by forces from outside it, such as human intervention, it becomes an artifact and loses its naturalness (Preston 2012a: 191). But this account is relatively unhelpful for evaluating climate geoengineering, because almost every portion of the globe has already been influenced in some way by human beings and thus has already lost its "naturalness." Moreover, it is unclear what is wrong with losing the earth's naturalness in this sense.

Steven Vogel (2003) has developed an alternative account of artifacts that provides the basis for a more compelling account of what is wrong with altering the earth's "natural" climate system. Vogel affirms that human artifacts can still display "naturalness," because a "gap" always remains between what the artificer intends and the manner in which the artifact actually behaves. As Preston (2012a) notes, this account of artifacts highlights the fact that no human endeavor goes precisely as planned. Therefore, it drives home the point that climate geoengineering would leave humanity with grave responsibilities that we have never faced before. As Preston puts it, "Wild nature has been the place people have gone to escape the pressing responsibilities of the human world" (2012a: 197). However, if we chose to geoengineer the climate, "There would be no place on earth – or under the sky – where anxiety-producing questions such as 'Are we succeeding?' could be avoided" (2012a: 197). Thus, the "intrinsic" argument that we should not turn earth into a human artifact is perhaps best cast as an "extrinsic" argument, based on the realization that our efforts at climate geoengineering are unlikely to go as we plan and that it is unwise to take on such a momentous responsibility.

This anxiety over taking responsibility for the climate highlights a second important ethical principle that needs to be examined in the climate geoengineering context: the precautionary principle (PP). According to this principle, decision makers should take precautionary measures to avoid creating grave threats for human health or the environment, even when the scientific information about those threats is incomplete (e.g., Fisher et al. 2006). At first glance, this principle seems to be a perfect guideline for addressing climate geoengineering; it appears to counsel decision makers to avoid schemes that could generate serious hazards for humans or the environment. Unfortunately, more detailed analysis indicates that the ramifications of the PP are less obvious than they initially appear. This is partly because the principle is ambiguous. Without further specification, it is not clear which threats are serious enough to merit precautionary action, or how much information about the threats is necessary to justify action, or precisely which precautionary actions ought to be taken (Sandin 1999). With this in mind, it may be most fruitful to think of the PP as a family of related principles, some of which demand more aggressive precautionary action than others. Thus, when evaluating the ramifications of the PP for climate geoengineering, one needs to consider which type of climate geoengineering is under consideration and which form of the PP is being applied to it (Elliott 2010a).

An even more serious problem is that at least some forms of the PP may end up being "self-defeating" when directed at climate geoengineering (Elliott 2010a). In other words, the PP could be used both for criticizing climate geoengineering and for criticizing the avoidance of it. A number of scholars have previously challenged the PP because of its potential to have these sorts of paralyzing effects (see e.g., Sunstein 2005). While these challenges are arguably exaggerated in many cases, it may have more purchase in the case of climate geoengineering (Elliott 2010a: 246). For example, if policy makers faced evidence in the future that catastrophic consequences of climate change were imminent, the PP would presumably call for effective steps to prevent disaster. But it is conceivable that the only truly effective steps to prevent at least some of these catastrophic consequences would involve climate geoengineering, which could plausibly pose serious threats of its own. Thus, the PP might simultaneously seem to call for engaging in climate geoengineering and banning it.

Additional forms of argumentative analysis could potentially mitigate some of this confusion. For example, it might be helpful to compare the uncertainties associated with engaging in climate geoengineering with those associated with failing to engage in it. Hansson (2016) argues that these uncertainties can be evaluated to determine their relevance and weight for decision making. This form of analysis could potentially help decision makers determine which precautionary actions are most important to prioritize (see Brun and Betz 2016; Betz 2016).

A third ethical concept that needs to be clarified in the context of climate geoengineering is the notion of a moral hazard (NRC 2015; Betz 2012: 479, Royal Society 2009: 39). Roughly, the concern is that researching geoengineering or actually engaging in it could give people a sense of complacency and make them less likely to mitigate or adapt to climate change. It might seem that efforts to determine whether climate geoengineering poses a moral hazard lie in the domain of the social sciences. However, a careful analysis by Ben Hale (2012) illustrates that argumentative analysis can be exceedingly valuable for addressing these uncertainties.

Hale (2012) identifies at least 16 different versions of the moral hazard argument. Without going into all the details, it is enlightening to see some of the broad categories into which these arguments fall. For example, he points out that some versions focus on the concern that climate geoengineering will encourage people to continue with "business as usual" rather than changing their behaviors. According to other versions, performing research on climate geoengineering is a moral hazard, because it could encourage people to go ahead and implement it. Still other versions express the worry that climate geoengineering could incite us to act in ways that are riskier than we have behaved in the past (Hale 2012: 119–122). Hale also points out that various versions of the moral hazard argument appeal to different moral principles. Some versions appeal to the worry that climate geoengineering will act like an insurance policy that causes people to act inefficiently, whereas other versions focus on the concern that it will encourage people to shirk their responsibilities for changing their behavior, while still others express the concern that climate geoengineering will encourage vicious character traits (e.g., overindulgence or hubris) (Hale 2012: 116–118). Given all this complexity, Hale argues that moral hazard arguments are largely unhelpful unless they are elaborated into very specific moral concerns. Argumentative analysis can play a valuable role in helping to provide this sort of clarification, as illustrated by Brun and Betz (2016).

A fourth ethical issue that needs to be clarified is whether climate geoengineering can be defended based on a sort of "lesser of two evils" argument. Stephen Gardiner (2010) has provided an influential analysis of this argument, pointing out that if substantial progress on emission reductions does not occur soon, humanity may face a choice between engaging in geoengineering or experiencing catastrophic effects of climate change. Thus, it is tempting to justify research on climate geoengineering, despite its morally worrisome characteristics, as a way of equipping society in case it were forced to opt for this "less bad" alternative. Gardiner argues that this argument faces significant difficulties (see also Betz 2012). Perhaps most importantly, it fails to take account of the moral corruption involved in placing future people in a situation where they have to choose between catastrophic climate change and climate geoengineering. He suggests that even if climate change were to become so severe in the future that climate geoengineering were to become the "lesser" of two evils, it might still "mar" the lives of those who were forced to engage in it. Moreover, if we failed to take appropriate actions to address climate change, thereby forcing others into such a marring evil, he argues that our own lives would be irredeemably blighted (Gardiner 2010: 300–301). Thus, Gardiner insists that we should think twice before blithely continuing with our "business as usual" approach to climate change and simultaneously calling for climate geoengineering research.

Kyle Powys Whyte (2012a) identifies a further problem with the lesser of two evils argument. He notes that it can play the role of silencing opposing perspectives, especially from traditionally disadvantaged groups such as indigenous peoples. Whyte points out that this form of argumentation has been used over and over in the face of moral dilemmas as a means of justifying harmful activities that are challenged by indigenous peoples. Once non-indigenous groups have failed to take the necessary steps to avoid these moral dilemmas (such as the choice between catastrophic climate change and climate geoengineering), they set aside typical requirements for consent and deliberation because of the perceived urgency or immediacy of the situation (Whyte 2012a: 70–71). In response, Whyte calls for a process of deliberation about climate geoengineering research and implementation that secures the permission of indigenous peoples in accordance with principles of free, prior, and informed consent (FPIC). He insists that this process should take place even before early research on climate geoengineering technologies is initiated, lest the research generates a technological trajectory that would be rejected by indigenous communities.

The plight of indigenous communities illustrates the importance of distributive justice, which is a fifth ethical concept that can guide decision makers in responding to the uncertainties associated with climate geoengineering. Numerous authors have warned that climate change is likely to impact already vulnerable populations in a disproportionate fashion, and climate geoengineering has the potential to make these inequalities even worse. According to Christopher Preston (2012c), for example, many of the world's poorest people live in geographic regions such as Asia and Africa where they are likely to experience particularly severe impacts from climate change, and they have limited economic resources for dealing with these impacts. These distributive inequities are exacerbated by the fact that these poor regions were responsible for very little of the greenhouse gas emissions that have contributed to climate change. While climate geoengineering might be thought to be an important avenue for alleviating these impacts on the world's poorest countries, Preston points out that these same regions are also predicted to bear the brunt of potential climatic disruptions associated with climate geoengineering (Preston 2012c: 81). And once again, their lack of economic resources will make it difficult to adapt to any potential impacts. Finally, Preston notes that these poor regions of the world are likely to play a very limited role in the research and development process for climate geoengineering, and they will have limited political power for deciding how to implement it (Preston 2012c: 82).

Preston (2012c) suggests that one of the best solutions to this unjust distribution of the risks associated with climate geoengineering is to promote the involvement of disadvantaged groups in public engagement concerning climate geoengineering research. He notes that numerous reports, including those by the Royal Society (2009) and the Solar Radiation Management Governance Initiative (SRMGI) (2011), call for public engagement in the early stages of climate geoengineering research. However, he worries that those reports do not adequately emphasize the special importance of engaging with vulnerable peoples (Preston 2012c: 88). He claims that there are special normative justifications for including these groups in engagement efforts, given that they face unique threats from climate change and further risks from climate geoengineering strategies for reversing those threats. He acknowledges that it will not be easy to incorporate the perspectives of these groups, especially because climate geoengineering research is typically not being performed in the most disadvantaged countries. Nevertheless, he suggests that international research teams could be formed, including specialists from vulnerable populations. These experts could include not only scientists but also scholars of law, ethics, and social science. Moreover, participatory technology assessment methods could provide avenues for incorporating perspectives from other members of vulnerable groups, even if they do not have technical expertise (Preston 2012c: 91–92).

5 Public Consent and Governance

A final set of issues that can be addressed by argumentative analysis involves procedural questions about how to generate and maintain an appropriate governance scheme for climate geoengineering. While many of these procedural questions could also be classified among the ethical issues discussed in the previous section, these procedural questions are so extensive and significant that it is helpful to highlight them in a section of their own. This section focuses on two ways in which argumentative analysis can help to address these issues: (1) it can help to elucidate the range of issues that need to be addressed as part of climate geoengineering governance schemes; and (2) it can generate critical reflection about the procedures needed for making legitimate governance decisions. Admittedly, this discussion provides only a brief introduction to the issues that need to be considered; there are a host of thorny questions that require further analysis. In fact, the profound difficulty of adequately addressing these issues may be a reason to remain skeptical about the justifiability of engaging in climate geoengineering.

Consider four issues that would need to be addressed as part of an adequate governance scheme for climate geoengineering. One issue, which was highlighted by Dale Jamieson in one of the first philosophical evaluations of climate geoengineering, is what would constitute an adequate "mandate for action" (1996: 330). Jamieson wonders whether all nations would have to agree to a climate geoengineering scheme or whether a majority would be sufficient or whether a decision by the United Nations Security Council would be adequate. But these issues quickly become more complicated, because some climate geoengineering strategies (like sulfur aerosol emissions) are so inexpensive that they could be initiated unilaterally. Thus, a second important governance issue is to determine how to prevent or respond to unapproved, unilateral efforts at climate geoengineering (Preston 2012b: 5).

A third set of governance issues concerns the appropriate means for regulating climate geoengineering. The preceding sections have already established that different countries would be likely to experience different levels of adverse effects from climate geoengineering, and they would also face different levels of harm from climate change. Thus, mechanisms for deliberating about how aggressively to geoengineer would need to be developed. In fact, as we have already seen, many of these governance mechanisms may already be needed in order to regulate research on climate geoengineering (Preston 2013: 27). Furthermore, if climate geoengineering schemes were ever implemented, those in control would probably need ongoing international guidance as they engaged in the "continual adjustments" that would presumably be needed to keep the climate in an appropriate trajectory (Preston 2013: 33). Finally, given that catastrophic warming could occur if SRM schemes were suddenly halted or changed, a fourth issue is to determine how to maintain sufficiently stable regimes that could maintain a chosen climate geoengineering scheme. All these issues are exceedingly complex and require extended reflection about how to partition them into a manageable set of distinct but related decisions (see Hirsch Hadorn 2016).

Argumentative analysis can also generate critical reflection about the procedures needed for making legitimate governance decisions about climate geoengineering. For example, as discussed in the previous section, some ethicists argue that obtaining consent from affected parties would be needed in order to justify a climate geoengineering scheme (e.g., Whyte 2012a). But it is not clear how to achieve adequate consent, because all people (as well as non-human organisms) have a stake in the earth's climate system. Moreover, future people and organisms have a stake in the climate as well. The international community currently depends heavily on negotiations between nation states as a means for obtaining consent to global decisions, but there are significant problems with this approach. First, as discussed in the previous section, some of the countries that are likely to experience the most severe effects from both climate change and climate geoengineering also have the least political power on the international stage (Preston 2012c; Corner and Pidgeon 2010). Second, nation states frequently fail to represent the interests of all their citizens in an effective manner. For example, they may ignore or downplay the interests of indigenous peoples within their borders (Whyte 2012b). Third, international negotiations tend to move very slowly and are thus limited in their ability to influence fast-moving efforts to develop and study climate geoengineering technologies.

For many of these reasons, Sven Ove Hansson (2006) has argued that it is misguided to try to apply the informed consent concept to public decisions about issues like climate geoengineering. He points out that this concept developed in the field of medical ethics as a way to give individuals "veto powers" against attempts by society to violate their rights. But requiring unanimous consent from every affected individual before making decisions about social issues like climate geoengineering makes it very difficult to move forward. Hansson (2006) also points out that the concept of informed consent has traditionally been employed when individuals need to choose whether to accept one or more courses of action that have already been selected by experts. This hardly seems like an appropriate model for addressing social issues where the public should be involved in framing the decision problem from the beginning.

There might be room for rethinking the concept of informed consent so that it can be applied to social decision making (Elliott 2010b; Shrader-Frechette 1993; Wong 2015), but perhaps it will be more fruitful to shift to a different concept, such as public engagement. Adam Corner and Nick Pidgeon (2010) point out that a number of novel approaches for promoting public engagement have been garnering increasing attention for assessing transformative technologies like climate geoengineering. Citizens' juries, panels, focus groups, deliberative workshops, scenario analyses, and various multi-stage methods could all be used for promoting "upstream public engagement" in the earliest stages of research on climate geoengineering. Corner and Pidgeon argue that citizens' juries and deliberative workshops in particular could provide valuable opportunities for select groups of citizens to become educated about the technology and to express their perspectives on the social and ethical issues that it raises. Moreover, these approaches need not be limited solely to small groups of citizens in a single locale. The World Wide Views on Global Warming project of September, 2009, engaged 4400 citizens in 38 countries in discussions about the UN climate negotiations in Copenhagen (Corner and Pidgeon 2010: 34).

Unfortunately, public engagement is not without problems of its own. Difficult questions remain about how to structure engagement efforts, how to frame the presentation of background information for participants, how to obtain the best possible representation of the full range of interested and affected parties (perhaps including nonhuman living organisms), and how to feed the results of these exercises into the international policy process. In other words, there is urgent need for *diagnosing* the most appropriate forms of deliberation and engagement for particular decision contexts (Elliott 2011: 109). Some of these issues are empirical (e.g., determining the extent to which particular engagement exercises meet particular criteria), but argumentative analysis is needed to determine what criteria should be used for evaluating public engagement exercises and how those criteria should be applied. Thus, argumentative analysis is crucially important both for determining the issues that need to be addressed as part of geoengineering governance schemes and for evaluating the procedures used for making decisions.

6 Conclusion

The climate geoengineering case provides a particularly vivid illustration of the value of argumentative analysis. It is a classic example of a decision under great uncertainty, in the sense that there are profound uncertainties of various kinds about the problems that climate geoengineering is designed to address, the extent to which it can effectively address those problems, its potential side effects, the future political context in which it will embedded, and the ethical and political principles that should guide decision makers. Given all these wide-ranging uncertainties, it would be foolhardy to base decisions about climate geoengineering solely on formal analyses of its costs and benefits. Rather, it becomes important to explore analogous decision scenarios, to attempt to break down the decision into more manageable components, and to develop principles that can provide guidance under great uncertainty.

In this chapter, it was only possible to scratch the surface of the many issues in this case that can be addressed through argumentative analysis. Three general topics were analyzed: (1) the terminology and framing of the decision problem; (2) the ethical principles that have been applied to the climate geoengineering case; and (3) issues of public consent and governance. Most of the issues discussed in the chapter could not be settled here, but were merely highlighted as deserving of further attention. Nevertheless, a number of lessons can be drawn from this analysis. First, the basic terms used in this case (including 'geoengineering' itself, as well as 'carbon dioxide removal' and 'solar radiation management') should not be treated as unproblematic. Even though they will probably continue to be used, decision makers should recognize that they frame the decision problem in ways that merit ongoing scrutiny and clarification. Second, it is probably misleading to refer to climate geoengineering as an "insurance policy" or even as a form of "compensation"; perhaps it would be more appropriate to regard it as a "technological fix." Third, efforts to justify *research* on geoengineering while continuing to challenge its *implementation* should be treated with a great deal of caution. Even if the two can be distinguished conceptually in many cases, efforts to perform research on climate geoengineering are likely to have a significant impact on the future implementation of it, and many of the same political issues arise for both activities.

Argumentative analysis is also helpful for clarifying a number of ethical issues that are at stake. First, while ethical concerns about turning nature into an artifact do not appear to be very compelling, they do highlight the burden of responsibility that we would be accepting by manipulating nature in such a pervasive manner. Second, while the precautionary principle appears to be an ideal moral principle for addressing an issue like this one, it probably does not provide the guidance needed by decision makers without further specification. Similarly, while it may be fruitful to conceptualize climate geoengineering as a moral hazard, further analysis is needed to clarify the precise sense in which this concept is being used. Argumentative analysis also indicates that efforts to justify climate geoengineering via "lesser of two evil" arguments should be evaluated with great care. Finally, it is very important to create venues for deliberating about climate geoengineering that incorporate traditionally marginalized and disadvantaged groups, lest we fall into traditional patterns of exploitation.

Lastly, the final section of this chapter indicated that argumentative analysis is desperately needed both to identify the issues that need to be addressed as part of geoengineering governance schemes and to evaluate the procedures used for making governance decisions. As part of developing a climate geoengineering scheme, it is necessary to determine what would constitute an adequate "mandate for action," what would be an appropriate procedure for responding to unapproved climate geoengineering efforts, and how climate geoengineering efforts could be maintained and regulated over an extended period of time. Procedurally, it is not clear whether the concept of informed consent is the appropriate goal when addressing a global issue of this sort. The concept of public engagement may be more helpful, but further work is needed to specify criteria for adequate engagement.

Faced with such a difficult problem, it is important to find reasonable ways to break down the decision into more manageable issues (Hirsch Hadorn 2016). We saw that the distinction between research and implementation is more porous than it initially appears, so deliberations about research on climate geoengineering cannot be divorced from considerations about how they might affect later decisions about implementation. But there may be other ways to break down the decision, such as by distinguishing different forms of climate geoengineering technologies or different forms of research. Thus, while climate geoengineering represents a terribly difficult decision problem, it provides an excellent example of the ways that argumentative analysis can prove helpful in cases where more formal approaches to decision analysis are inadequate.

Recommended Readings

Gardiner, S., Caney, S., Jamieson, D., & Shue, H. (Eds.). (2010). *Climate ethics: Essential readings*. New York: Oxford University Press.

National Research Council. (2015). *Climate intervention: Reflecting sunlight to cool earth*. Washington, DC: National Academies Press.

Preston, C. (Ed.). (2012). *Engineering the climate: The ethics of solar radiation management*. Lanham: Lexington Books.

Royal Society. (2009). *Geoengineering the climate: Science, governance, and uncertainty*. Royal Society Policy Document 10/09.

References

Betz, G. (2012). The case for climate engineering research: An analysis of the "arm the future" argument. *Climatic Change, 111*, 473–485.

Betz, G. (2016). Accounting for possibilities in decision making. In S. O. Hansson & G. Hirsch Hadorn (Eds.), *The argumentative turn in policy analysis. Reasoning about uncertainty* (pp. 135–169). Cham: Springer. doi:10.1007/978-3-319-30549-3_6.

Bipartisan Policy Center. (2011). *Geoengineering: A national strategic plan for research on the potential effectiveness, feasibility, and consequences of climate remediation technologies*. http://bipartisanpolicy.org/wp-content/uploads/sites/default/files/BPC%20Climate%20Remediation%20Final%20Report.pdf. Accessed 1 June 2015.

Blackstock, J., & Long, J. (2010). The politics of geoengineering. *Science, 327*, 527.

Brun, G., & Betz, G. (2016). Analysing practical argumentation. In S. O. Hansson & G. Hirsch Hadorn (Eds.), *The argumentative turn in policy analysis. Reasoning about uncertainty* (pp. 39–77). Cham: Springer. doi:10.1007/978-3-319-30549-3_3.

Caldeira, K. (2007, October 24). How to cool the globe. *New York Times*. http://www.nytimes.com/2007/10/24/opinion/24caldiera.html?_r=0. Accessed 14 Apr 2015.

Cicerone, R. (2006). Geoengineering: Encouraging research and overseeing implementation. *Climatic Change, 77*, 221–226.

COMEST (World Commission on the Ethics of Science and Technology). (2005). *The precautionary principle*. Paris: United Nations Educational, Scientific, and Cultural Organization.

Corner, A., & Pidgeon, N. (2010). Geoengineering the climate: The social and ethical implications. *Environment, 52*, 24–37.

Crutzen, P. (2006). Albedo enhancement by stratospheric sulfur injections: A contribution to resolve a policy dilemma? *Climatic Change, 77*, 211–219.

Cullen, J., & Boyd, P. (2008). Predicting and verifying the intended and unintended consequences of large-scale ocean iron fertilization. *Marine Ecology: Progress Series, 364*, 295–301.

Elliott, K. (2009). The ethical significance of language in the environmental sciences: Case studies from pollution research. *Ethics, Place & Environment, 12*, 157–173.

Elliott, K. (2010a). Geoengineering and the precautionary principle. *International Journal of Applied Philosophy, 24*, 237–253.

Elliott, K. (2010b). Hydrogen fuel-cell vehicles, energy policy, and the ethics of expertise. *Journal of Applied Philosophy, 27*, 376–393.

Elliott, K. (2011). *Is a little pollution good for you? Incorporating societal values in environmental research*. New York: Oxford University Press.
Elliott, K., & Dickson, M. (2011). Distinguishing risk and uncertainty in risk assessments of emerging technologies. In T. B. Zülsdorf, C. Coenen, A. Ferrari, U. Fiedeler, C. Milburn, & M. Wienroth (Eds.), *Quantum engagements: Social reflections of nanoscience and emerging technologies* (pp. 165–176). Heidelberg: AKA Verlag.
Elliott, K., & Resnik, D. (2015). Scientific reproducibility, human error, and public policy. *BioScience, 65*, 5–6.
Fisher, E., Jones, J., & von Schomberg, R. (Eds.). (2006). *Implementing the precautionary principle: Perspectives and prospects*. Northampton: Edward Elgar.
Gardiner, S. (2010). Is "arming the future" with geoengineering really the lesser evil? Some doubts about the ethics of intentionally manipulating the climate system. In S. Gardiner, S. Caney, D. Jamieson, & H. Shue (Eds.), *Climate ethics: Essential readings* (pp. 284–312). New York: Oxford University Press.
Grüne-Yanoff, T. (2016). Framing. In S. O. Hansson & G. Hirsch Hadorn (Eds.), *The argumentative turn in policy analysis. Reasoning about uncertainty* (pp. 189–215). Cham: Springer. doi:10.1007/978-3-319-30549-3_8.
Hale, B. (2012). The world that would have been: Moral hazard arguments against geoengineering. In C. Preston (Ed.), *Engineering the climate: The ethics of solar radiation management* (pp. 113–131). Lanham: Lexington Books.
Hansson, S. O. (2006). Informed consent out of context. *Journal of Business Ethics, 63*, 149–154.
Hansson, S. O. (2016). Evaluating the uncertainties. In S. O. Hansson & G. Hirsch Hadorn (Eds.), *The argumentative turn in policy analysis. Reasoning about uncertainty* (pp. 79–104). Cham: Springer. doi:10.1007/978-3-319-30549-3_4.
Hansson, S. O., & Hirsch Hadorn, G. (2016). Introducing the argumentative turn in policy analysis. In S. O. Hansson & G. Hirsch Hadorn (Eds.), *The argumentative turn in policy analysis. Reasoning about uncertainty* (pp. 11–35). Cham: Springer. doi:10.1007/978-3-319-30549-3_2.
Hirsch Hadorn, G. (2016). Temporal strategies for decision making. In S. O. Hansson & G. Hirsch Hadorn (Eds.), *The argumentative turn in policy analysis. Reasoning about uncertainty* (pp. 217–242). Cham: Springer.
Hoffert, M., Caldeira, K., Benford, G., Criswell, D., Green, C., & Wigley, T. (2002). Advanced technology paths to global climate stability: Energy for a greenhouse planet. *Science, 298*, 981–987.
Jamieson, D. (1996). Ethics and intentional climate change. *Climatic Change, 33*, 323–336.
Katz, E. (1992). The big lie: Human restoration of nature. *Research in Philosophy and Technology, 12*, 231–241.
Keith, D., Parson, E., & Granger Morgan, M. (2010). Research on global sun block needed now. *Nature, 463*, 426–427.
Kellogg, W., & Schneider, S. (1974). Climate stabilization: For better or for worse? *Science, 186*, 1163–1172.
National Research Council (NRC). (1996). *Understanding risk: Informing decisions in a democratic society*. Washington, DC: National Academies Press.
National Research Council (NRC). (2015). *Climate intervention: Reflecting sunlight to cool earth*. Washington, DC: National Academies Press.
Parker, W. (2011). When climate models agree: The significance of robust model predictions. *Philosophy of Science, 78*, 579–600.
Pierrehumbert, R. (2015). *Climate hacking is barking mad*. http://www.slate.com/articles/health_and_science/science/2015/02/nrc_geoengineering_report_climate_hacking_is_dangerous_and_barking_mad.html. Accessed 14 Apr 2015.
Preston, C. (2012a). Beyond the end of nature: SRM and two tales of artificity for the Anthropocene. *Ethics, Policy & Environment, 15*, 188–201.

Preston, C. (2012b). The extraordinary ethics of solar radiation management. In C. Preston (Ed.), *Engineering the climate: The ethics of solar radiation management* (pp. 1–11). Lanham: Lexington Books.

Preston, C. (2012c). Solar radiation management and vulnerable populations: The moral deficit and its prospects. In C. Preston (Ed.), *Engineering the climate: The ethics of solar radiation management* (pp. 77–93). Lanham: Lexington Books.

Preston, C. (2013). Ethics and geoengineering: Reviewing the moral issues raised by solar radiation management and carbon dioxide removal. *WIREs Climate Change, 4*, 23–37.

Robock, A. (2008). 20 reasons why geoengineering may be a bad idea. *Bulletin of the Atomic Scientists, 64*(14–18), 59.

Robock, A., Bunzl, M., Kravitz, B., & Stenchikov, G. (2010). A test for geoengineering? *Science, 327*, 530–531.

Royal Society. (2009). Geoengineering the climate: Science, governance, and uncertainty. Royal Society Policy document 10/09. http://royalsociety.org/policy/publications/2009/geoengineering-climate/. Accessed 14 Apr 2015.

Sandin, P. (1999). Dimensions of the precautionary principle. *Human and Ecological Risk Assessment, 5*, 889–907.

Schneider, S. (2001). Earth systems engineering and management. *Nature, 409*, 417–421.

Scott, D. (2012). Insurance policy or technological fix? The ethical implications of framing solar radiation management. In C. Preston (Ed.), *Engineering the climate: The ethics of solar radiation management* (pp. 151–168). Lanham: Lexington Books.

Shrader-Frechette, K. (1993). Consent and nuclear waste disposal. *Public Affairs Quarterly, 7*, 363–377.

Solar Radiation Management Governance Initiative (SRMGI). (2011). Solar radiation management: The governance of research. http://www.srmgi.org/files/2012/01/DES2391_SRMGI-report_web_11112.pdf. Accessed 14 Apr 2015.

Sunstein, C. (2005). *Laws of fear: Beyond the precautionary principle*. Cambridge: Cambridge University Press.

Tuana, N., Sriver, R., Svodoba, T., Olson, R., Irvine, P., Haqq-Misra, J., & Keller, K. (2012). Towards integrated ethical and scientific analysis of geoengineering: A research agenda. *Ethics, Policy & Environment, 15*, 136–157.

Vogel, S. (2003). The nature of artifacts. *Environmental Ethics, 25*, 149–168.

Whyte, K. P. (2012a). Indigenous people, solar radiation management, and consent. In C. Preston (Ed.), *Engineering the climate: The ethics of solar radiation management* (pp. 65–76). Lanham: Lexington Books.

Whyte, K. P. (2012b). Now this! Indigenous sovereignty, political obliviousness and governance models for SRM research. *Ethics, Policy & Environment, 15*, 172–187.

Wilde, G. (1998). Risk homeostasis theory: An overview. *Injury Prevention, 4*, 89–91.

Wong, P.-H. (2015). Consenting to geoengineering. *Philosophy & Technology*. doi:10.1007/s13347-015-0203-1.

Chapter 14
Synthetic Biology: Seeking for Orientation in the Absence of Valid Prospective Knowledge and of Common Values

Armin Grunwald

Abstract Synthetic biology seeks employing technology to shape living systems, possibly up to creating artificial life. This obviously raises the issue of responsibility. However, at this stage, there is almost no valid prospective knowledge available, neither about specific innovation paths and products based on research in synthetic biology nor about consequences and impacts of production, use, side-effects and disposal of such products. So, the traditional consequentialist approach to providing orientation by analysing and assessing prospective knowledge about anticipated consequences cannot be applied. Today's responsibility debate on synthetic biology consists of narratives about future developments such as visions, expectations, fears, concerns and hopes. A hermeneutic analysis of this debate can tell us something about ourselves, our contemporary expectations and concerns, diagnoses and judgments, hopes and fears. A better understanding of this mental, cultural, or philosophical background helps for better imbedding arguments in the absence of valid prospective knowledge and common values.

Keywords Prospective knowledge • Visionary narratives • Hermeneutic orientation • Consequentialist approach • Synthetic biology

1 The Dependency of Responsibility on the Quality of Knowledge

The goal of synthetic biology is to employ technology to influence and shape living systems to a greater degree compared to existing types of biotechnology and genetic engineering. It even offers the perspective of becoming able to create artificial life in some future. The question whether and under which conditions such developments can be regarded morally responsible has frequently been raised in recent years. Several ELSI studies (ethical, legal, and social implications) on risks and benefits

A. Grunwald (✉)
Institute for Technology Assessment and Systems Analysis (ITAS), Karlsruhe, Germany
e-mail: armin.grunwald@kit.edu

S.O. Hansson, G. Hirsch Hadorn (eds.), *The Argumentative Turn in Policy Analysis*, Logic, Argumentation & Reasoning 10, DOI 10.1007/978-3-319-30549-3_14

of synthetic biology have already been performed (see Sect. 2). Synthetic biology became a focal item of the emerging field of RRI "Responsible Research and Innovation" (see Grunwald 2012: 191–226).

While RRI is focusing on procedural aspects and participation, taking the notion of responsibility mostly as a self-explanatory phrase, a theoretical debate on how to understand responsibility in this context is still lacking (there are only few papers in this direction, e.g. Grinbaum and Groves 2013; Grunwald 2012). First reflections based on earlier concepts within the ethics of responsibility showed, however, that the notion of responsibility is far more complex than being a merely ethical term. Responsibility comprises at least three dimensions (e.g. Grunwald 2014a):

- The *empirical dimension* of responsibility considers the attribution of responsibility as a social act done by specific actors and affecting others. Attributing responsibility therefore must involve issues of accountability, distributed governance, and power. It is a social process which needs a clear picture of the empirical social and political constellation (actors, decision-makers, stakeholders, people affected etc.) in the respective field.
- The *ethical dimension* of responsibility concerns asking for *criteria and rules* for judging actions and decisions as responsible or irresponsible (e.g. Jonas 1984), or for helping to find out how actions and decisions could be designed to be (more) responsible.
- The *epistemic* dimension is about the quality of knowledge about the subject of responsibility. This is crucial in particular in fields showing a high degree of uncertainty. Because "mere possibility arguments" (Hansson 2006) are difficult to deal with (Betz 2016; Hansson 2016) the uncertainty about the available knowledge must be critically reflected.

In many RRI fields it quickly became clear that responsibility analyses, statements, and attributions are difficult or even impossible to provide in a knowledge-based, unanimous and consensual way. The familiar approach of discussing responsibilities of agents is to consider future consequences of their actions (e.g. the development and use of new technologies) and then to reflect on these consequences from an ethical point of view (e.g. with respect to the acceptability of technology-induced risk). In the field of synthetic biology (and also other developments called NEST – newly emerging science and technology), a crucial precondition of this approach is not fulfilled. Because of the early stage of development, there is almost no valid prospective knowledge available, neither about specific innovation paths and products based on synthetic biology nor about consequences and impacts of the production, use, side-effects and disposal of such products (Sect. 2.3).

Thus, the epistemic dimension of responsibility becomes decisive in the field of synthetic biology. The ethical debate on synthetic biology consists of narratives about future developments involving visions, expectations, fears, concerns and hopes, which can hardly be assessed with respect to their validity, or even their epistemic possibility. This renders the traditional consequentialist approach to providing orientation by assessing future consequences impossible, but also ethical arguments referring to epistemic possibilities (Betz 2016; Hansson 2016). Exactly this observation is the conceptual point of departure of this Chapter, related with the questions:

- What kind of reflection on/analysis of today's debate on synthetic biology is appropriate in the absence of valid knowledge about consequences of synthetic biology?
- What kind of orientation is provided by reflecting on/analyzing today's debate on synthetic biology?

It will be shown that a *hermeneutic approach* (Grunwald 2014b), i.e. a method for understanding the meaning of narratives and further expressions,[1] changes the perspective on the debate in order to provide a different kind of orientation in this deficient situation of uncertain knowledge. If it is not appropriate to provide orientation by looking at the narratives of synthetic biology within the consequentialist paradigm we could try to explore these narratives in a different, non-consequentialist way. A system of different modes of providing orientation will be presented to illustrate this change (Sect. 3) which then will be exemplified by focusing on two narratives of synthetic biology (Sect. 4). The *argumentative turn* shows itself in this field as a change of perspective (Sect. 5).

2 ELSI Reflections on Synthetic Biology Facing Lack of Knowledge

The basic idea of this Section is to present roughly the concept of synthetic biology (Sect. 2.1) and to give a brief overview about recent ELSI (ethical, legal, social implications) activities in this field (Sect. 2.2) in order to prepare the ground for a more specific analysis of the epistemic dimension of responsibility (Sect. 2.3).

2.1 Synthetic Biology as a NEST Field

Synthetic Biology entered the visionary NEST field rather late, after nanotechnology and human enhancement technologies. It has only recently turned into a vibrant field of scientific inquiry (Grunwald 2012: 191–226). Synthetic biologists hope, both by employing off-the-shelf parts and methods already used in biology and by developing new tools and methods, e.g. based on informatics, to hasten the advent

[1] See Brun and Betz (2016) for the principles of the hermeneutic method and their application in reconstructing arguments.

of far-ranging promises (Synth-Ethics 2011). Various suggestions have been made for definitions describing synthetic biology as:

- Synthetic biology focuses on producing engineered cells, microbes, and biological systems to perform new, useful functions. It aims to develop technologies, methods, and biological components that will make the engineering of biology safer, more reliable, more predictable and, ultimately, standardized (Synthetic Biology Institute 2015).
- The design and synthesis of artificial genes and complete biological systems, and the modification of existing organisms, aimed at acquiring useful functions (COGEM 2006).
- The engineering of biological components and systems that do not exist in nature and the re-engineering of existing biological elements; it is determined by the intentional design of artificial biological systems, rather than by the understanding of natural biology (Synbiology 2005).

A characteristic feature of each of these definitions is the turn to artificial forms of life – whether they will be newly constructed or produced via the redesign of existing life – each of which is associated with an expectation of a specific utility. The knowledge provided by Synthetic Biology can be used to produce new functions in living systems (Pade et al. 2014) by modifying bio-molecules or the design of cells, or designing artificial cells. The promises of synthetic biology go far beyond those of traditional biotechnology (e.g. in the field of GMO - genetically modified organisms) regarding the depth of intervention into living systems. The traditional self-understanding of biology in the framework of natural sciences aiming at *understanding* natural processes is reinterpreted by synthetic biology (Ball 2005) as a *new invention* of nature and as the creation of artificial life on the basis of our knowledge about 'natural' life. This transforms biology into an engineering science of a new type (de Vriend 2006).

There are some relations between synthetic biology and nanotechnology in the field of nanobiotechnology (Grunwald 2012; Schmid et al. 2006). The combination of engineering with biology promises to make it possible to fulfill many of the goals which have been expected of nanotechnology in earlier times in an even easier fashion: while nanotechnology involves the development of materials, processes and structures at the nanoscale, synthetic biology builds on the insight that nature already employs components and methods for constructing materials, processes and structures at very small scales.

These expectations are grounded in the observation that basic life processes take place on a nanoscale because this is precisely the size of life's essential building-blocks. Nanobiotechnology is expected to make it possible to control biological processes by means of nanotechnology. Molecular "factories" (mitochondria) and "transport systems" can – precisely because they play an essential role in cellular metabolism – be models for controllable bio-machines. Thus, nanotechnology could make it possible to engineer cells.

2.2 ELSI Activities on Synthetic Biology

The second World Conference on Synthetic Biology in 2006 brought about first interest among CSOs (civil society organisations) (ETC Group 2007). In view of the fact that, compared to traditional gene technology, synthetic biology leads to a further increase in the depth of man's interventions in living systems and that the pace of innovation continues to increase, discussions on precautionary measures (Paslack et al. 2012) and on the responsibility of scientists and researchers emerged and manifested itself mainly in the form of several ELSI activities so far.

Issues of bio-safety and bio-security have frequently been discussed (see already de Vriend 2006). The moral dimension touches questions such as: how safe is safe enough, what risk is acceptable according to which criteria, and is it legitimate to weigh up expected benefits with risks, or are there knock-out arguments morally forbidding cost/benefit comparisons? Furthermore, the production of new living things or technically strongly modified ones by synthetic biology will raise the question of their moral status. And, even metaphysical questions entered the game. In synthetic biology, man moves stronger from being a modifier of what is present to a creator of something new, compared to earlier stages of biotechnology, at least according to the visions of some biologists: "In fact, if synthetic biology as an activity of creation differs from genetic engineering as a manipulative approach, the Baconian *homo faber* will turn into a creator" (Boldt and Müller 2008: 387). In 2005 a high-level expert group on behalf of the European Commission called it likely that work to create new life forms will give rise to fears, especially that of human hubris and synthetic biologists "playing God" (Dabrock 2009).

Several ELSI and some TA (technology assessment) studies in this field have already been performed or are still ongoing. Funding agencies and political bodies early recognized the importance to get insight into possible ethical challenges and possible conflict situations with the public. Some examples are:

Ethical and regulatory challenges raised by synthetic biology – Synth-Ethics

Synth-Ethics, funded by the European Commission, was among the first ELSI projects on synthetic biology. It applied a special focus on biosafety and biosecurity and on notions of life. It also analyzed early public debates around these issues and identified challenges for current regulatory and ethical frameworks. Finally, it formulated policy recommendations targeted at the synthetic biology community, at EU policy-makers, at NGOs and the public (see www.synthethics.eu).

Engineering life

This project was funded by the German ministry on education and research. Its objectives were (1) to investigate whether synthetic biology would enable humans to create life and what this would mean in ethical respect; (2) to analyze the rhetoric phrase of 'Playing God' from a theological perspective; (3) to explore risks and chances of synthetic biology in a comprehensive manner; and (4) to scrutinize legal boundary conditions for research in synthetic biology (see www.egm.uni-freiburg.de/forschung/projektdetails/SynBio(ELSA)?set_language=en).

Synthetic Biology

This project was commissioned by the German *Bundestag* and conducted by its Office of Technology Assessment. Main issues are – in addition to the scientific-technological aspects – ethics, safety and security, intellectual property rights, regulation (or governance), public perception, and adequate and early communication about chances and risks (see https://www.tab-beim-bundestag.de/en/research/u9800.html).

SYNENERGENE – Synthetic Biology Engaging with New and Emerging Science and Technology in Responsible Governance of the Science and Society Relationship

The aim of the EU funded SYNENERGENE project is to initiate various activities with a view to stimulating and fostering debate on the opportunities and risks of synthetic biology. Among other things, it monitors developments in synthetic biology, identifies critical aspects, experiments with diverse participation formats – from citizen consultations to theatrical debates – and engages stakeholders from science, the arts, industry, politics, civil society and other fields in the debate about synthetic biology (see https://www.itas.kit.edu/english/iut_current_coen13_senergene.php).

Presidential Commission

The Presidential Commission on Bioethics (2010) advising the U.S. President explored potential benefits of synthetic biology, including the development of vaccines and new drugs and the production of biofuels that could someday reduce the need for fossil fuels. It also addressed the risks possibly posed by synthetic biology, including the inadvertent release of a laboratory-created organism into nature and the potential adverse effects of such a release on ecosystems. The Commission urged the policy level to enhance coordination and transparency, to continuously perform risk analysis, to encourage public engagement and to establish ethics education for researchers.

This quick look on some ELSI activities gives a more or less coherent picture and allows for concurrent conclusions:

- The focus of the considered activities varies according to the respective setting; however, the issues addressed show considerable overlap. Some issues such as biosafety and biosecurity appear in all of the studies.
- Understanding the novelty of synthetic biology, of its promises and challenges is a significant part of all the studies.
- There is no consensual system of values to be applied in assessments – to the contrary, values are diverse, controversial and contested.
- Lack of knowledge about innovation paths and products based on synthetic biology as well as on possible consequences of their use was reported in all of the studies.

The latter point will be examined in greater detail in the next Section.

2.3 Lack of Prospective Knowledge

Thus, as stated by almost all of the ELSI and TA studies available so far there is lack of knowledge about foreseeable consequences of synthetic biology. The responsibility debate so far is based on mere assumptions about future developments without a clear epistemic status of e.g. being an epistemic possibility or having a certain probability. This debate consists mostly of narratives including visions, expectations, fears, concerns and hopes. An example is the debate on the possible risk to biosecurity. This is a typical field of "unclear risk" (Wiedemann and Schütz 2008) with the basic preconditions of applying familiar approaches such as cost benefit analysis not being fulfilled: no availability of (ideally quantitative) data on probabilities, the extent of possible damage or of expected benefits. Rather, stories about synthetic biology as "Do It Yourself-Technologies" and bio-hacking are told. Avoiding the danger of fallacies based on "mere possibility arguments" (Hansson 2006, 2016) would imply renouncing drawing any simple conclusion from those stories. The following quote taken from a visionary paper of synthetic biology hits the crucial point – probably not by intention:

> Fifty years from now, synthetic biology will be as pervasive and transformative as is electronics today. And as with that technology, the applications and impacts are impossible to predict in the field's nascent stages. Nevertheless, the decisions we make now will have enormous impact on the shape of this future. (Ilulissat Statement 2007: 2)

This statement is an ideal illustration of what the editors of this Volume write in their Introduction: "In some decisions we are even unable to *identify the potential events that we would take into account if we were aware of them*" (Hansson and Hirsch Hadorn 2016). It expresses (a) that the authors expect synthetic biology will lead to deep-ranging and revolutionary changes, (b) that our decisions today will have high impact on future development, but (c) we have no idea what that impact will be. In this situation of 'great uncertainty' (according to the classification given in Hansson and Hirsch Hadorn (2016)[2] there would be no chance of assigning responsibility; even speaking about responsibility would no longer have a reasonable purpose. It is indeed a 'great uncertainty' showing most of the characteristics mentioned in the Introduction: "insufficient information about options, undetermined or contested demarcation of the decision, lack of control over one's own future decision, multiple values and goals, combination problems when there are several decision-makers, etc." (see Hansson and Hirsch Hadorn 2016). The quote also shows the characteristics of uncertainty of consequences, unknown possibilities and disagreement among experts which legitimates the diagnosis of 'great uncertainty' (Hansson 1996).

[2] The term "great uncertainty" is used for "a situation in which other information than the probabilities needed for a well-informed decision is lacking" (Hansson and Hirsch Hadorn 2016). The term "risk" is used to characterise a decision problem, if "we know both the values and the probabilities of these outcomes" (Hansson and Hirsch Hadorn 2016).

The challenge to responsibility reflections and assessments is further deepened by the absence of a set of common values as normative basis of the assessment (Möller 2016). As a typical example, values in favour of the technological advance conflict with more precautionary attitudes (Synth-Ethics 2011). And values protecting life as it is or as is seen as our heritage from either evolution or creation conflict with values emphasizing human's transcending nature and the emancipatory character with regard to nature. Thus, the absence of common values coincides with the absence of valid prospective knowledge – in terms of decision theory probably the worst case. I would like to demarcate this as the coincidence of high cognitive uncertainty with high normative uncertainty (Grunwald 2012).

This coincidence prevents drawing simple conclusions for today's decision-making but rather would allow decisions in arbitrariness (Hansson 2006). Any ethics of responsibility would be obsolete because of an unclear or even missing subject what could be scrutinized with respect to its responsibility (Bechmann 1993). This would make reflections on the desirability or acceptability of those future developments impossible; or would make completely arbitrary any conclusions on today's attribution of responsibility (for the field of risk assessment see also Shrader-Frechette 1991; Rescher 1983).

A first conclusion could be: okay, it simply might be too early to seriously think about chances and risks of synthetic biology. Let the researchers do their work and come back to the field as soon as better knowledge will be available – and then provide orientation in the familiar consequentialist manner. Nordmann's criticism on the so-called speculative nano-ethics (Nordmann and Rip 2009; Nordmann 2007) might be interpreted in this sense. But in spite of the early stage of development of synthetic biology there are good arguments not to wait for better times regarding preconditions of consequence-regarding reflection to become fulfilled (Grunwald 2010).

While futuristic narratives often appear somewhat fictitious in content, it is a fact that such narratives can and will have real impact on scientific and public discussions (Grunwald 2007). We must distinguish between the degree of facticity of the *content* of the narratives and the fact that they are used in genuine communication processes *with their own dynamics*. Even a narrative without any facticity at all can influence debates, opinion-forming, acceptance and even decision-making. E.g. visions of new science and technology can have a major impact on the way in which political and public debates about future technologies are currently conducted, and will probably also have a great impact on the results of such debates – thereby considerably influencing the pathways to the future in two ways at least:

- Futuristic narratives are able to change the perception of present and possible future developments. The societal and public debate about the chances and risks of new technologies will revolve around these narratives to a considerable extent, as was the case in the field of nanotechnology (see Schmid et al. 2006) and as is currently the case in Synthetic Biology. Futuristic narratives motivate and fuel public debate. E.g. negative visions and dystopias could mobilise resistance to specific technologies while positive ones could create acceptance and fascination.

- Visionary narratives have a particularly great influence on the scientific agenda (Nordmann 2004) which, as a consequence, partly determines which knowledge will be available and applicable in the future. Directly or indirectly, they influence the views of researchers, and thus ultimately also have a bearing on political support and research funding. Visionary communication therefore influences decisions about the support and prioritisation of scientific progress and are an important part of the governance of knowledge (Selin 2008).

The factual power of futuristic narratives in public debate and for decision-making on funding is a strong argument in favour of carefully and critically analysing and assessing them in early stages of development. But how can conclusions be drawn from epistemologically completely unclear narratives?

3 Hermeneutic Mode of Orientation Beyond Consequentialism

Thus we seemingly end up in an aporetic situation. Orientation in the field of Synthetic Biology is needed but cannot be provided because of lack of knowledge and of common values. This diagnosis – which is similar in other NEST-type debates (Nordmann 2014) – was the reason to think more fundamentally about the possibilities for providing orientation out of techno-futures. Recently three modes of providing orientation according to differing epistemic quality of the respective prospective knowledge have been distinguished (Grunwald 2013, see Table 14.1):

- *Mode 1* (i.e., *prognostic*) *orientation*: The prognostic imagination of future technologies and their consequences is supposed to produce a reliable basis for decision-making. For instance, possibilistic knowledge about future developments may be taken in this mode as information on boundary conditions within the Rational Choice paradigm in order to optimize decisions. Experience and theoretical analyses have shown, however, that as a rule this mode does not work in considering the consequences of technology (e.g. Grunwald 2009a). Instead of hoping for certain knowledge about the future substantial uncertainty in different kind is the rule (see below).
- *Mode 2* (i.e., *scenario-based*) *orientation*: Scenarios have become the established means in many areas of prospective analyses, e.g., in sustainability studies (e.g., Heinrichs et al. 2012). In this mode we reflect systematically on a future that is in principle open and thus cannot be prognosticated. The necessary precondition for mode 2 orientation to be applicable is the existence of well-founded corridors of the envisaged future development, or at least an imagination of such corridors agreed upon by relevant persons or groups. Frequently, the space of plausible futures is imagined between a ‘worst case’ and a ‘best case’ scenario.

Table 14.1 Rough sketch of the three modes of orientation

	Prognostic	Scenario-based	Hermeneutic
Approach to the future	The most probable future	Corridor of possible futures	Open space of futures
Spectrum of futures	Determining the best as ideal	Bounded diversity	Unbounded divergence
Preferred methodology	Quantitatively model-based	Quantitatively or qualitatively model based; participatory deliberation	Narrative
Knowledge used	Causal and statistical knowledge	Scientific models and results, knowledge of stakeholders	Associative knowledge, qualitative arguments
Role of normative issues	Low (at least in the self-understanding of the resp. communities)	Depends on case	High
Orientation provided	Decision-making support for optimization	Robust action strategies	Self-reflection and contemporary diagnostics of embeddings of the problem

Source: Grunwald 2013, modified

- *Mode 3* (i.e., *hermeneutic*) *orientation*: This mode comes into the play in case of overwhelming uncertainty, by which is meant that the knowledge of the future is so uncertain or the images of the future are so strongly divergent that there are no longer any valid arguments for employing scenarios to provide orientating structure of the future, which corresponds to great uncertainty (Hansson 1996, 2006). For this situation rendering any form of consequentialism non-applicable – which is the case in the field of synthetic biology as has been shown above – a *hermeneutic turn* was proposed (Grunwald 2014b). The change of perspective consists of raising the question what could be learned by analyzing the visionary narratives *about the contemporary situation*. The techno-visionary narratives could be examined for what they mean and under which diagnoses and values they originated. Understanding by means of a hermeneutic approach how the problem for decision – in this case research on synthetic biology – is embedded in various broader perspectives held by different groups is of help in clarifying the more specific different framings of problems, e.g. in ELSI activities (see Sect. 2, Grüne-Yanoff 2016). Understanding the different positions and the reasons for their differences might be of substantial help in public deliberation.

The three modes of orientation do not exclude each other logically. They provide different kinds of orientation and require knowledge of different epistemological quality that ranges from even certain knowledge or mostly probabilistic knowledge (mode 1) to full ignorance (mode 3). In addition, knowledge on different parts of the complex problem might be of different quality. So the distinguished modes of orientation may be combined in accordance with the purposes and the quality of knowledge at hand.

In particular, it becomes obvious that a hermeneutic perspective is not restricted to the mode 3 case. Questions of meaning and the attribution of meaning are also of interest at least in the mode 2 but might even be worthwile to apply to mode 1 type approaches. Scenario-building – the mode 2 approach – is a *constructive* process: scenarios cannot be 'derived' from present-day's knowledge alone. Thus, qualitative assumptions about 'plausible' developments or 'best-case' and 'worst-case' developments have to be made in order to build a set of scenarios which can be expected to orientate decision-making in the respective field. Those assumptions could be made subject to the hermeneutic approach, e.g. by reconstructing the arguments applied in defining the scenarios, analogously to the consideration of the narratives in mode 3.

Thus the result is a kind of hierarchy: while the hermeneutic approach could be applied to all the modes in order to improve understanding, there is nothing else than this approach applicable in mode 3. More or less warranted arguments support orientation in mode 1 and mode 2 – but not in mode 3.

4 Techno-visionary Narratives of Synthetic Biology

In the debate on synthetic biology neither the mode 1 nor the mode 2 approach is applicable (see Sect. 2). Therefore we have to focus on the hermeneutic mode (3) and ask for opportunities to provide orientation by understanding the various perspectives on how the problem is embedded. Coming back to the field of synthetic biology two narratives will be recalled which might be promising subjects to a more in-depth hermeneutic consideration.

Techno-visionary narratives are present in the debate on Synthetic Biology at different levels (Synth-Ethics 2011). They include "official" visions provided and disseminated by scientists and science promoters, and visions disseminated by mass media including negative visions up to dystopian views as well. They include stories about great progress solving the energy problem or contributing to huge steps in medicine but also severe concerns about a possible non-controllability of self-organising systems (Dupuy and Grinbaum 2004) or the already mentioned narrative of humans "Playing God". As stated above there is epistemologically no chance to clarify today whether these narratives do tell us something sensible about the future or not. Therefore we can only take the narratives (including their origins, the intentions and diagnoses behind them, their meanings, their dissemination and the impacts) as the empirical data and ask for their role in contemporary debates, renouncing on any attempt of anticipation (Nordmann 2014).

For example, take the debate on "Playing God". Independent from that there is no argument behind this debate (Dabrock 2009) it should be scrutinized seriously, especially since playing God is one of the favorite buzzwords in media coverage of synthetic biology. A report by the influential German news magazine *Der Spiegel* (following Synth-Ethics 2011) titled "Konkurrenz für Gott" (Competing with God). This is a reference to a statement by the ETC Group ("For the first time, God has competition", 2007). The introduction states that the aim of a group of biologists is to reinvent life, thereby raising fears concerning human hubris. The goal of understanding and fundamentally recreating life would, according to the article, provoke fears of mankind taking over God's role and that a being such as Frankenstein's monster could be created in the lab. This narrative is a dystopian version of the Baconian vision of full control over nature. The hermeneutic approach means to understand what such debates with unclear epistemic status or even without any epistemic claims could tell us, e.g. by reconstruction of the arguments and their premises used in the corresponding debates, or by a historical analysis of the roots of the narratives used.

In the following I will refer to two narratives relevant to synthetic biology with diverging messages (based on Grunwald 2012). Because a comprehensive reconstruction and exploration of these is beyond the scope of this Chapter the presentation shall mainly serve the purpose of illustration of the argumentation. A concise hermeneutic consideration would need a much more in-depth investigation which cannot be given here.

4.1 *The 'Nature as Model' Narrative*

Many visions of Synthetic Biology tell well-known stories about the paradise-like nature of scientific and technological advance. Synthetic Biology is expected to provide many benefits and to solve many of the urgent problems of humanity. These expectations concern primarily the fields of energy, health, new materials and a more sustainable development. The basic idea behind these expectations is that solutions which have developed in nature could directly be made useful to human exploitation by Synthetic Biology:

> Nature has made highly precise and functional nanostructures for billions of years: DNA, proteins, membranes, filaments and cellular components. These biological nanostructures typically consist of simple molecular building blocks of limited chemical diversity arranged into a vast number of complex three-dimensional architectures and dynamic interaction patterns. Nature has evolved the ultimate design principles for nanoscale assembly by supplying and transforming building blocks such as atoms and molecules into functional nanostructures and utilizing templating and self-assembly principles, thereby providing systems that can self-replicate, self-repair, self-generate and self-destroy. (Wagner 2005: 39)

In analysing those solutions of natural systems and adopting them to human needs the traditional border between biotic and abiotic systems could be transgressed. It is one of the visions of Synthetic Biology to become technically able to design and construct life according to human purposes and ends (Pade et al. 2014, see Sect. 2).

While this objective is widely agreed upon there are diverging understandings of what this would mean:

1. Humans take full control over nature following the Baconian idea (see Sect. 4.2 for this interpretation), and

2. Humans regard nature as a model and go for technologies following this model expecting a reconciliation of technology and nature

In the first-mentioned understanding the term of nano-bionics is used in order to apply a particular perspective on Synthetic Biology. Bionics attempts, as is frequently expressed metaphorically, to employ scientific means to learn from nature in order to solve technical problems (von Gleich et al. 2007). The major promise of bionics is, in the eyes of the protagonists, that the bionic approach will make it possible to achieve a technology that is more natural or better adapted to nature than is possible with traditional technology. Examples of desired properties that could be achieved include adaptation into natural cycles, low levels of risk, fault tolerance, and environmental compatibility.

In grounding such expectations, advocates refer to the problem-solving properties of natural living systems, such as optimization according to multiple criteria under variable boundary conditions in the course of evolution, and the use of available or closed materials cycles (von Gleich et al. 2007: 30ff.). According to these expectations, the targeted exploitation of physical principles, of the possibilities for chemical synthesis, and of the functional properties of biological nanostructures is supposed to enable synthetic biology to achieve new technical features in hitherto unachieved complexity, with nature ultimately serving as the model.

These ideas refer to traditional bionics which aimed (and aims) at learning from nature (e.g. animals or plants) at a macroscopic level. Transferred to the micro- or even nano-level it gets an even more utopian character. If humans become able to act following nature as the model at the level of the "brick-stones" of life an even more "nature-friendly" or nature-compatible technology could be expected. Philosophically this reminds the idea of the German philosopher Ernst Bloch who proposed an "alliance technology" (*Allianztechnik*) in order to reconcile nature and technology. While in the traditional way of designing technology nature is regarded as a kind of "enemy" which must brought under control by technology Bloch proposes to develop future technology in accordance with nature in order to arrive at a status of peaceful co-existence of humans and the natural environment.

Thus, this narrative related with "Synthetic biology" is not totally new but goes back to earlier philosophical concerns about the dichotomy between technology and nature. But the postulate related to this narrative would not work straight forward. It suffers from the fallacy of naturalness, which takes naturalness as a guarantee against danger (Hansson 2016). In addition, it is easily possible to tell a narrative of Synthetic Biology in the opposite direction, based on the same characteristics of Synthetic Biology (see below).

4.2 The "Dominion Over Nature" Narrative

Based on a completely different philosophical background, namely on traditional Baconism, Synthetic Biology could be regarded as the fulminant triumph of

Bacon's "dominion over nature" utopia. The idea of controlling more and more parts of nature continues basic convictions of European Enlightenment in the Baconian tradition. Human advance includes, in that perspective, to achieve more and more independence from any restrictions given by nature or by the natural evolution and to enable humankind to shape its environment and living conditions according to human values, preferences and interests to maximum extent.

The cognitive process of Synthetic Biology attempts to gather knowledge about the structures and functions of natural systems from *technical intervention*, not from contemplation or via distanced observation of nature. Living systems are not of interest *as such*, for example in their respective ecological or aesthetical context, but are analyzed in the *relationship of their technical functioning*. Living systems are thus interpreted as *technical systems* by Synthetic Biology. This can easily be seen in the extension of classical machine language to the sphere of the living. The living is increasingly being described in techno-morph terms:

> Although it can be argued that synthetic biology is nothing more than a logical extension of the reductionist approach that dominated biology during the second half of the twentieth century, the use of engineering language, and the practical approach of creating standardized cells and components like in an electrical circuitry suggests a paradigm shift. Biology is no longer considered "nature at work," but becomes an engineering discipline. (de Vriend 2006: 26)

Living systems are examined within the context of their technical function, and cells are interpreted as machines – consisting of components, analogous to the components of a machine which have to co-operate in order to fulfil the overall function. For example, proteins and messenger molecules are understood as such components that can be duplicated, altered or recombined in new ways by synthetic biology. A "modularisation of life" is thereby made as well as an attempt to identify and standardise the individual components of life processes. In the tradition of technical standardisation, gene sequences are saved as models for various cellular components of machines. Following design principles of mechanical and electrical engineering, the components of living systems are regarded as having been put together according to a building plan in order to obtain a functioning whole. The recombination of different standardised bio-modules (sometimes called "bio-bricks") allows for the design and creation of different living systems. With the growing collection of modules, out of which engineering can develop new ideas for products and systems, the number of possibilities grows exponentially.

Thus the main indicator of the relevance of this understanding of Synthetic Biology and its meaning is the use of language. Examples of such uses of language are referring to hemoglobin as a vehicle, to adenosine triphosphate synthase as a generator, to nucleosomes as digital data storage units, to polymerase as a copier, and to membranes as electrical fences. From this perspective, Synthetic Biology is linked epistemologically to a technical view of the world and to technical intervention. It carries these technical ideas into the natural world, modulates nature in a techno-morph manner, and gains specific knowledge from this perspective. Nature is seen as technology, both in its individual components and also as a whole.

> This is where a natural scientific reductionist view of the world is linked to a mechanistic technical one, according to which nature is consequently also just an engineer . . .. Since we can allegedly make its construction principles into our own, we can only see machines wherever we look — in human cells just as in the products of nanotechnology. (Nordmann 2007: 221)

Instead of eliciting a more natural technology per se as promised by a bionic understanding of Synthetic Biology (see above) the result of this research signifies a far-reaching technicalization of what is natural. Learning from nature for technical problem solving must of necessity already take a technical view of nature. *Prior to* considering Synthetic Biology from the perspective of technology ethics or societal debate and assessment, it appears sensible to ask if and how such changes in the use of language and such re-interpretations aiming at a different understanding modify the relationship between technology and life or modify our view of this relationship.

4.3 *Open Questions: What Could Be Learned in the Hermeneutic Mode?*

The presentation of the two narratives of Synthetic Biology showed unanimously the completely diverging nature of the underlying convictions and images of the relation between technology and nature. This divergence is not about a consequentialist weighing of chances against risks or about performing cost-benefit analyses. It is also not about specific innovation paths, products or services based on progress in synthetic biology. Instead, the following questions might be raised facing the situation sketched above:

- What are the underlying convictions, attitudes and pictures of the relations between humans and nature or between nature and technology? What could be done to make them as explicit as possible?
- What does it mean that, after a period of more humility concerning human's relation to nature, now the "dominion over nature" narrative comes back and seems to dominate the debate?
- How does this situation relate to the earlier debate on GMOs, and what does a possible shift tell us about a changing contemporary situation?
- In what way could the tension, even the contradiction, between the two narratives presented be made fruitful for the further debate on Synthetic Biology?
- Moreover, both narratives and their normative presumptions and pre-occupations might be inadequate or might show severe shortcomings. This suspicion calls for more in-depth philosophical inquiry.
- How could it be possible to realize the expectation "Argumentative analysis is a means for better substantiating deliberation to achieve democratic legitimacy of decisions" (Hansson and Hirsch Hadorn 2016) facing this situation?

Searching for answers to this (and related) question does need a hermeneutic approach by which the *meaning* of the patterns, notions, arguments, attitudes and convictions in the debate on synthetic should be investigated (Grunwald 2014b). Methodologically, this hermeneutic approach would draw from different disciplines and adopt different methods, tailor-made to the type of question to be answered. If we take the example of the narratives on more or less speculative techno-futures a hermeneutic investigation could view at the 'biography' of those narratives: who are the authors, what were their intentions and points of departure, what are the cultural, philosophical and historical roots of their thoughts, how are these narratives communicated, debated, and perceived, which consequences and reactions could be observed etc. (Grunwald 2014b).

To answer questions about the biography of techno-futures and the consequences of their diffusion and communication, an interdisciplinary procedure employing various types of methods appears sensible. The empirical social sciences can contribute to clarifying the communication of techno-futures by using media analyses or sociological discourse analysis and generate, for example, maps or models of the respective constellations of actors. Political science, especially the study of governance, can analyze the way in which techno-futures exert influence on political decision-making processes (Grunwald 2014b). Philosophical inquiry could deliver reconstructions and assessments of arguments brought forward (Betz 2016; Hansson 2016), in particular concerning the different legitimisation and justification strategies behind the narratives. Philosophy of the arts could provide insights into the meaning of movies or other pieces of art which play a strong role in the debate on Synthetic Biology.

The question, however, remains: what can specifically be learned from such an investigation? The examples presented show clearly that a direct support to decision-makers in the sense of a classical decision-support cannot be expected. If a specific research field of Synthetic Biology would be challenged in terms of whether proceeding with it would be responsible at all, hermeneutic considerations would provide a clear indication. It could only contribute to a better understanding of the mental, cultural, or philosophical background of the field under consideration, the options and arguments presented, and the narratives disseminated and debated in its context. Though this will not allow deriving a clear conclusion with respect to the responsibility of the field under consideration it could help in an indirect sense. Making implicit backgrounds of alternatives and narratives explicit may contribute to better and more transparent embedding the options under consideration into their – philosophical, cultural, ethical – aura. It serves rational reasoning and debates in deliberative democracy by providing the 'grand picture' more comprehensively and thus allows for giving the field under consideration a place in the broader picture.

This means that insights provided by a hermeneutic approach may be expected which do not directly support decision-making but which could help better framing the respective challenge by embedding it into the broader picture mentioned above (Grüne-Yanoff 2016). This broader picture would include a transparent picture of all the uncertainties and areas of ignorance involved, of the diverse and possibly diverging values affected by the research under consideration and of moral conflicts or normative uncertainties possibly involved. By considering this broader picture instead of a more narrowed description of the challenge there should be a better basis to search for agreed research goals or for defining temporal strategies to work into the direction of those goals and to foresee specific, e.g. anticipatory or regulatory, measures to approach the future.

5 Lessons Learned: The Hermeneutic Side of the Argumentative Turn

In the absence of valid prospective knowledge and common values about the future of synthetic biology and its impacts and consequences for society and humankind the argumentative turn has to include a hermeneutic perspective: Instead of trying to derive orientation from prospective knowledge in the sense of consequentialism (as is the usual business of technology assessment and applied ethics) we have to consider the more or less speculative narratives as elements of current debates and try to learn more about ourselves by better understanding their origin, their expression, their content, their normative backgrounds, their cultural traditions, their ways of spreading, and so forth within a *hermeneutic approach* (Grunwald 2014b).

The hermeneutic approach to visionary narratives of synthetic biology aims at: (1) understanding the processes by which *meaning* is attributed to developments in synthetic biology by using narratives about the future, (2) understanding the contents and backgrounds of the communicated futures, and (3) understanding their reception, communication, and consequences in the social debates and political decision-making processes. By analysing these narratives we will probably be able to learn something about our contemporary situation by "making the implicit explicit". All this serves then as a basis to reconstruct and assess the argumentations put forward in this debate.

> We can use argumentation analysis for instance to better understand the uncertainties involved in decisions, to prioritize among uncertain dangers, to determine how decisions should be framed, to clarify how different decisions on interconnected subject-matter relate to each other, to choose a suitable time frame for decision-making, to analyze the ethical aspects of a decision, to systematically choose among different decision options, and not least to improve our communication with other decision-makers in order to co-ordinate our decisions. (Hansson and Hirsch Hadorn 2016)

Applying the hermeneutic approach would on the one hand help clarifying current debates as well as prepare for coming debates in which it could then, for example, be about concrete technology design. Within this context, a "vision assessment" (Grunwald 2009b) would study the cognitive as well as evaluative content of tech-based visions and their impacts. They would be the fundamental building blocks of a cognitively informed and normatively oriented dialogue – a dialogue, for example, between experts and the public or between synthetic biology, ethics, research funding, the public and regulation.

Thus it becomes obvious that the argumentative turn involves an additional perspective which is not accounted for in traditional policy analysis and technology assessment, namely towards a hermeneutic approach to narratives of the future of synthetic biology. This turn opens up a new type of thinking of using visionary narratives in NEST debates and a new field of methods to investigate this field. The subjects of a hermeneutic investigation are not only narratives as texts but also pieces of art used in those fields. Research fields such as philosophical or sociological discourse analysis, linguistics, media research and philosophy of arts might enter the field of investigating visionary futures in NEST debates.

Recommended Readings

Grunwald, A. (2012). *Responsible nanobiotechnology. Philosophy and ethics.* Singapore: Pan Stanford Publishing.

Nordmann, A. (2014). Responsible innovation, the art and craft of future anticipation. *Journal of Responsible Innovation, 1*, 87–98.

Wiedemann, P., & Schütz, H. (Eds.). (2008). *The role of evidence in risk characterization.* Weinheim: WILEY-VCH Verlag.

References

Ball, P. (2005). Synthetic biology for nanotechnology. *Nanotechnology, 16*, R1–R8.

Bechmann, G. (1993). Ethische Grenzen der Technik oder technische Grenzen der Ethik? In Studiengesellschaft für Zeitgeschichte und politische Bildung (Ed.), *Geschichte und Gegenwart. Vierteljahreshefte für Zeitgeschichte, Gesellschaftsanalyse und politische Bildung* (12th ed., pp. 213–225). Graz: Studiengesellschaft für Zeitgeschichte und politische Bildung.

Betz, G. (2016). Accounting for possibilities in decision making. In S. O. Hansson & G. Hirsch Hadorn (Eds.), *The argumentative turn in policy analysis. Reasoning about uncertainty* (pp. 135–169). Cham: Springer. doi:10.1007/978-3-319-30549-3_6.

Boldt, J., & Müller, O. (2008). Newtons of the leaves of grass. *Nat Biotechnol, 26*, 387–389.

Brun, G., & Betz, G. (2016). Analysing practical argumentation. In S. O. Hansson & G. Hirsch Hadorn (Eds.), *The argumentative turn in policy analysis. Reasoning about uncertainty* (pp. 39–77). Cham: Springer. doi:10.1007/978-3-319-30549-3_3.

COGEM (2006). *Synthetische biologie. Een onderzoeksveld met voortschrijdende gevolgen.* COGEM signalering CGM/060228-03. See: www.cogem.net/index.cfm/nl/publicaties/publicatie/synthetische-biologie-een-onderzoeksveld-met-voortschrijdende-gevolgen. Accessed 3 May 2015.

Dabrock, P. (2009). Playing God? Synthetic biology as a theological and ethical challenge. *Syst Synth Biol, 3*, 47–54.

de Vriend, H. (2006). *Constructing life. Early social reflections on the emerging field of synthetic biology*. The Hague: Rathenau Institute.

Dupuy, J.-P., & Grinbaum, A. (2004). Living with uncertainty: Toward the ongoing normative assessment of nanotechnology. In J. Schummer & D. Baird (Eds.), *Nanotechnology challenges: Implications for philosophy, ethics and society* (pp. 287–314). Singapore: World Scientific Publishing Co. Pte. Ltd.

ETC – The Et-cetera Group (2007). *Extreme genetic engineering. An introduction to synthetic biology.* http://www.etcgroup.org/sites/www.etcgroup.org/files/publication/602/01/synbioreportweb.pdf. Accessed 3 May 2015.

Grinbaum, A., & Groves, C. (2013). What is 'responsible' about responsible innovation? In R. Owen, J. Bessant, & M. Heintz (Eds.), *Responsible innovation: Managing the responsible emergence of science and innovation in society* (pp. 119–142). West Sussex: Wiley.

Grüne-Yanoff, T. (2016). Framing. In S. O. Hansson & G. Hirsch Hadorn (Eds.), *The argumentative turn in policy analysis. Reasoning about uncertainty* (pp. 189–215). Cham: Springer. doi:10.1007/978-3-319-30549-3_8.

Grunwald, A. (2007). Converging technologies: Visions, increased contingencies of the Conditio Humana, and search for orientation. *Futures, 39*, 380–392.

Grunwald, A. (2009a). Technology assessment: Concepts and methods. In A. Meijers (Ed.), *Philosophy of technology and engineering sciences* (Vol. 9, pp. 1103–1146). Amsterdam: Elsevier.

Grunwald, A. (2009b). Vision assessment supporting the governance of knowledge – The case of futuristic nanotechnology. In G. Bechmann, V. Gorokhov, & N. Stehr (Eds.), *The social integration of science. Institutional and epistemological aspects of the transformation of knowledge in modern society* (pp. 147–170). Berlin: Edition Sigma.

Grunwald, A. (2010). From speculative nanoethics to explorative philosophy of nanotechnology. *NanoEthics, 4*, 91–101.

Grunwald, A. (2012). *Responsible nanobiotechnology. Philosophy and ethics.* Singapore: Pan Stanford Publishing.

Grunwald, A. (2013). Modes of orientation provided by futures studies: Making sense of diversity and divergence. *European Journal of Futures Studies, 15*, 30. doi:10.1007/s40309-013-0030-5.

Grunwald, A. (2014a). Synthetic biology as technoscience and the EEE concept of responsibility. In B. Giese, C. Pade, H. Wigger, & A. von Gleich (Eds.), *Synthetic biology. Character and impact* (pp. 249–266). Heidelberg: Springer.

Grunwald, A. (2014b). The hermeneutic side of responsible research and innovation. *Journal of Responsible Innovation, 1*, 274–291.

Hansson, S. O. (1996). Decision-making under great uncertainty. *Philos Soc Sci, 26*, 369–386.

Hansson, S. O. (2006). Great uncertainty about small things. In J. Schummer & D. Baird (Eds.), *Nanotechnology challenges – Implications for philosophy, ethics and society* (pp. 315–325). Singapore: World Scientific Publishing Co. Pte. Ltd.

Hansson, S. O. (2016). Evaluating the uncertainties. In S. O. Hansson & G. Hirsch Hadorn (Eds.), *The argumentative turn in policy analysis. Reasoning about uncertainty* (pp. 79–104). Cham: Springer. doi:10.1007/978-3-319-30549-3_4.

Hansson, S. O., & Hirsch Hadorn, G. (2016). Introducing the argumentative turn in policy analysis. In S. O. Hansson & G. Hirsch Hadorn (Eds.), *The argumentative turn in policy analysis. Reasoning about uncertainty* (pp. 11–35). Cham: Springer. doi:10.1007/978-3-319-30549-3_2.

Heinrichs, D., Krellenberg, K., Hansjürgens, B., & Martínez, F. (Eds.). (2012). *Risk habitat megacity.* Heidelberg: Springer.

Ilulissat Statement. (2007). *Synthesizing the future. A vision for the convergence of synthetic biology and nanotechnology.* See: https://www.research.cornell.edu/KIC/images/pdfs/ilulissat_statement.pdf. Accessed 3 May 2015.

Jonas, H. (1984). *The imperative of responsibility.* Chicago: The University of Chicago Press. German version: Jonas, Hans. 1979. *Das Prinzip Verantwortung. Versuch einer Ethik für die technologische Zivilisation.* Frankfurt/M.: Suhrkamp.

Möller, N. (2016). Value uncertainty. In S. O. Hansson & G. Hirsch Hadorn (Eds.), *The argumentative turn in policy analysis. Reasoning about uncertainty* (pp. 105–133). Cham: Springer. doi:10.1007/978-3-319-30549-3_5.

Nordmann, A. (2007). If and then: A critique of speculative NanoEthics. *Nanoethics, 1*, 31–46.

Nordmann, A. (2014). Responsible innovation, the art and craft of future anticipation. *Journal of Responsible Innovation, 1*, 87–98.
Nordmann, A. (2004). *Converging technologies – Shaping the future of European societies.* European Commission. See www.ec.europa.eu/research/social-sciences/pdf/ntw-report-alfred-nordmann_en.pdf. Accessed 3 May 2015.
Nordmann, A., & Rip, A. (2009). Mind the gap revisited. *Nat Nanotechnol, 4*, 273–274.
Pade, C., Giese, B., Koenigstein, S., Wigger, H., & von Gleich, A. (2014). Characterizing synthetic biology through its novel and enhanced functionalities. In B. Giese, C. Pade, H. Wigger, & A. von Gleich (Eds.), *Synthetic biology. Character and impact* (pp. 71–104). Heidelberg: Springer.
Paslack, R., Ach, J., Lüttenberg, B., & Weltring, K.-M. (Eds.). (2012). *Proceed with caution. Concept and application of the precautionary principle in nanobiotechnology.* Münster: LIT.
Presidential Commission for the Study of Bioethical Issues (2010). *New directions: The ethics of synthetic biology and emerging technologies.* See www.bioethics.gov/synthetic-biology-report. Accessed 3 May 2015.
Rescher, N. (1983). *Risk. A philosophical introduction to the theory of risk evaluation and management.* Lanham: University Press of America.
Schmid, G., Ernst, H., Grünwald, W., Grunwald, A., et al. (2006). *Nanotechnology – Perspectives and assessment.* Berlin: Springer.
Selin, C. (2008). The sociology of the future: Tracing stories of technology and time. *Sociology Compass, 2*, 1878–1895.
Shrader-Frechette, K. S. (1991). *Risk and rationality. Philosophical foundations for populist reforms.* Berkeley: University of California Press.
Synbiology (2005). *SYNBIOLOGY – an analysis of synthetic biology research in Europe and North America.* http://www2.spi.pt/synbiology/documents/SYNBIOLOGY_Literature_And_Statistical_Review.pdf. Accessed 3 May 2015.
Synth-Ethics (2011). *Homepage of the EU-funded project ethical and regulatory issues raised by synthetic biology.* http://synthethics.eu/. Accessed 3 May 2015.
Synthetic Biology Institute (2015). *What is synthetic biology*? See www.synbio.berkeley.edu/index.php?page=about-us. Accessed 3 May 2015.
von Gleich, A., Pade, C., Petschow, U., & Pissarskoi, E. (2007). *Bionik. Aktuelle Trends und zukünftige Potentiale.* Berlin: Universität Bremen.
Wagner, P. (2005). Nanobiotechnology. In R. Greco, F. B. Prinz, & R. Lane Smith (Eds.), *Nanoscale technology in biological systems* (pp. 39–55). Boca Raton: CRC Press.
Wiedemann, P., & Schütz, H. (Eds.). (2008). *The role of evidence in risk characterization.* Weinheim: WILEY-VCH Verlag.

Appendix

Ten Core Concepts for the Argumentative Turn in Policy Analysis

Sven Ove Hansson and Gertrude Hirsch Hadorn

Abstract Ten core concepts for the argumentative turn in uncertainty management and policy analysis are explained and briefly defined. References are given to other chapters in the same book where these concepts are introduced and discussed more in depth. The 10 concepts are argument analysis, argumentative approach, fallacy, framing, rational goal setting and goal revision, hypothetical retrospection, possibilistic arguments, scenario, temporal strategy, and uncertainty.

In this appendix we provide brief definitions of some of the concepts that are most important for characterizing the argumentative turn in policy analysis and the methods that it employs. References are given to the chapters in the book where these concepts are introduced and discussed more extensively and used to develop methods and tools for policy analysis.

Argument Analysis

When we provide reasons for or against a claim, we argue. More precisely, an *argument* consists of an inference from one or several premises to a conclusion. Often, we combine several arguments into a more complex *argumentation*. Argument analysis can be defined in a narrow and a wide sense: "Argument analysis, understood in a wide sense, involves two basic activities: reconstruction and evaluation of argumentation and debate" (Brun and Betz 2016:42). Each of these activities – reconstruction and assessment – includes several tasks, one of which is argument analysis in a narrow sense. By this is meant a process in which complex argumentation is broken down into its component arguments and their relations. For example,

S.O. Hansson (✉)
Department of Philosophy and History, Royal Institute of Technology, Stockholm, Sweden
e-mail: soh@kth.se

G. Hirsch Hadorn
Department of Environmental Systems Science, Swiss Federal Institute of Technology, Zurich, Switzerland
e-mail: hirsch@env.ethz.ch

S.O. Hansson, G. Hirsch Hadorn (eds.), *The Argumentative Turn in Policy Analysis*, Logic, Argumentation & Reasoning 10, DOI 10.1007/978-3-319-30549-3 we can "identify attack and support relations between arguments, or distinguish 'hierarchical' argumentation in which one argument supports a premise of another argument, from 'multiple' argumentation, in which several arguments support the same conclusion" (Brun and Betz 2016:42). The reconstruction and evaluation of argumentation is best performed in an iterative fashion so that each of these methods can be applied several times before the analysis has been completed. *Argument maps* are a means to structure and visualize attack and support relations between the single arguments of a complex argumentation. Argument maps serve as a reasoning tool: "the argument map identifies the questions to be answered when adopting a position in the debate, and merely points out the implications of different answers to these questions" (Brun and Betz 2016:62). In policy analysis, arguments that speak for or against given policy options are scrutinized. In philosophy, arguments for or against policy options are called *practical arguments*. "Such 'practical' arguments have a normative – more precisely, prescriptive – conclusion: they warrant that certain policy options are obligatory (ought to be taken), permissible (may be taken), or prohibited (must not be taken)" (Betz 2016:140).

Argumentative Approach

The standard approach in policy analysis is *expected utility maximization*. It requires that we calculate the expected (probability-weighted) value of each option in the decision. A rational decision-maker is assumed to choose an option that has maximal aggregated expected utility, as compared to the other available options. The application of this method requires that the options for choice, the probabilities of the outcomes, and the values of these outcomes are well determined or determinable. In real life we often have to make decisions although we lack much of this information. The *argumentative approach* to decision-making provides means to systematize our deliberation about decisions under such, more difficult conditions. It is "a *widened rationality approach* that scrutinises inferences from what is known and what is unknown in order to substantiate decision-supporting deliberations. It includes and recognises the normative components of decisions and makes them explicit to help finding reasonable decisions with democratic legitimacy" (Hansson and Hirsch Hadorn 2016:11). The argumentative approach includes a large and open-ended range of methods and strategies to tackle the various tasks that come up with the analysis of a decision problem. It is a pluralistic and flexible approach that does not try to squeeze all decision problems into a uniform format.

Fallacy

A fallacy is "a deceptive or misleading argument pattern" (Hansson 2016:80). Most fallacies that are known from other contexts can also be encountered in the context of decision-making. But there are also some types of fallacious reasoning that are specific for arguments on decision-making. Some examples are the fallacies of disregarding unquantifiable effects, disregarding indetectable effects, cherry-picking uncertainties, disregarding scientific uncertainty, and treating uncertain probability estimates as certain. Most of the decision-related fallacies have in common that they induce us to programmatically disregard certain types of decision-relevant information. They can therefore be subsumed under a joint larger category, *the fallacies of programmatically excluding decision-relevant information*. Obviously, in each particular decision the decision maker should focus on the most important information, but the types of information that can in practice be only cursorily attended to will differ between different decisions. There are, for instance, decisions in which the scientific uncertainty can be disregarded, but there are other decisions in which it is a crucial consideration. Decision rules or decision behaviour that excludes certain types of information from all decision-making can lead us seriously wrong.

Framing

The concept of a "decision frame" was introduced as "the decision maker's conception of the acts, outcomes, and contingencies associated with a particular choice… controlled partly by the formulation of the problem, and partly by the norms, habits, and personal characteristics of the decision maker" (Tversky and Kahneman 1981:453). In the classical cases, framing refers to how one can describe one and the same outcome in different but logically equivalent ways – e.g. describing a glass as half full or half empty. In psychological laboratory studies, the choice of different, but logically equivalent, descriptions of an option has been shown to have a large impact on the decisions made by the experimental subjects. This has often been seen as a sign of irrationality, but other interpretations are also possible. Framing effects are important in policy analysis for at least three reasons. "First, they are used to caution about various elements of uncertainty that are introduced through framing into policy interventions. Second, framing is often referred to in order to justify certain policy interventions, as framing effects are often seen as sources of irrationality in need of correction. Third, framing effects are often used as instruments for policy-making, as they are seen as effective ways to influence behaviour" (Grüne-Yanoff 2016:189).

Goal Setting and Goal Revision

In decision analysis, goals (ends) are typically taken as given and stable, while rationality refers to means-ends relations. Arguments for and against goal revision go beyond this instrumental perspective. Goals guide and motivate actions. They need to have a certain stability "to fulfil their typical function of regulating action in a way that contributes to the satisfaction of the agent's interests in getting what she wants [. . .] . Frequent goal revision not only makes it difficult for the agent to plan her activities over time; it also makes it more difficult for the agent to coordinate her actions with other agents upon whose behaviour the good outcome of her plans and actions is contingent" (Edvardsson Björnberg 2016:172). Therefore, frequent reconsideration of one's goals is not in general commendable. However, there are situations when goal revision is an option that should be seriously considered, in particular situations when the agent has found reasons to revise her beliefs about the achievability of some of her goals and/or the desirability of achieving them.

Hypothetical Retrospection

In our everyday decision-guiding deliberations we often try to apply a future temporal perspective. We ask ourselves how the decision we are going to make will be judged by ourselves (and others) in the future. In some cases, this is easy to find out. For instance, some of the consequences of drinking excessively tonight can, for practical purposes, be regarded as foreseeable. In other cases, in particular those concerning societal decisions under great uncertainty, it will be necessary to carefully think through several possible future developments, often conceptualized as "branches" of the future. The performance of this argumentative strategy has been called *hypothetical retrospection*, and guidelines for its performance have been developed (Hansson 2007, 2016). At least as a first approximation, its aim is to ensure that whatever such "branch" of the future materializes, we will not in the future come to the conclusion that what we do now was wrong (given what we now know). The goal of hypothetical retrospection can also be described as a kind of decision-stability: Our conviction that the decision is right should not be perturbed by information that reaches us after the decision.

Possibilistic Arguments

When precise probabilities of the various potential outcomes are available, they form an important part of the information on which we should base our decisions. But justified choices of policy options can also be made when we lack such information. For that purpose, argumentative methods can be used that consider what is possible according to the state of our background knowledge. Decision relevant possibilities fall into two categories: those which are shown to be consistent with the background knowledge and those which are articulated without that being demonstrated. As the background knowledge changes, arguments based on possibilities may have to be revised. Previous possibilities may, for example, turn out to be inconsistent with the novel background beliefs (Betz 2016: Sect. 4). Important types of practical arguments that account for articulated possibilistic hypotheses are: arguments from best and worst cases, from robustness and from risk imposition. "The fine-grained conceptual framework of possibilistic foreknowledge does not only induce a differentiation of existing decision criteria, it also allows us to formulate novel argument schemes for practical reasoning under deep uncertainty, which could not be represented in terms of traditional risk analysis. These novel argument schemes concern the various options' potential of surprise" (Betz 2016:162).

Scenario

By a scenario we can mean "a postulated or projected situation or sequence of potential future events" (Oxford English Dictionary). In the decision sciences a scenario is a narrative summarizing a particular future development that is held to be possible. In decision-making under (great) uncertainty, multiple scenarios can be used to make sure that various future possibilities are taken into account. In general, only a small selection of the possible future developments can be developed into a scenario. It would therefore be fallacious to infer that some future event is impossible just on the grounds that it doesn't figure in any scenario one has explicitly considered so far (Betz 2016; Hansson 2016). Scenarios have often been used in technology assessment in order to ensure that several different potential developments of a technology and its social embedding are considered. The climate change scenarios developed by the IPCC have a central role in the integration of science from different fields that provides the background knowledge necessary both for international negotiations on emission limitation and in national policies for climate mitigation and adaptation.

Temporal Strategy

Temporal strategies for decision making are "plans to extend decisions over time, such as delaying decisions (postponement), reconsidering provisional decisions later on (semi-closure), or partitioning decisions for taking them stepwise (sequential decisions)" (Hirsch Hadorn 2016:217). The purpose of temporal strategies is to open opportunities for learning about, evaluating and accounting for uncertainty in taking decisions. In many cases, temporal strategies enable the application of argumentative methods in order to systematize deliberation on policy decisions. For proper uses of temporal strategies one has to focus on those uncertainties that need to be clarified more and to consider whether it is feasible to achieve these improvements with a particular temporal strategy. To prevent the problem from worsening in the course of a temporal strategy or decision-makers eschewing the decision problem, it is also necessary to consider trade-offs that may arise from following the temporal strategy instead of taking a definitive decision, and – not least! – to assure appropriate governance of the temporal strategy across time.

Uncertainty

"The case traditionally counted as closest to certainty is that in which at least some of our options can have more than one outcome, and we know both the values and the probabilities of these outcomes. This is usually called *decision-making under risk*… The next step downwards in information access differs from the previous case only in that we do not know the probabilities, at least not all of them. This is usually called *decision-making under uncertainty*" (Hansson and Hirsch Hadorn 2016:16). But although uncertainty and risk are usually defined in this way, as two mutually exclusive concepts, the term "uncertainty" is often also used to cover both concepts, so that risk is seen as a form of uncertainty. The term *great uncertainty* is used for a situation in which other information than the probabilities needed for a well-informed decision is lacking (Hansson 2004). Great uncertainty covers a wide range of types of uncertainties, including uncertainty of demarcation, of consequences, of reliance, and of values. In the same vein, *deep uncertainty* refers to situations when "decision-makers do not know or cannot agree on: (i) the system models, (ii) the prior probability distributions for inputs to the system model(s) and their interdependencies, and/or (iii) the value system(s) used to rank alternatives" (Lempert et al. 2004:2). The terms "great uncertainty" and "deep uncertainty" can for most purposes be treated as synonyms. *Value uncertainty* "may be both about *what* we value – e.g. freedom, security, a morning cup of coffee – and about *how much* value we assign to that which we value" (Möller 2016:107). This can preferably be interpreted broadly, pertaining not only to uncertainty explicitly expressed in terms of values, but also to uncertainty expressed in terms of preferences, norms, principles or (moral or political) theories. Value uncertainty has an important role in many decisions, and special argumentative strategies to deal with it are often needed.

References

Betz, G. (2016). Accounting for possibilities in decision making. In S. O. Hansson & G. Hirsch Hadorn (Eds.), *The argumentative turn in policy analysis. Reasoning about uncertainty* (pp. 135–169). Cham: Springer. doi:10.1007/978-3-319-30549-3_6.

Brun, G., & Betz, G. (2016). Analysing practical argumentation. In S. O. Hansson & G. Hirsch Hadorn (Eds.), *The argumentative turn in policy analysis. Reasoning about uncertainty* (pp. 39–77). Cham: Springer. doi:10.1007/978-3-319-30549-3_3.

Edvardsson Björnberg, K. (2016). Setting and revising goals. In S. O. Hansson & G. Hirsch Hadorn (Eds.), *The argumentative turn in policy analysis. Reasoning about uncertainty* (pp. 171–188). Cham: Springer. doi:10.1007/978-3-319-30549-3_7.

Grüne-Yanoff, T. (2016). Framing. In S. O. Hansson & G. Hirsch Hadorn (Eds.), *The argumentative turn in policy analysis. Reasoning about uncertainty* (pp. 189–215). Cham: Springer. doi:10.1007/978-3-319-30549-3_8.

Hansson, S. O. (2004). Great uncertainty about small things. *Techne, 8*, 26–35.

Hansson, S. O. (2007). Hypothetical retrospection. *Ethical Theory and Moral Practice, 10*, 145–157.

Hansson, S. O. (2016). Evaluating the uncertainties. In S. O. Hansson & G. Hirsch Hadorn (Eds.), *The argumentative turn in policy analysis. Reasoning about uncertainty* (pp. 79–104). Cham: Springer. doi:10.1007/978-3-319-30549-3_4.

Hansson, S. O., & Hirsch Hadorn, G. (2016). Introducing the argumentative turn in policy analysis. In S. O. Hansson & G. Hirsch Hadorn (Eds.), *The argumentative turn in policy analysis. Reasoning about uncertainty* (pp. 11–35). Cham: Springer. doi:10.1007/978-3-319-30549-3_2.

Hirsch Hadorn, G. (2016). Temporal strategies for decision making. In S. O. Hansson & G. Hirsch Hadorn (Eds.), *The argumentative turn in policy analysis. Reasoning about uncertainty* (pp. 217–242). Cham: Springer. doi:10.1007/978-3-319-30549-3_2.

Lempert, R. J., Nakicenovic, N., Sarewitz, D., & Schlesinger, M. (2004). Characterizing climate-change uncertainties for decision-makers. An editorial essay. *Climatic Change, 65*, 1–9.

Möller, N. (2016). Value uncertainty. In S. O. Hansson & G. Hirsch Hadorn (Eds.), *The argumentative turn in policy analysis. Reasoning about uncertainty* (pp. 105–133). Cham: Springer. doi:10.1007/978-3-319-30549-3_5.

Oxford English Dictionary Online. (2015, August). *"scenario"*. Oxford University Press. http://dictionary.oed.com/. Accessed 14 Aug 2015.

Tversky, A., & Kahneman, D. (1981). The framing of decisions and the psychology of choice. *Science (New Series), 211*, 453–458.